SECOND HOME TOURISM IN EUROPE

Second Home Tourism in Europe

Lifestyle Issues and Policy Responses

Edited by

ZORAN ROCA
Universidade Lusófona de Humanidades
e Tecnologias, Lisbon, Portugal

Routledge
Taylor & Francis Group

LONDON AND NEW YORK

First published 2013 by Ashgate Publishing

Published 2016 by Routledge
2 Park Square, Milton Park, Abingdon, Oxfordshire OX14 4RN
711 Third Avenue, New York, NY 10017, USA

First issued in paperback 2016

Routledge is an imprint of the Taylor & Francis Group, an informa business

British Library Cataloguing in Publication Data
Second home tourism in Europe : lifestyle issues and policy responses.
   1. Second homes – Europe. 2. Vacation homes – Europe. 3. Tourism – Europe.
   4. Real estate development – Europe.
   I. Roca, Zoran.
   306.4'819–dc23

The Library of Congress has cataloged the printed edition as follows:
Roca, Zoran.
   Second home tourism in Europe : lifestyle issues and policy responses / by
   Zoran Roca.
      p. cm.
   Includes bibliographical references and index.
   ISBN 978–1–4094–5071–9 (hardback : alk. paper)
   1. Tourism – Europe. 2. Second homes – Europe. 3. Housing policy – Europe.
   4. Europe – Social conditions.
   I. Title.
   G155.E8R64 2013
   333.33'8–dc23   012038426

ISBN 13: 978-1-138-27958-2 (pbk)
ISBN 13: 978-1-4094-5071-9 (hbk)

Contents

PART I   OWNING SECOND HOMES: FROM TRANSNATIONAL CRISIS TO PLACE ATTACHMENT

PART II   BACK TO NATURE: BETWEEN URBAN SPRAWL AND COUNTRYSIDE IDYLL

List of Figures

List of Tables

List of Abbreviations

CPER	Centre for Planning and Economic Research
CReST	Centre for Research and Studies on Tourism, University of Calabria (Italy)
CSO	Central Statistics Office (Ireland)
DCLG	Department of Communities and Local Government (England and Wales)
GFC	Global Financial Crisis
HAS	Hellenic Statistical Authority (Greece)
INE	National Institute for Statistics (Portugal/Spain)
INSEE	Institut National de la Statistique et des Études Économiques (France)
ISNART	National Institute for Research on Tourism (Italy)
ISTAT	Italian National Institute for Statistics (Italy)
IU	Izquierda Unida (Spain)
LMA	Lisbon Metropolitan Area (Portugal)
MEECC	Ministry of the Environment, Energy and Climate Change (Greece)
NCSR	National Centre for Social Research (Greece)
NTUA	National Technical University (Greece)
NUTS	Common Nomenclature of Territorial Units for Statistics
OLIR	Oslo Leisure Influence Region (Norway)
PNPOT	National Programme for Spatial Planning Policy
PP	Partido Popular (Spain)
PSOE	Partido Socialista Obrero Español (Spain)
RRS	Seaside resort scheme (Ireland)
SEGREX	Second home expansion and spatial development planning in Portugal
SHE	Survey of English Housing
SRS	Rural renewal scheme (Ireland)
TERCUD	Territory Culture and Development Research Centre, Lusófona University (Portugal)
USP/UPSP	Unidos por Santa Pola (Spain)
UTh	University of Thessaly (Greece)
WTO	World Tourism Organization

List of Contributors

Tor Arnesen: Eastern Norway Research Institute, Lillehammer, Norway.

Paul Claval: Université de Paris I – Sorbonne, Paris, France.

Elena Delgado Laguna: Universidad de Alicante, Alicante, Spain.

Birgitta Ericsson: Eastern Norway Research Institute, Lillehammer, Norway.

C. Michael Hall: University of Canterbury, Christchurch, New Zealand.

Mervi J. Hiltunen: University of Eastern Finland, Joensuu, Finland.

José A. Hurtado: Universidad de Alicante, Alicante, Spain.

Olga Iakovidou: Aristotle University of Thessaloniki, Greece.

Olga Karayiannis: University of the Aegean, Greece.

Tomás Mazón: Universidad de Alicante, Alicante, Spain.

Dieter K. Müller: Umeå University, Umeå, Sweden.

Maria de Nazaré Oliveira Roca: Universidade Nova de Lisboa, Lisbon, Portugal.

Tatyana Nefedova: Institute of Geography, Russian Academy of Sciences, Moscow, Russia.

José António de Oliveira: Universidade Lusófona de Humanidades e Tecnologias, Lisbon, Portugal.

Chris Paris: CHURP, University of Adelaide, Adelaide, Australia.

Judith Pallot: Oxford University, Oxford, United Kingdom.

Antonella Perri: Università della Calabria, Arcavacata di Rende, Italy.

Kati Pitkänen: University of Eastern Finland, Joensuu, Finland.

Zoran Roca: Universidade Lusófona de Humanidades e Tecnologias, Lisbon, Portugal.

Tullio Romita: Università della Calabria, Arcavacata di Rende, Italy.

Paris Tsartas: University of the Aegean, Greece.

Mia Vepsäläinen: University of Eastern Finland, Joensuu, Finland.

Jean-Marc Zaninetti: Université d'Orléans, Orléans, France.

Acknowledgments

The editor and publisher would like to thank all entities (Archives, Government Boards, Municipalities and Publishers) for permission to reprint the illustrations, figures and other material used in this book. Every effort has been made to trace copyright holders of materials reproduced in this book. Any rights not acknowledged here will be acknowledged in subsequent printings if notice is given to the publisher.

The editor is indebted to many for their advice and assistance at various stages in the preparation of this book. Particular recognition goes to Isabel Canhoto of ULHT – Universidade Lusófona de Humanidades e Tecnologias, Lisbon, for her proficiency and devotion throughout the process of producing the final typescript, as well as to Valerie Rose of Ashgate Publishers for her encouragement in developing this publication.

The editor also acknowledges with gratitude the financial assistance from the FCT – Foundation for Science and Technology, Lisbon, to the project 'SEGREX – Second Home Expansion and Spatial Development Planning In Portugal', implemented in the period 2008–2012 by TERCUD – Territory, Culture and Development Research Centre of ULHT in cooperation with e-GEO Research Centre for Geography and Regional Planning of Universidade Nova de Lisboa, Lisbon. This Project's rescarch experience and findings, as well as international networking and discussions, inspired the production of this book.

Introduction

Zoran Roca

As Paul Claval eloquently tells us in his closing essay to this book, second homes have long been part and parcel of the evolving form of population mobility and settlement in Europe. Their ubiquitous presence and continued expansion over the last several decades, however, have turned second homes into a phenomenon which, first in a handful of countries and nowadays across the entire continent, has gained a prominent place on research agendas in a number of academic fields, become the subject of public policy concern and raised challenges to economic development options at all scales, from local to transnational.

The complexity of the driving forces across diverse geographic contexts have resulted in countless types of second homes – ranging from old to modern buildings and from modest to opulent dwelling units, from isolated locations to contiguous developments and from single to multi-storey properties, from aggressive to harmonious integration in natural or cultural landscapes, within or outside of tourist resorts, etc. – as well as in numerous motives to own, purposes of use and frequency of occupancy of second homes. This complexity has made it rather difficult to achieve a generalised international consensus about the overall significance of second homes. In fact, the term 'second home' itself is actually a surrogate for the English translation of what has become widely accepted as the common denominator to a phenomenon that is in fact culturally bound by a multiplicity of national, regional and local specificities.

Moreover, the dynamic character of the phenomenon – particularly the changing relationship between 'first' and 'second' homes, and the evolving concept of 'multiple dwelling homes' – often makes it difficult to distinguish second residences in research and planning contexts, as well as to target them in public policies. In the sizable world literature on second homes this is mirrored in terminological diversity, hence in the myriad of social, economic, cultural, environmental, legal and other meanings attributed to the phenomenon. For example, to mention only the terms used in this book, second homes in rural and/ or urban parts of European countries are referred to as 'holiday homes', 'vacation homes', 'seasonal homes', 'summer homes', 'weekend homes', 'secondary homes', 'seasonal second homes', 'seasonal migration homes', 'residential tourism homes', 'vacation properties', 'leisure homes', 'recreational cottages', or just as 'cottages', 'cabins', '*pieds-à-terre*' and 'dachas'.

The now classic dilemma identified by Coppock in the late 1970s – 'second homes, curse or blessing?' – still summarises well the ongoing debate on how to grasp the socio-spatial impacts of the second home phenomenon. In fact, there are considerable gaps between the academic, political and media discourses on the one hand, and, on the other, reality in terms of the effective integration of second-home expansion into housing policies and real-estate regulations, land use and landscape protection plans, as well as in development programmes and projects, especially in economic activities related to tourism, in specific geographical settings such as rural, peri-urban, coastal, mountainous, and other areas rich with natural and cultural heritage and/or other amenities, etc. Increasing evidence from different countries shows that to narrow this gap research into the second home phenomenon cannot rely solely on the definitions and interpretations provided by the available statistical data sources, but, rather, on thorough first-hand scientific inquiry – in which fieldwork has to have pivotal importance – and on multidisciplinary explanations of the phenomenon as involving a mix of lifestyle and policy issues.

One way of closing the gap between discourse and reality has been, as evidenced in the literature, through the convergence between social science research, spatial planning and public policy concerns regarding second homes and the evolving functional and morphological features of tourism as a steady driving force behind increased geographical mobility, on the one hand, and the various motives for establishing second homes, on the other. By connecting second home expansion and tourism development, the integration of desk-based macroscopic analyses, based on statistics, cartography and/or political options, with field-based studies, relying on stakeholder analyses and participatory diagnostics and prognoses, can be ensured. This in turn should lead to the effective integration of top-down and bottom-up perspectives in defining second home tourism policy options, with corresponding spill-over effects in a wide range of related areas such as housing construction and real estate markets, natural and cultural heritage protection, as well as landscape and land use planning.

This book aims at bringing attention to recent insights about the spatial diversity and social complexity inherent in second home expansion in Europe in the context of mobility patterns, largely induced by tourism in Europe. The book contains 12 thematic chapters ranging from theoretical research essays to detailed empirical studies in 12 countries by 22 authors and co-authors from all parts of the continent, from Scandinavia to the Mediterranean and from the British Isles to Russia. As befits the overall conception of the book as a compendium of current second home research, planning and policy issues, the book endorses (i) multidisciplinary approaches to the second home phenomenon as an expression of the 'leisure class' mobility and recreation-based lifestyles, as well as a constitutive element of post-productivist land-use patterns and landscape change, (ii) socio-economic and territorial development planning and policy-related perspectives on social change and spatial re-organisation provoked by the expansion of second home tourism in times of prosperity and crisis, and (iii) trans-European coverage of impacts of second homes that have been similar across the continent and those which, by being recorded as country, region and/or place specific, can inspire comparative assessments and good use of the lessons-learned.

The chapters are clustered in three sections that revolve around key dimensions of the second home problematique – owners' conditions, perceptions and roles; location factors, criteria and patterns; and, expansion drivers and public policy choices – followed by Paul Claval's concluding essay. Each chapter is now given a brief *précis* of its main ideas within this overall framework

Part I – Owning Second Homes: From Transnational Crisis to Place Attachment

Chris Paris examines recent developments in transnational second home ownership through a comparative analysis of the UK and Ireland. Second home ownership continued to grow in the UK over the last years domestically, through inward investment and overseas purchase of country estates. This contributed to 'de-coupling' the prime London housing market from the rest of the UK and to the post-productivist character of the British rural settlement. By way of contrast, the global financial crisis accelerated the collapse of overheated housing markets in Ireland, especially in areas with high share of second homes, leading to large surpluses of empty unsold and unfinished dwellings. Whereas the growth of second home ownership in Britain has been associated with systematic gentrification of high amenity country and coastal areas, this was absent in Ireland due to a very different history of rural development and a *laissez-faire* planning regime.

Tomás Mazón, Elena Delgado Laguna and José A. Hurtado bring forward opinions of key social actors about the state of tourism in the municipalities on the coasts of Southern Alicante Region, ranging from the perception of the development policies established when these tourist destinations originally emerged to the possibly adverse future perspectives if urgent and adequate corrective actions are not taken. Social actors also express their views on the tourism model and the type of tourists that come to these territories, especially the issue of the overall dominance of residential tourism and 'mortgaged tourists', i.e., hyper-indebted second home owners, in relation to the almost symbolic hotel offer, as well as of the impacts generated by the current economic crisis and environmental problems associated with growth of occupancy rate and the management patterns practiced by the municipalities.

Based on the results of an extensive survey conducted in Southern Italy, Antonella Perri shows that there is a close link between tourism and residential roots tourism, or return tourism, commonly referred to as the movement of people who come to stay in places where they themselves and/or their families were born, and where they lived before migrating to other places of their permanent residence. The main pull motives are cultural and emotional ties with the places of origin. This phenomenon needs to be closely studied because it may represent a key resource for economic development and social integration, especially in rural and inland areas marked by intensive emigration flows, and which still continue to lose population. Although tourists travelling to their roots do not need advertising or promotional activities for such destinations, it is essential to satisfy their needs and aspirations, and to know their intentions about possible return, as well as tourist behaviour patterns.

Maria de Nazaré Oliveira Roca elucidates on the issue of the relationship between place attachment among second home owners and their role as local development stakeholders. Based on field survey of owners in the Oeste Region, about 100 km north-west of the Lisbon Metropolitan Area, she established levels of 'topophilia' (the affective bond between people and place, or setting), and 'terraphilia' (the experience-based affection between people and a territory that spurs them into local development intervention). Findings point to substantial differences between the two main groups of second home owners: Lisbon residents, who praise escape-related features for enjoying owing a second home in the Oeste, and Portuguese emigrants who honour their previous bonds to the place in this region. These two groups differ in the intensity of relationships with the local community, and motives to participate, or not, in activities promoting local development.

Part II – Back to Nature: Between Urban Sprawl and Countryside Idyll

Arguing that summer residences, *dachas*, are a potential agent of suburbanisation in contemporary Russia, Tatyana Nefedova and Judith Pallot explain how *dacha* is a phenomenon which involves all strata of the population, from the poor to the richest. The former have little more than garden sheds on their allotments to which they repair at weekends for recreation and food production, while the super-rich repair to their 'castles' to 'perform' the country life behind high fences while protected by private security guards. What they have in common is that both are heirs to the tradition of the summer retreat from the city, each pursuing their own version of the rural idyll. The extraordinary expansion in the number of people who take part in the weekend and summer exodus to the countryside since the collapse of the Soviet regime has fuelled an enlargement and simultaneous fragmentation of the spaces of '*dacha* colonisation', as illustrated in the case of metropolitan region of Moscow.

Based on ideas of complementary spaces and multiple dwelling, according to which households with access to second homes are able to divide activities between different homes, Dieter K. Müller analyses the intersection of second home use and nature-based tourism. The research question asked to what extent outdoor recreation among second home owners differs from patterns recorded for other recreationists. The hypothesis that second home owners use their leisure residence for engaging in outdoor recreation and thus they are more active than other recreationists, since their visits to rural areas are more regular, is tested using national survey data gathered by the Swedish research program Outdoor Recreation in Change in 2008. The results indicate that second home owners are indeed more active outdoor recreationists. Second homes are thus not complementary spaces, but rather they reinforce this activity pattern of the everyday environment.

Jean-Marc Zaninetti describes how in France only a minority of regions have specialised in second home development since the 1990s. Regional contrasts are likely to become ever more accentuated because second homes are actually tourist accommodation along the coastlines and in mountainous ski areas, where they are mostly an investment that brings income to the owner. In contrast, a countryside summer home or an urban *pied-à-terre* is unlikely to generate rents, except perhaps in Paris *intra-muros*. For this reason, the second home stock is likely to decrease in many regions. As the French society is becoming more polarised, and the disposable income gap *vis-à-vis* the affluent upper class is widening, middle class households are less likely to retain a second home ownership title that does not generate income. In the vicinity of urban centres, many second homes are now being sold in the market and returning to the condition of main residences.

Mervi J. Hiltunen, Kati Pitkänen, Mia Vepsäläinen and C. Michael Hall explain the ways in which second homes in Finland are an integral part of Finnish lifestyle, cultural identity, and the relationship between people and nature. It is estimated that over half of Finns have access to a second home, and it is widely believed that the popularity and growth of second home tourism is continuous. However, many environmental and social questions challenge the limits to the growth of the phenomenon. The future of second home tourism in Finland emphasising the ecological and community impacts of potential developments are discussed around three distinctive themes – dual dwelling, internationalisation and regional differentiation – indicating changes in the use, user groups, forms and locations of second homes, thus pointing to the underlying factors and processes behind the impacts of these trends.

Part III – Leisure Housing Expansion: Driving Forces and Policy Choices

Olga Karayiannis, Olga Iakovidou and Paris Tsartas present the reality and particularities of second homes in Greece. After explaining the historical evolution of this phenomenon in the light of the post-war boom of the construction industry at national level, largely under the legislation of building 'off-plan', the spatial concentration of second homes on the insular complex of Cyclades is analysed and discussed. The traditionally high dependency of the Greek economy on the local construction sector, combined with the current international financial crisis and the crucial fiscal deficit of the country, are expected to further hold back the adoption of necessary legislative measures to enable the preparation and implementation of a comprehensive national land-use plan, hence clearly obscuring the presently fragmented and short-term approach to national space and commons.

José António Oliveira explores the conceptual and methodological controversies related to the significance of the second home phenomenon, and the relationship between second homes and tourism in the context of the Portuguese territorial planning policies. An attempt is made to clarify these controversies within a theoretical framework anchored in three analytical domains in which territorial management instruments can be applied effectively in regulating land occupation and use in Portugal: the evolution of the housing market and its relationship with a fragile economy; the increased adoption of new consumption habits by a growing urban population; and, the actions of public authorities, from municipal to central governments, which since the 1960s have put high stakes on tourism as an activity generating employment, wealth and capitalisation of public accounts.

According to Tullio Romita, 'undetected tourism' in Italy refers to a significant part of residential tourism which has created spaces of 'spontaneous tourist contexts' without geographic and/or administrative dimension, or boundaries, but which start and end wherever social, cultural and economic development is based on the ability to organise independently and beyond the formal rules, and to discover a shared space and time at the local community level. In this context, the main actors are the do-it-yourself tourists and other related actors who actually enable them to take on this role. They are special social figures, not just tourists, in terms of the ability to decide for themselves, to foster the creation of tourism spaces and to influence their evolution. This kind of tourism has overtaken official tourism in many parts of Italy, but remains difficult to appreciate in terms of the tourism industry management logic.

Tor Arnesen and Birgitta Ericsson explain how, in pace with the accelerating general growth in welfare in Norway in the post-war period, the quality of leisure housing evolved from the plain cabins to increasingly high standard second homes developed in rural landscapes rich with amenities. Also, multi-house home use has emerged as a complementary lifestyle, typically with one house for work and daily life functions in an urban area, and the other house for recreation in a well-equipped rural hinterland. This development has been partly met by a number of both proactive and reactive policy responses at central and local government levels. The policy aims have mainly been to regulate land use from a nature protection perspective as well as to counteract displacement effects. As recreation-based or prioritised lifestyles seem to advance to a new level – such as expressed by the concept of multi-house homes – new policy measures are called for.

Part IV – Conclusion

In the concluding chapter, Paul Claval begins by offering a comprehensive panoramic overview of changing geographical-historical contexts in which second homes and tourism in Europe have evolved. He then goes on to explain the contemporary distribution of second homes in light of the rural legacy of pre-industrial Europe and of the effects of industrialisation and modernisation, as well as elucidate how nowadays 'building second homes has ceased to result mainly from individual initiatives and the work of local craftsmen. It is increasingly the job of property developers.' In his transversal analysis of the twelve chapters in this book, Claval pinpoints their common denominators, namely the diversity and complementarily of authors' approaches; problems of definition and measurement of the second homes phenomenon; the importance of geography of amenities and of historical and cultural dimensions, as well as of the role of heritage in the distribution, age and use of second homes; today's naturalistic and holiday utopias and their eco-social impacts, ever more evident in the 'landscapes of second homes'. Drawing on the book's focus on second homes from the tourism perspective, Claval concludes that 'second home tourism has become such an important sector of economy that it is no more possible to let it develop freely: it is the source of new forms of social deprivation; it generates residential economies that are particularly sensitive to the economic cycle; it often impairs beautiful landscapes and increases human pressure on natural environments. As a result, it is one of the major physical planning stakes of touristic areas.'

By bringing together a broad range of views, approaches and contexts, this collection adds new empirical depth to the contemporary discussion about second homes and their wider effects. The hope is that in doing so it may also inspire further research on a topic that encapsulates many of the dilemmas of contemporary lifestyles and their policy implications.

PART I

Owning Second Homes: From Transnational Crisis to Place Attachment

Second Home Ownership since the Global Financial Crisis in the United Kingdom and Ireland

Chris Paris

Introduction

This chapter examines recent developments in second home ownership through a comparative analysis of the United Kingdom and Ireland. Whereas the growth of second home ownership in Britain was associated with systematic gentrification of high amenity country and coastal areas, this was absent in Ireland due to a very different history of rural development and a *laissez-faire* planning regime. Second home ownership in the UK and Ireland was extremely low by European standards at the start of the 1990s, but grew rapidly until the onset of the global financial crisis (GFC) in 2007 (Gallent et al. 2005, Norris et al. 2010, Paris 2011). The main drivers of growing second home ownership in both the UK and Ireland had been growing affluence and mobility, combined with a strong belief in the utility of investment in property as a core element of households' investment and consumption strategies. The promotion of second home ownership was a booming business in the UK and Ireland, illustrated by property pages of major newspapers in both countries with extensive advertising of second homes, airline in-flight magazines promoting part-time living overseas, and many websites aimed at prospective second home buyers at home and abroad.

The UK is distinctive in terms of transnational second home ownership in two ways. Firstly, as second home ownership has grown, a rapidly increasing share has been in other countries rather than at home. Secondly, prime residential areas of London and country estates in southern England are sites of buoyant transnational second home ownership by non-citizens, demonstrating the increasingly global nature of property markets. Whereas growing affluence in Ireland also stimulated in growth of transnational second home ownership by Irish citizens overseas, there has been little evidence of overseas buyers in prime residential areas of Dublin or elsewhere in Ireland, especially since the GFC.

The growth and impacts of second homes have been largely ignored by housing scholars, apart from some studies of rural housing, with most scholarly coverage in other academic areas, especially planning, rural studies and leisure studies (Hall and Müller 2004, Gallent et al. 2005, McIntyre et al. 2006). But second homes are not just items of leisure investment in rural areas; they are significant elements of urban housing markets. The growing ownership of up-market housing in London and SE England by rich overseas-based owners, moreover, has significant impacts on metropolitan housing markets with ripple effects both locally and across the UK.

Official data sources and series are quite good for the UK, especially for England. In addition to census data every ten years, there have been regular questions in housing surveys, especially the Survey of English Housing (SEH), which ran to 2008, and the current English Housing Survey, which integrates the former SEH with a survey of housing conditions. In addition, other data on second homes in England derive from the registration of second homes with local authorities in order to obtain council tax rebates.[1] There are no equivalent official housing surveys in other UK countries, though some data derive from house conditions surveys. Official data on second homes are very poor in Ireland. There have not been any systematic housing surveys and local authorities had no reason to collect data on second homes as local property taxes were abolished in 1977. We must rely primarily on census data, available every five years, industry estimates and expert commentary and analysis. There is good evidence that construction of dwellings targeted at the second homes market was a significant feature of the Irish housing boom, though it is harder to tease out evidence relating to second home ownership in Irish cities. Other data sources on second home ownership in both countries include housing and real estate industry research and commentary, media commentary and countless web-based sources.

The poor quality of Irish data on second homes has parallels in many other countries, except where such data are collected for tax purposes. But even in those countries, as well as in countries like the UK, Ireland and Australia, there are good reasons for considering that official estimates routinely under-estimate the number of second homes (Paris 2011).

National censuses and surveys in the UK and Ireland typically distinguish a 'usual place of residence' (or 'primary residence') from 'second homes' (UK) or 'holiday homes' (Ireland). Most commentators use the term 'second home' to refer to dwellings used primarily by their owners,[2] and families and friends, for leisure purposes or as *pieds-à-terre* (*not* dwellings deemed to be private rental investments or mainly rented as 'holiday lets'). Most researchers accept Coppock's (1977) argument that second homes are not a discrete type of dwelling. The changing use of dwellings over time, moreover, undermines attempts to fix categorical definitions: some are used often and solely by their owners for leisure purposes, but others are used less often and/or sometimes let out on commercial basis. Seasonal variations in type of use compound problems of trying to count dwellings as 'primary' or 'second' homes, and the distinction may be stretched

---

1   Council tax is the local property tax levied on residential dwellings in England, Wales and Scotland; in Northern Ireland the local property tax is called 'rates'.

2   A small proportion of second homes are rented in some countries, though the overwhelming majority are owned by their users.

beyond breaking point as a growing number of affluent households own more than one 'second home'; it may then be preferable to think in terms of 'multiple homes' (Paris 2012).

There is no official specification of 'second homes' for tax purposes in the UK or Ireland, though various tax breaks in the latter encouraged the development of many dwellings in seaside and rural areas. One implication of the general tax treatment of second homes is that they have constituted an attractive form of investment for affluent households, combining free leisure or other use with the possibility of untaxed capital gains, as one element of household investment and consumption strategies. The opportunity to organise and reorganise residential circumstances on retirement means that such investments may be especially attractive at the end of a working life as a legitimate way to minimise tax liabilities.

Second Home Ownership in the UK and Ireland before the GFC

There was continuous growth in the number of English households with second homes in the ten years leading up to the GFC. Figure 1.1 (below), based on official UK government data, shows that the total number of English-owned second homes increased from around 338,000 second homes in 1996/97 to 525,000 in 2006/07, an overall increase of 55 per cent.[3] Around two-thirds of English-owned second homes were in England and the rest of the UK in 1996/97 but much growth of English second home ownership after 2001/02 was outside the UK, fuelled by a long period during which the pound had strong purchasing power against the euro and mobility was enhanced by the growth of budget airlines. Spain and France were the main European destinations, accounting for 33 per cent and 24 per cent respectively in 2006/07, when the main reasons for having second homes were 'long-term investment' (47 per cent) and 'holiday home' (31 per cent). Recent research in Northern Ireland also showed growth in the ownership of overseas second homes and strong interest, before the GFC at least, in further overseas purchases (Paris 2011).

Savills (2007a) identified a high level of relatively young buyers of overseas second homes, with incomes of twice the UK average, with growing second homes purchase facilitated by access to large cash reserves as well as equity from primary residences. Other factors driving overseas second home ownership included perceived differences between 'home' and 'overseas' countries: more attractive locations for holidays and/or future retirement, lower house prices in many countries, better opportunities for part-time letting opportunities and/or capital growth, as well as new opportunities to access mortgage finance across more liberalised banking systems (Mintel 2006, 2009, Savills 2007a, 2008).

---

3   The DCLG notes that the total number of households with second homes is lower, as some have more than one second home or second homes in more than one location.

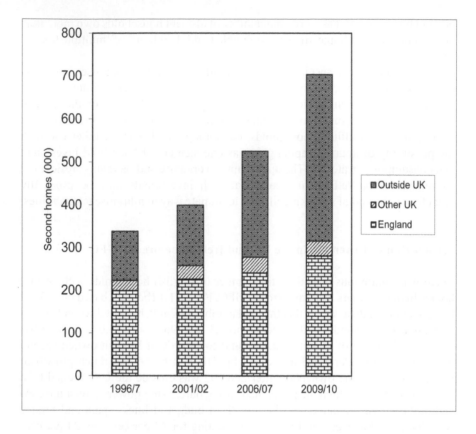

Figure 1.1   English-owned second homes, 1996/97 to 2009/10.

Note: The data use three-year moving averages because annual survey figures fluctuate considerably. The total includes properties owned by households with more than one second home.

Sources: DCLG 2008, 2011.

One factor enabling the growth of second home ownership in the UK has been the strong growth in the number of households owning their primary homes outright, thus increasing the pool of potential second home purchasers: the total increased by 1.5m households in England alone between 1991 and 2001 to a total of 5.6m. Whereas most outright home owners are elderly, growth in outright ownership in the 1990s was highest among younger home owners (Paris 2008). The number of households in England that owned their homes outright had increased to 6.8 million in 2007 (DCLG 2011), suggesting that there was considerable additional scope for future growth in second home ownership

Second home ownership in the UK differs from most other EU countries because its highly restrictive planning regime has constrained housing development generally, and especially so for second homes. Strong demand for second homes, therefore, has been met by purchasing existing dwellings, thus contributing to widespread gentrification of high amenity coastal and countryside areas, as well as stimulating diverse and innovative forms of second home provision (Paris 2011). In Ireland, by way of contrast, a *laissez-faire* planning regime was combined with a series of generous tax breaks for home owners and property investors during the Celtic Tiger boom, resulting in explosive growth of newly-built second homes, as shown in Figure 1.2 (Fitz Gerald 2005, Norris and Winston 2010, Norris et al. 2010).

Figure 1.2   Seaside second homes in County Donegal, Ireland.

Source: Gardiner Mitchell (reprinted by permission).

Two area-based tax incentives were especially important in stimulating construction outside the main population centres: the seaside resort scheme (SRS) and the rural renewal scheme (RRS). The SRS operated between 1995 and 1998 in designated coastal centres, and the RRS between1998 and 2006 in five predominantly inland counties in northwest Ireland. The local impacts of these schemes were examined in two of Norris and Winston's (2009) case studies of the growth of second (and vacant) homes in rural Ireland: Courtown (County Wexford) and Schull (County Cork) both experienced rapid growth of housing development, much of which was rarely if ever occupied. A third case study, Schull (County Cork), was a longer-established second homes destination and, although no area-based tax incentives applied, Norris and Winston argued that other nationwide tax incentives were utilised to construct holiday villages.

It is difficult to estimate the extent of second home ownership in Ireland up to 2007, as there is no official data source on the topic, though the census records some dwellings as 'holiday homes' and others classified as 'vacant' are almost certainly second homes. The proportion of all dwellings counted as holiday homes increased from one to three per cent between 1991 and 2006, with over 90 per cent located in rural, peripheral and coastal areas, typically of high landscape value (Norris and Winston 2009). Census data also indicated that Ireland had an extremely high vacancy rate in 2006, which was similar to southern European countries, especially Spain, Portugal and Italy, all of which also have large numbers of second homes officially described as 'vacant' (Norris and Winston 2009). Nearly three quarters of all dwellings listed as permanently vacant in 2006 were in the same areas as holiday homes, leading Norris and Winston to argue that many were second homes that were unoccupied on the night of the census rather than vacant and never occupied. There is no systematic evidence on Irish second home ownership overseas, though anecdotal evidence, primarily from media and industry sources, suggests that growth in the ownership of overseas second homes was at levels very similar to England.

The evidence overall points to a strong growth of second home ownership across Ireland, especially in attractive coastal and countryside areas, driven by booming affluence and property wealth during the long period of economic expansion from the early 1990s to 2007. Supposedly reputable commentators from leading Irish banks reviewed the growth of incomes and assets in glowing terms. The Irish National Bank published a report, *The Emerald Isle: The Wealth of Modern Ireland*, which claimed that Irish household wealth had doubled in the previous five years, and that the new wealth was 'not an illusion of the property market, but is reflected in our now substantial holdings of financial assets' (O'Toole and Callan 2008: 10). The Bank of Ireland Private Banking Ltd predicted 'strong growth over the remainder of this decade', to follow a year in which 'Irish household wealth per capita increased from €168,000 to €196,000 at the end of last year' (O'Sullivan 2007: 1–2).

Such predictions have proven to be utterly unfounded, but they reveal the mood of optimism and the expectation that the boom would go on and on as home owners were 'encouraged to take advantage of the equity in their homes to scale-up or purchase second or holiday homes or release equity to enable their children to get onto the housing ladder' (Kitchen at al. 2010: 36). And the typical second homes of Dublin's affluent elite were often big: 'larger than their main properties – three or four-bedroom detached houses on at least an acre of land in a coastal location' (Devane 2008).

Second Homes in Britain: Gentrification and Diversification

The growth of British second home ownership has involved extensive gentrification and diversification. Some of the first reported conflicts over second homes in Britain in the 1960s and 1970s concerned the purchase and rehabilitation of abandoned or dilapidated dwellings in areas of depopulation (Coppock 1977). By 2009, most homes in attractive areas had been renovated and gentrified (Gallent and Tewdwr-Jones 2001, Gallent et al. 2005, Paris 2011). Figure 1.3 depicts an example of a gentrified village in the Cotswolds. British second home ownership was often depicted as damaging rural and coastal areas by driving up prices and making housing unaffordable for 'locals', increasing social polarisation and creating 'ghost villages' (for example Monbiot 1998, 2006, 2009).

Figure 1.3   Cotswolds village, England.
Source: Natural England (reprinted by permission).

Many studies examining the impacts of second homes on housing affordability, however, concluded that they were rarely the *main* cause of rural change or house prices increasing beyond the reach of local working people. Rather, many factors in combination had changed local communities: declining employment opportunities in primary industries, growing numbers of dwellings purchased to let to holiday makers, with affluent incoming retirees and long-distance commuters buying properties, as well as growing numbers of dwellings used as second homes (Paris 2011). Thus country towns and villages were increasingly 'post-productivist' as the restructuring of rural industries and greater personal mobility had resulted in the out-migration of primary workers and their replacement by residents whose incomes were generated elsewhere (Mather et al. 2006). The post-productivist idea of the countryside is less as a place of work and more as a setting for residence, leisure, or a sanitised scenic backdrop to affluent everyday life.

Although most British commentators discussed second homes in countryside and coastal areas, some of the largest numbers of second homes in England are in urban areas, especially London boroughs. Most may be considered *pieds-à-terre*, but their owners by definition also have another dwelling purely for their own use, often in rural areas. Many define their city home as the 'second' home and claim council tax discounts there because discounts are higher in cities than in country areas. Data from the Department for Communities and Local Government (DCLG) identifies the spatial distribution of second homes, for households that applied for council tax discounts (see Figure 1.4). Most of the local authorities with the highest percentages of second homes are in coastal and country areas, but three of the top ten in 2007/08 were in London. The City of London has the highest concentration of second homes, at 26 per cent of all dwellings (Savills 2007b). The top five local authorities in London together had a combined total of 25,000 second homes, including established upmarket Westminster, Camden, and Kensington and Chelsea, plus formerly run-down Tower Hamlets, which was transformed by the London Docklands regeneration and extensive gentrification. Overall, half of all declared English second homes were in districts classified as 'predominantly urban'. Thus second home ownership has been an important but overlooked element of gentrification in urban as well as in country and coastal areas. Moreover, the data in table 1 under-state the extent of urban second home ownership as they do neither include second homes where discounts were not claimed, nor count second and multiple homes with overseas-based owners.

The Savills Director of Research suggested that house price growth was resulting in growing second home ownership as 'the investment motive is boosted by expectations of high capital value growth which results in higher levels of second home buying as well as buying to let' (Savills 2007b: 3). Direct Line Insurance (2005) predicted strong growth in the 'work based' sector of the second homes market in major provincial cities such as Newcastle, Liverpool, Glasgow, as well as central London; also many affluent parents purchase houses and apartments for the use of their children while they are living away from home attending university.

The post-war British planning system stopped most building in the countryside and regulated all housing developments more tightly. Planning restrictions are tightest in National Parks or Areas of Outstanding Natural Beauty, which together covered around 30 per cent of the land area in England and Wales in 2009. Most rural and coastal councils with high concentrations of second homes shown in Figure 1.4 are within or adjacent to these areas, especially in Devon, Cornwall, Norfolk, and the Lake District. Growing demand for second homes spilled over into areas further from designated areas as second home owners purchased dwellings from lower income owner occupiers or landlords seeking to capitalise on increasing property values. Many coastal and country villages became gentrified leisure sites, notably Padstow in Cornwall, ironically called 'Padstein' to reflect

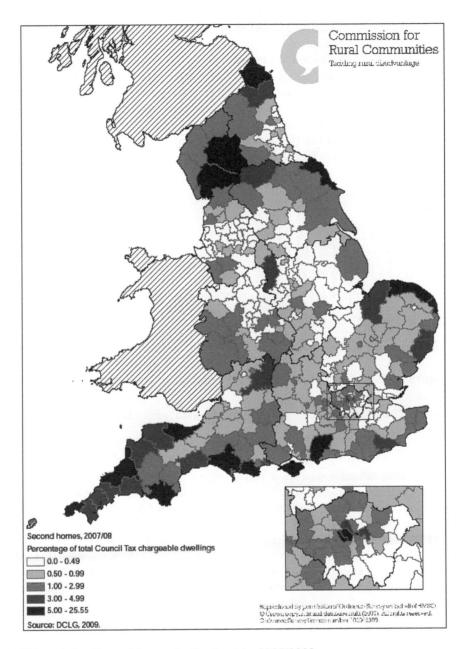

Figure 1.4　Second homes in England in 2007/2008.

Source: Commission for Rural Communities (reprinted by permission).

the influence of the celebrity TV chef and restaurateur, Rick Stein. There has been widespread diversification of developments as well as growth of developments aimed at second home buyers just outside zones of severe development constraint, including expensive developments as well as mixtures of property types and prices oriented towards mass consumption. Many recent up-market developments are in existing built-up areas having obtained planning permission from local authorities eager for prestigious new development, on the basis of existing use rights or replacement of like for like (for example, replacing an old hotel with new holiday apartments, as shown in Figure 1.5).

Figure 1.5   Redevelopment of former hotel as up-market second homes in Cornwall, England.

Source: Coastal Partnerships, with permission from John Goodman (reprinted by permission).

Growing affluence enabled a wider range of households to acquire second homes and the increased mobility of household members has enabled them routinely to enjoy the pleasures of weekends and holidays by the coast or in the countryside. New forms of mobile and static prefabricated dwellings are widespread, often carefully packaged and marketed for different market segments, with complex overlaps between housing and leisure markets. Some new planned developments cater for upmarket second home ownership within high quality landscaped settings based on principles of ecological sustainability with stunning architect-designed houses and apartments.

Planning has also influenced the location of many developments by prioritising the re-use of 'brownfield' sites; policies relating to sand and gravel extraction in river valleys produced sites which were then allocated 'brownfield' status as former 'industrial' areas; so housing developers could obtain planning permission to develop holiday homes that may not be occupied as permanent residences. The sites were regenerated as attractive settings of mixed woodland and lakes for up-market second homes, especially around disused gravel pits in the Cotswolds

Water Park area, some 150 kilometres from central London. Lower Mill Estate is one prestigious development of luxurious second homes in a private estate with high environmental standards and abundant wildlife (Pearman 2007). European beavers re-introduced into the Thames tributary running through the estate gave birth to Britain's first beaver kits in 500 years. Other nearby developments also target affluent second home buyers. Watermark claimed that it was 'the most established second home developer in the Cotswolds' (http://www.watermarkclub. co.uk). The Lakes, developed by Yoo Design Studio with interior designs by Jade Jagger, advertised second homes in January 2012 with prices starting at £870,000 (http://www.thelakesbyyoo.com); its marketing emphasised nature and the surrounding area was sold as part of the package with organic food producers, old pubs and artisanal industries including glass blowing and saddle making.

The Cotswolds is one of the most expensive housing areas of England, with extremely high house price to income ratios. But the development of up-market second homes in the Cotswold Water Park did not involve gentrification directly or replace locals with incomers: one attraction of these 'communities' it is that there *are* no locals! These are a distinctively British type of second homes development, crucially influenced by the planning system, depending on high levels of affluence and mobility of households with incomes generated far away. They are delightful sites for hyper-consumption: exclusive, gated leisure enclaves within a post-productivist countryside marketed through commodified concepts of 'nature', 'community' and 'the Cotswolds'.

Concerns about the impact of second homes on local housing affordability were rarely articulated in Ireland, though some commentators argued that second home ownership was an important factor in driving up land and house prices (National Economic and Social Council 2004) and placing excess pressure on infrastructure in 'summer boomtowns' which became largely deserted in winter (Finnerty et al. 2003). Norris et al. (2010), however, argued that second home developments did not create significant problems of housing affordability in Ireland due to the *laissez-faire* planning system and widespread self-building within rural communities. Their case study interviews did not identify any strong feelings of antipathy towards second homes development and the literature generally on Irish second homes confirmed their view that a strong belief in the importance of home ownership, including second home ownership, made respondents very reluctant to criticise any types of housing development.

Overall housing production had increased enormously in Ireland from under 20,000 a year in the early 1990s to 93,000 in 2006, in stark contrast to static levels of new housing construction across the Irish Sea in Britain. Thus whilst the high proportion of vacant dwellings recorded in the Irish 2006 census was a sign of increased second home ownership, it also reflected 15 years of continuously-rising housing production with many recently-constructed dwellings remaining unoccupied. To some commentators, however, high vacancy levels suggested that housing production had over-shot demand and that the housing boom was due for a significant downward correction (Kelly 2007).

Second Homes within Global Property Markets

It was clear by 2007 that Britain was an extremely attractive site for second and multiple home purchases by overseas-based buyers, who owned uncounted numbers of dwellings in London and SE England. Housing researchers have been largely silent on the impact of globalisation on British housing markets, despite extensive commentary in the media and widespread publicity and commentary by real estate and leisure industries. Increasingly, however, affluent global elites own multiple dwellings in many countries, in both iconic leisure environments such as the Swiss Alps and French Riviera, and also global cities such as London, Paris and New York. One magazine devoted to this subject, *International Homes*, available in the first and business class lounges of international airports, emphasises exclusivity and lifestyle: 'We reach the most desirable high net worth audience with strong spending power and impressive profiles' (http://www.international-homes.com, 20 December 2009). The property sections of major Sunday newspapers also contain many pages devoted to luxury dwellings for leisure use in dozens of overseas countries.

Britain was a well-established 'haven for the international super-rich' during the period leading up to the GFC, as the prices of London mansions increasingly reflected the buying capacity of players in global housing markets rather than any developments within the UK economy as a whole (Woods 2007). The world's super-rich were 'queuing up to buy in the capital, with billionaires from India and China joining the Russian oligarchs and the oil sheikhs who have overheated the market in the past five years' (Partridge 2006). Many of the London mansions owned by non-resident foreigners are rarely if ever visited by their owners; these are investments or possible bolt-holes should political or other problems emerge in their own, typically oil-rich countries (Caesar 2008). Almost half of the country homes in SE England over £5m were being bought by foreigners in 2007 as 'Russian oligarchs and tycoons from Asia and the Middle East...emulate the lifestyle of Britain's landed gentry' and up to 75 per cent over £10m were bought by overseas investors (Gader and Davies 2007).

These developments highlighted the consolidation of an international property market within which 'real estate prices at the center of New York City are more connected to prices in central London or Frankfurt than to the overall real estate market in New York's metropolitan areas' (Sassen 2006: 10). Just before the GFC and during the emerging US sub-prime mortgage debacle it was reported that 'international estate agents maintain that the market for mansions should be relatively unscathed (because) the high end of the market operated under different rules' as the 'boom for top London properties seems to go on for ever' (Shearer 2007).

Second Home Ownership in the UK and Ireland since the GFC

Surging growth in affluence has been checked since 2007 in the UK and Ireland as well as many other countries, during the most widespread recession since the 1930s. Despite the general housing market malaise following the global financial crisis, there is clear evidence of continued growth of second home ownership by English households and evidence of growth of overseas ownership of dwellings as second (3rd, 4th etc.) homes in the UK, especially in London. Extensive purchase of luxury homes by the global super-rich, for a variety of reasons and uses, has contributed to a 'de-coupling' of the prime London housing market from the rest of the UK, with spatial ripple effects across London and SE England. Albeit to a lesser extent, growing overseas purchase of country estates also has contributed to the post-productivist character of British rural settlement. By way of contrast, there is no evidence of continued growth of second home ownership in Ireland or of overseas purchasers of up-market homes in Dublin.

There are three interwoven stories of second home ownership in the UK and Ireland since the GFC, all substantially influenced by wider trends in the two economies since 2007, and very different trends in the two housing systems. The next section provides an overview of housing system developments in the UK and Ireland, before exploring the three dimensions of second home ownership in the two countries: continued strong growth of second home ownership by people based in the UK; a remarkable surge of inward investment by overseas-based purchasers of UK housing, particularly in London; and, the lack of evidence of continued growth in Irish second home ownership.

A Tale of Two Housing Systems

The early signs of possible over-production of housing in Ireland were confirmed after the summer of 2007 when the speculative housing, land and investment boom came to an end. The Irish economy and housing market were massively hit by the GFC, exacerbated by wild bank lending practices, leading to extensive job losses, deteriorating public finances and massive falls in housing and other construction. Unemployment reached 14 per cent in December 2010 and net outmigration in 2010 'was at its highest since 1989 at the end of the last recession' (Drudy and Collins 2011: 10). Housing completions fell dramatically to below 15,000 in 2009 and high vacancy rates were markers of wildly over-optimistic speculative building projects, with thousands of unoccupied and/or unfinished new dwellings in hundreds of 'ghost estates' across the country (Drudy and Collins 2011, Kitchen et al. 2010). Thousands of new homes remained unsold in early 2012; negative equity was widespread among home buyers who purchased after 2003 and house prices had fallen by 50 per cent from their peak of 2007.

The Irish government has taken draconian measures, cutting public sector pay, imposing job caps, raising new taxes and creating a new institution to take over much of the debt of banks and property developers at taxpayers' expense. There may be much more pain to come as it remains uncertain whether measures taken so far will resolve problems of the Irish fiscal crisis, economy, banking and housing systems. The sharp about-turn in Irish house prices is clearly illustrated in Figure 1.6 which also shows that there was no significant difference between trends house prices in Dublin and the rest of the country.

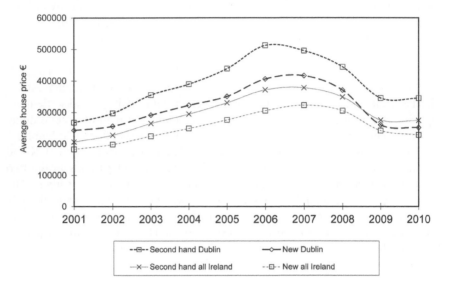

Figure 1.6   Average house prices in the Republic of Ireland 2001–2010.
Source: Central Statistics Office Live Tables.

The housing system in the UK has not been affected as dramatically by the GFC as in Ireland, though the wider economy has been shaken with rising unemployment, higher inflation and widespread cuts in public expenditure. Shortly before the GFC, Ferri (2004: 21) had argued that that 'for those born in 1970, prolonged economic dependence and steeply rising house prices relative to incomes mean that home ownership has become an unattainable goal for many'. But then house prices fell between 2007 and 2009 in most of the UK, though not as much as in Ireland. After 2009, however, house prices increased in most regions, as shown in Figure 1.7. Prices in England have been consistently higher than Scotland, Wales and Northern Ireland. A brief housing boom in Northern Ireland between 2004 and 2007, stimulated by speculative investment by Irish buyers and fuelled by inflated expectations, turned into bust with both house prices and new building collapsing more like the Republic of Ireland than the rest of the UK.

Despite falling house prices between 2007 and 2009 across the UK, home ownership has remained out of the reach of the majority of first time buyers, largely through much tighter restrictions on access to loans for purchase and extremely low levels of new building, as the GFC had a greater impact on housing construction than on house prices. Annual housing construction in the UK had only increased slightly from already-low levels of the late 1990s to a peak at 219,000 in 2006/07. New building fell dramatically after 2007, as in Ireland. Dwelling completions

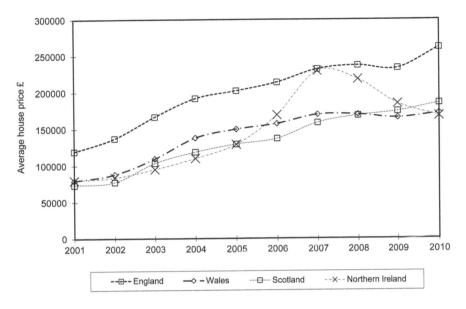

Figure 1.7   Average house prices in the UK 2001–2010.

Source: DCLG Live Table 511.

fell by nearly 40 per cent between 2006/07 and 2010/11 to just 135,500;[4] dwelling starts fell over 50 per cent from 234,000 in 2005/06 to 111,000 in 2008/09, though they rose slightly to 131,000 in 2010/11. Thus UK house building recently has been running at the lowest level since the 1920s, despite increased housing need fuelled by household formation and record net in-migration.

There also have been significant regional variations in English house prices, as shown in Figure 1.8 (below), consistently highest in London, followed by the South East and East regions. All other regions were below the average for England, with the South West, in fourth position, consistently just below the English average. House prices in some parts of the midlands and northern England were falling during 2010/11 (Land Registry 2011) and major lenders predicted static or falling prices over the coming two years; thus first-time buyers who are able to obtain mortgage finance are wary of possible falls in house prices (Milner 2011). By way of contrast, the number of homes sold for over £1m massively out-performed the rest of the market during the first half of 2011, rising 'to its highest level since the peak of the housing market in 2007' (Lloyds 2011: 1). The great bulk of the £1m sales in England were in London (64 per cent) and SE England (22 per cent). The East and South West regions together accounted for another 10 per cent, thus 96 per cent of the £1m sales were in southern England (op cit.). The number of million

---

4   Data on UK dwelling starts and completions are from CLG Live Tables 208 and 209 accessed on 6 September 2010 at http://www.communities.gov.uk/.

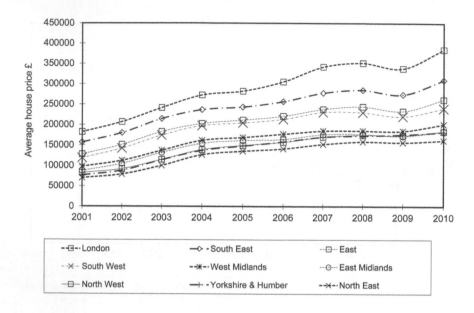

Figure 1.8   Average house prices in English regions 2001–2010.
Source: DCLG Live Table 511.

pound sales had risen across southern England, by 12 per cent in London which contributed around 80 per cent of the overall increase (op cit.). The housing stock in southern England is thus *hugely* more valuable in early 2012 than in other parts of the UK, with over 80 per cent of homes valued above £2m in London, especially Kensington, Westminster and Camden (with 40, 20 and 13 per cent respectively of the total) (Budworth 2012). Such regional differentials highlight growing inequalities in income and wealth, maintaining the relatively privileged position of existing home owners in London and SE England, in part driven by inward investment by overseas purchasers of second and multiple homes in the same areas.

British Second Home Ownership after the GFC

The GFC has had negligible impact of the growth of second home ownership by English households (see Figure 1.1); rather, the story of British second home ownership since the GFC has been one of increasing socio-economic polarisation, in a country which already had the greatest extent of socio-economic polarisation in the EU (Irvin 2008). The rate at which second home ownership has grown increased rapidly from the mid-1990s to the GFC: by 18 per cent from 1996/97 to 2001/02, then by another 32 per cent to 2006/07. The total continued to increase at a faster rate during the three years since 2006/07, with an additional net 180,000 households owning second homes to above 700,000 in 2009/10. That is the equivalent of between three and four per cent of all English households. The rate of increase slackened slightly after 2007, but the most that could be argued is that the GFC may have moderated the trend of increasing second home ownership by English households. It clearly did not reverse the trend.

The total number of second homes owned by UK households is considerably higher than the total for England alone. Research in Northern Ireland has demonstrated strong growth in second home ownership both locally and overseas, during the period up to 2007 (Paris 2008) and there are no reasons for supposing that the situation was significantly different in Scotland and Wales. Levels of wealth and assets were much higher in London and SE England than in other parts of the UK, but the cost of second homes in other regions would also be correspondingly lower. There were about 4.5 m households in the rest of the UK in 2010 and so, if their levels of second home ownership were similar to England, they owned around 158,000 second homes, bringing the UK total to around 850,000. But this is almost certainly a substantial under-estimate due to poor recording of second homes in many countries and to households' reluctance in many instances to admit to such ownership. For example, whilst official data on English-owned second homes provides useful consistent time-series data, other evidence suggests that the number of British-owned homes in Spain may have been at least twice as high as official survey-based estimates (Paris, 2011: 152–5). There are good reasons, therefore, for believing that the total number of second homes owned by UK households is well over a million.

As well as indicating overall growth, Figure 1.9 (below) also highlights rapid growth in the ownership of second homes overseas. In response to an observed increase in the number of second homes owned abroad, the 2003/03 SEH included a new question about where such homes were located. At that time around 100,000 English-owned overseas second homes were located in Spain and France, together representing almost 60 per cent of the total number of overseas second homes. Most of the net growth over the last ten years has been further afield. Figure 1.6 shows that the proportion of overseas second homes in France and Spain together fell to around half in 2009/10, by which time the total had more than doubled. The number of second homes in other European countries had grown from around 32,000 in 2003/04 to 114,000 in 2009/10 whilst the number outside Europe increased from 40,000 to over 100,000.

The market research company Mintel (2006) had reported strong growth in second home ownership abroad before the GFC. In 2009 it argued that the recession was not having any effect on people's desire to buy a second home overseas and that strong growth would continue into the future (Mintel 2009). The most recent official data indicate that this prediction was coming true. In the period 2006/07 to 2009/10 there was *much* stronger growth in the purchase of second homes overseas than in the UK as the number increased by over 140,000 compared to a net increase of just 40,000 in England.

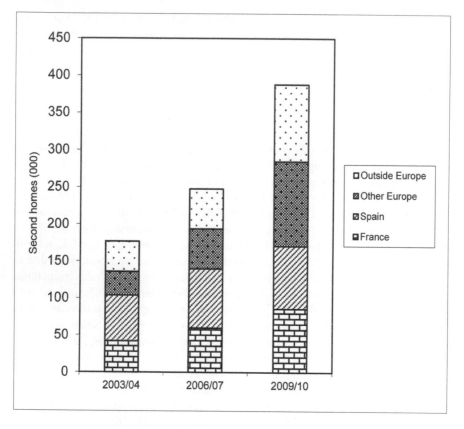

Figure 1.9   English second homes abroad, 2006/7 and 2009/10.
Source: DCLG 2006, 2008, 2011.

Some commentators had suggested that there could be a shift back towards a preference for owning second homes within Britain (Direct Line Insurance 2005) but there is no evidence at present that this has occurred. There are clearly greater risks associated with the purchase of second homes abroad, many of which have been well publicised in the media, especially problems facing second home owners in Spain, with many cases of dramatic failures of building projects, widespread negative equity and loss of properties that had been constructed unlawfully. Concerns about volatility in overseas second homes markets more generally have been widely reported in the media, together with scandals of miss-selling in Bulgaria and Turkey. Other factors also may come increasingly into play, including a weaker pound and the growing cost and inconvenience of air travel, but there was no indication in the most recent data that the trend towards more overseas second home ownership was being reversed.

Despite the relatively slow growth of domestic second home ownership by English households, some industry commentators have argued that demand for newly built second homes in England 'has bucked the trend of slowing activity seen in the mainstream housing market since the financial crisis hit in 2007' (Knight Frank 2011). Demand was especially strong for new *luxury* second homes, especially in Devon and Cornwall, stimulated by desire for 'staycations' as well as a weaker pound making investment 'at home' better value than buying overseas (Knight Frank 2011). The higher incomes and greater assets of households in London and the rest of southern England ensure that this region is the source of the bulk of continuing demand for additional luxury second home ownership.

Meanwhile, by way of contrast, social housing organisations and the homelessness charity Shelter have also argued that second home ownership is continuing to fuel house price increases in many rural areas. The peak body representing British housing associations, the National Housing Federation, launched a 'Save Our Villages' campaign in 2009 to highlight the loss of affordable housing in rural areas and to advocate change in housing policy and expenditure. It reported rapid increase in second home ownership in many parts of England, with demand for second homes spilling over from the main concentrations shown in Figure 1.1 into adjacent areas where house price increases have been much faster than regional averages. It identified ten 'new second homes hotspots' within which the total number of second homes increased by over 50 per cent between 2004 and 2009. Shelter (2011) advocated the abolition of council tax rebates for second homes, arguing that this was effectively a subsidy encouraging under-use of the stock.

The overlap between housing and leisure markets is a key factor in determining demand for second homes in England, as much of the interest in new build second homes comes from investors who plan both to use the properties at times themselves as well as letting them to holiday makers at other times. Strong demand for up-market second homes and continuing gentrification in country and coastal areas have been evident since the GFC, whilst construction more generally was falling dramatically, sales of primary residences were sluggish and house prices also were falling in many other areas.

Transnational Home Ownership within the UK since the GFC

During the same period that British second home purchasers were increasingly buying overseas, large amounts of up-market housing in the UK was being purchased by affluent overseas-based buyers. Prime real estate in London, in particular, was in demand as it increasingly became the second homes capital of the world! Many such mansions are rarely occupied by their owners, though security and other staff are typically in residence most of the year. Thus large swathes of inner west London, like 'ghost villages' in the English countryside, are second homes enclaves which only become populated at night on the rare occasions when their absentee owners grace them with their presence. Overseas-based property owners are not required to pay UK tax on their overseas income and many avoid liability for paying future sales tax by registering their UK properties to companies based in offshore tax havens. Critical commentators have argued that apartments in some developments such as the 'One Hyde Park' complex in Knightsbridge are little more than 'an empty shell for overseas investors to park their assets' (Arlidge 2011: 23).

The growth of transnational housing purchase makes it increasingly less meaningful to conceptualise and analyse housing markets as sets of interactions constrained by national boundaries. Purchases and sales by global economic actors have dramatic impacts on prime property prices, as parts of city-regional and/or national housing markets are 'decoupled' from local and national economic and demographic factors driving housing markets (Knight Frank 2011). There has thus been a disconnection between purchases by the global super-rich and the dynamics of 'national' housing and/or leisure markets. In response to this growing international market, Savills (2011) recently launched its new 'global billionaire property index' focusing on the 'new global super class of real estate' in 10 'global' cities where mansions are marketed to super prime 'global billionaire' property investors.

The notion that the boom for top London properties seems to go on for ever (Shearer 2007) looked over-optimistic after June 2008, as the price of luxury housing in prime London locations fell by 10–15 per cent. From May 2009, however, there was 'a remarkable revival' in London's residential market' as the price of luxury houses began to surge once again (Knight Frank 2009). The strongest sector comprised properties selling between five and ten million pounds, with the London market 'benefitting (*sic*) from substantial inward investment from overseas buyers looking to take advantage of the weak pound and lower overall prices'. Rich buyers from Russia, Italy and the Middle East had been especially active, and 'the revival of the City economy has brought more traditional buyers from the banks, hedge funds and private equity houses back into the market'. Despite the announcement of a new tax aimed at banker's huge bonuses announced in the Pre-Budget Report, in the following week four of Knight Frank's central London offices 'saw contracts exchanged on 22 deals, with aggregate sales prices of over £60m, terms were agreed on a further £45m worth of sales' (op. cit.).

The boom in luxury London house sales accelerated further during 2011, as many super-rich overseas buyers did not simply buy one property in the most expensive areas, but wealthy families from the Middle East or India bought 'a cluster of houses or apartments for themselves, their children and small teams of personal staff' (Phadnis 2011). Mounting political unrest in Arab states and fear of fiscal collapse in some euro-zone countries added to the attraction of London as a safe haven for mobile affluent families. Estate agents reported an influx of rich Greeks relocating their assets and buying £3m-plus homes in London (Davies 2011).

In a recent review of London's housing market (Knight Frank 2011: 2) suggested that 'there is no escaping the fact that the prime London sales market has de-coupled from the wider UK residential market and economy'. Growth of prime central London house prices had been 10 per cent higher over the previous year than in the wider UK market, with exchange rates and currency movements favouring overseas buyers. Knight Frank (2011: 3) predicted that strong demand during 2011 would continue through 2012 as 'the strength of the market in London appears to be untroubled by the latest bout of the financial crisis'. Further evidence of de-coupling comes from official government data with the UK Land Registry house price index (2011) showing falling house prices across the whole UK at the county level, except in Greater London, where prices rose by one per cent. But prices in prime inner London Boroughs with super-rich second homes hot-spots, increased *much* more rapidly: seven per cent in Westminster and five to six per cent in Islington, Kensington and Chelsea, and Hammersmith and Fulham.

The same factors that have contributed to gentrification of British countryside and coasts are also operating to make London an attractive prospect for overseas second home ownership: an extremely restrictive planning regime underpinning house price increases by limiting any housing development, strong anti-development lobbies funded by affluent landowners and supported by better-off existing home owners; and strong inward migration contributing to ever-increasing housing demand. In addition, the British tax system allows overseas-based second owners to benefit from capital gains whilst avoiding taxes. Thus Knight Frank partners shared a £73m bonus pot in October 2011 'after the frenzied interest in luxury London homes helped push profits to pre-crunch levels' (Shah 2011).

Irish Second Home Ownership since the GFC

The situation in Ireland contrasts dramatically with the relative buoyancy of second home ownership in Britain. Irish data on second home ownership remains poor, so it is impossible to use official statistics to monitor developments in any meaningful way. But the housing market remains depressed, with falling prices and production and not the slightest sign of rich overseas buyers bidding up prices in Dublin or anywhere else in Ireland. Many commentators have argued that over-investment in housing and inflated valuations of properties, in a context of government-orchestrated speculation, were major elements in the collapse of Irish banks and developers:

> ...the Irish boom and bust arose due mainly to the failure to control domestic demand and in particular to manage an escalating property market. The significant reduction in capital gains tax and income tax, the persistence for many years (...) with generous property-based tax incentives and some government expenditure during the boom was ill-considered. (Drudy and Collins 2011: 14)

The early signs of over-production of housing in Ireland were confirmed after the summer of 2007, as high vacancy rates were seen increasingly less as an indication of buoyant second home ownership and more a marker of wildly over-optimistic speculative building projects. The GFC accelerated the collapse of over-heated housing markets in Ireland, especially in areas with high shares of second homes, leading to large surpluses of empty unsold and unfinished dwellings Housing construction ceased on many developments and completions fell from a peak of

93,400 in 2006 down to levels not seen since the recession of the late 1980s: just 14,600 in 2010 and 10,500 in 2010. Thousands of unoccupied and/or unfinished new dwellings were scattered across the country in hundreds of 'ghost estates' and hundreds of unfinished as well as uncounted numbers of unfinished or unsold one-off developments in the countryside, of which Figure 1.10 is an example (Drudy and Collins 2011, Kitchen et al. 2010).

Figure 1.10   Unsold second homes development in County Donegal, Ireland.
Source: Elma Lynn.

Irish house prices continued to fall during 2011, with no apparent end in sight. The government's Residential Property Price Index showed a national fall of over 15 per cent in the twelve months to November 2011 (Central Statistics Office 2011). House prices in Ireland had fallen overall by 46 per cent from their peak in 2007. The decrease in property prices was higher in Dublin than in the rest of the country: Dublin house prices had fallen by 52 per cent and apartments were down by 58 per cent. These price falls, together with the weakened economy, job losses and wage reductions had resulted in widespread negative equity and more than 10 per cent of mortgages being in arrears.

We can derive some indication of the extent to which developments aimed at second home buyers have been affected by the collapse of the housing construction boom from the spatial distribution of ghost estates. Kitchen et al. (2010) developed a standardized measure of the number of ghost estate per county in relation to the overall population of the county. This showed that the counties with the highest relative level of ghost estates were in the rural north west of the country and suggested that many of these were vacant holiday home developments. An intensive fieldwork in the Irish border counties has revealed many empty unsold new developments clearly targeted at second home buyers, especially in Co Donegal (Paris 2011: 103–4).

The massive falls in property values may offer a once-in-a-lifetime opportunity for prospective purchasers to buy a delightful second home in Ireland at a knock-down price. Many attractive potential second homes are for sale in early 2012 at less than half the price at which they were valued in 2007. But prospective purchasers should select their bargains carefully, and almost certainly avoid the ghost estates, as many unfinished and unsold properties may be knocked down literally, in order to prevent their falling further into decay.

Conclusions

Affluence and mobility, combined with a preference for investment in property, were the key drivers of the growth in second home ownership in the UK and Ireland. Household wealth increased considerably from the early the 1990s to 2007, especially in the form of domestic residential property and with an increasing number of households owning their homes outright (Paris 2011). Second home ownership in the UK and Ireland was relatively low but grew quickly in both cases after the early the 1990s. Enhanced mobility, especially with the growth of budget airlines, enabled second home buyers to look further afield into Europe and beyond.

British second home ownership was distinctive in a number of ways due to the tight planning regime which influenced the form of domestic second home ownership. It grew largely through gentrification and change of use of existing dwellings from primary residences to second homes. Growing demand for second home ownership, in the context of restrictive British planning, stimulated diverse and innovative forms of second home developments, as in the regeneration of former mineral working sites or the redevelopment of hotels as apartment blocks marketed to non-permanent residents. Such developments avoided the criticism often made of second home ownership in terms of displacing low income locals. The restrictive planning regime also helped to maximise the value of existing dwellings and thus encouraged the search for less expensive second homes overseas especially between 1997 and 2005 when the pound was relatively strong against the euro. Overseas second homes also typically benefitted from warmer climates and more sunshine, even if some of the risks involved were considerably greater than buying at home. London had become well established as a site for second and multiple home acquisitions by affluent non-domiciles by the 1970s, but there was strong growth in the number and scale of such purchases after the mid-1990s.

Second home ownership within Ireland mainly comprised the purchase of new or recently-constructed dwellings, with no significant gentrification, during a period of unprecedented prosperity, with incomes and house values growing strongly from the early 1990s to 2007. The full extent of the growth of Irish second home ownership is difficult to quantify due to poor official housing data sources, but there is strong evidence of rapid domestic growth and reasonable anecdotal evidence of growth in overseas purchases. The house building and development industry boomed in Ireland for over 15 years, as housing production to a peak of 93,000 completed new dwellings in 2006. Unlike the UK, however, there were few signs of any significant flow of inward purchase by overseas-based second home owners and the growth in house prices in Dublin, as elsewhere in the country, was largely a function of domestic demand.

The GFC had major impacts on housing systems in both countries, but was particularly devastating in Ireland, with systemic failure of financial institutions and the housing boom came to a shuddering halt. New building dropped enormously as the extent of over-production became increasingly clear, especially in 'ghost estates' in 'a haunted landscape' (Kitchen et al. 2010). The impact on Irish second home ownership is hard to assess, though we know that over-production was high in areas where second homes have been developed and there is clear evidence of falling demand as house prices had fallen by 40 or 50 per cent. In the starkest contrast to London, Dublin house prices fell more than the rest of the country.

The recent story of second home ownership in the UK has been very different from Ireland, though it is more a tale of socio-economic polarisation and deepening regional divisions, rather than of all-round prosperity. New housing construction, already relatively low compared to growth in demand, fell even lower, to levels not seen since the 1920s. House prices initially fell across the whole of the UK in 2007 and 2008 but subsequently moved in different directions in different places. Prices rose slightly in most regions in 2010 but fell again across most of the UK during 2011, especially in Northern Ireland which had been caught up in the contagion of the Irish property bubble.

House prices in London and SE England have continued to increase, especially prime London real estate which was popular with overseas-based second and multiple home buyers. The level of English second home ownership continued to increase strongly after the GFC, albeit at a slightly slower rate of growth than before 2007. As in the earlier period, there was much stronger growth in the purchase of second homes overseas rather than in England, in ever-further away locations. There was some growth of English domestic second home ownership, with clear evidence of second home gentrification spreading out from such areas into emerging second homes locations but strongest demand for up-market properties in prime second homes hot spots. Thus demand for luxurious new second homes was increasing whilst overall housing demand was limited by restricted mortgage availability and uncertainty in the market.

Are these developments structural or cyclical? The Irish economy has recently shown some signs of recovery, but it will probably take a generation to work through the consequences of wild speculative bank lending practices and widespread over-production of housing. It is almost certain that prime London real estate will continue to be de-coupled from the rest of the UK housing market whilst London remains a global financial centre and relatively safe haven for global capital. The rest of the British housing market is also likely gradually to move back towards 'normal' operation, until the next boom and bust come around.

The strong demand from affluent English, primarily southern English households for second homes shows no signs of abating at present, nor does their preference for buying overseas. There could be a shift back to more domestic-based second home ownership but that would require major reform of planning regulations, which would be ferociously opposed by strong anti-development lobbies. Under current policy settings it would probably mean wider spread of second home gentrification and redevelopment of existing leisure use sites, such as caravan parks, rather than additional green field construction.

The combined effects of the British planning and tax systems appear to be crucial drivers of some of these recent trends. Planning restricts new development, boosting the price of existing dwellings and preserving the British countryside for royalty, the aristocracy and rich households, including a growing stream of overseas-based country landowners. The tax system has almost entirely avoided taxing property wealth, and non-domiciled affluent owners of second home mansions or country estates do not have to pay any tax. The general malaise in house prices across most of the UK means that affluent second home buyers can pick and choose where and what to buy, whilst the flood of overseas money into prime London real estate has enabled the local merely-rich to look further afield for mansions or country homes. The ownership of second homes remains a minority pursuit even in affluent societies and, despite the GFC, the best may be still to come for affluent minorities.

References

Affordable Rural Housing Commission. 2006. *Final Report*. London: Affordable Rural Housing Commission.

Arlidge, J. 2011. Anybody home? *The Sunday Times* (Magazine), 20 November, 16–23.

Barker, K. 2004. *Review of Housing Supply, Delivering Stability: Securing our Future Housing Needs, Final Report – Recommendations*. London: HMSO.

Bramley, G. 2009. Meeting the demand for new housing, *Housing, Markets and Policy*, edited by P. Malpass and R. Rowlands. London: Routledge.

Budworth, D. 2012. Property agents selling stamp duty dodge despite crackdown, *The Times*, 3 March, 10–11.

Caesar, E. 2008. A street named desire, *The Sunday Times*, 22 June 2008. [Online] Available at: http://property.timesonline.co.uk/tol/life_and_style/property/article4164403.ece [accessed: 1 June 2009].

Central Statistics Office. 2011. *Residential Property Price Index*. [Online] Available at: http://www.cso.ie/en/media/csoie/releasespublications/documents/prices/2011/rppi_nov2011.pdf [accessed: 5 January 2012].

Champion, T., Coombes, M., Raybould, S. and Wymer, C. 2005. *Migration and socioeconomic change, a 2001 census analysis of Britain's larger cities*. York: Joseph Rowntree Foundation.

Conradi, P. 2009. Racing king buys piece of empire, *The Sunday Times*, 5 July, 6.

Coppock, J.T. 1977. Second homes in perspective, in *Second Homes, Curse or Blessing*, edited by J.T. Coppock. Oxford: Pergamon Press.

Davies, H. 2011 Greeks in UK homes odyssey. *The Sunday Times*, 23 October, 7.

DCLG. 2006. *Housing in England 2003/04*. London: DCLG.

DCLG. 2008. *Housing in England 2006/07*, London: DCLG.

DCLG. 2011. *English Housing Survey 2009–10*. [Online]. Available at: http://www.communities.gov.uk/housing/housingresearch/housingsurveys/englishhousingsurvey/ehspublications [accessed 20 November 2011].

Devane, M. 2008. Home for the holidays abroad, *The Sunday Times Business Post* [Online, 9 March]. Available at: http://archives.tcm.businesspost/2008/03/09/story31034.asp [accessed 19 March 2009].

Direct Line Insurance. 2005. *Direct Line Second Homes in the UK*. London: Direct Line.

Drudy, P., and Collins, M. 2011. Ireland: from boom to austerity. *Cambridge Journal of Regions, Economy and Society*, 1–16.

Ferri, L. 2004. Thirty-something: time to settle down? in *Seven Ages of Man and Women*, edited by I. Stewart and R. Vaitilingam, R. Swindon: ESRC.

Finnerty, J., Guerin, D., and O'Connell, C. 2003. Ireland, in *Housing in the European Countryside: Rural Pressure and Policy in Western Europe*, edited by N. Gallent et al. London: Routledge.

Fitz Gerald, J. 2005. The Irish housing stock: Growth in number of vacant dwellings. *Quarterly Economic Commentary*, 1–22.

Gader, D., and Davies, H. 2007. Shires fall to foreign land rush. *The Sunday Times*, 28 October 2007.

Gallent, N., Mace, A. and Tewdwr-Jones, M. 2005. *Second Homes, European Perspectives and UK Policies*. Aldershot: Ashgate.

Gallent, N. and Tewdwr-Jones, M. 2001 Second homes and the UK planning system. *Planning Practice and Research*, 16(1), 59–69.

Gkartzios, M. and Scott, M. 2005. Urban-generated rural housing and evidence of counterurbanisation in the Dublin city-region, in *Renewing Urban Communities: Environment, Citizenship and Sustainability in Ireland*, edited by N. Moore and M. Scott. Aldershot: Ashgate, 132–56.

Hall, C.M. 2005. *Tourism: Rethinking the Social Science of Mobility*. Edinburgh: Pearson.

Hall, C.M. and Müller, D.K. 2004. *Tourism, Mobility and Second Homes*. Clevedon Buffalo and Toronto: Channel View Publications.

Irvin, G. 2008. *The Super-Rich: The Rise of Inequality in Britain and the United States*. Cambridge: Policy Press.

Kelly, M. 2007. On the likely extent of falls in Irish house prices. UCD Centre for Economic Policy Research Working Paper 07/01. Dublin: University College. [Online] Available at: http://www.ucd.ie/economics/research/papers/2007/WP07.01.pdf [accessed 10 July 2009].

Kitchin, R., Gleeson, J., Keaveney, K. and O'Callaghan, C. 2010. *A Haunted Landscape: Housing and Ghost estates in Post-Celtic Tiger Ireland*. NISRA Working Paper 59, Maynooth: NUI Maynooth.

Knight Frank. 2009. London's luxury residential sector ends the year on a high. [Online] Available at: http://www.knightfrank.co.uk/news/London%E2%80%99s-luxury-residential-sector-ends-the-year-on-a-high-084.aspx [accessed 20 March 2010].

Knight Frank. 2011. *Autumn 2011 London Residential Review*. [Online] Available at: http://my.knightfrank.com/research-reports/the-london-review.aspx [accessed 20 November 2011].

Land Registry. 2011. *House price index November 2011*. [Online] Available at: http://www.landreg.gov.uk/house-prices [accessed 4 January 2012].

Lloyds TSB. 2011. Million pound homes sales rise hit four year high, *Lloyds TSB Million Pound Property Report*. [Online], broadcast 22 October at http://www.lloydsbankinggroup.com/media/pdfs/LTSB/2011/22_10_11_Lloyds_TSB_2011H1_Review_Million_Pound_SalesFINAL.pdf [accessed 10 January 2012].

McIntyre, N., Williams, D. and McHugh, K. 2006. *Multiple Dwelling and Tourism, Negotiating Place, Home and Identity*. Cambridge, MA: CABI.

Mather, A., Hill, G. and Nijnik, M. 2006. Post-productivism and rural land use: cul de sac or challenge for theorization? *Journal of Rural Studies*, 22, 441–55.

Milner, L. 2011. First-time buyers may be walking into an equity trap. *The Times*, 31 December, 6–7.

Mintel. 2006. *Home Ownership Abroad and Timeshare*. London: Mintel.

Mintel. 2009. *Holiday Property Abroad*. London: Mintel.

Monbiot, G. 1998. Ghost towns. *The Guardian* [Online, 18 July] Available at: http://www.monbiot.com/archives/1998/07/18/ghost-towns [accessed 10 June 2009].

Monbiot, G. 2006. Second-home owners are among the most selfish people in Britain. *The Guardian* [Online, 18 July] Available at: http://www.guardian.co.uk/commentisfree/2006/may/23/comment.politics3 [accessed 1 May 2007].

Monbiot, G. 2009. Flying over the cuckoo's nest. *The Guardian*. [Online, 13 January] Available at: http://www.monbiot.com/archives/1998/07/18/ghost-towns/ [accessed 10 June 2009].

National Housing Federation. 2010. Second home ownership rockets by up to 2000% in rural hotspots. [Online, 27 November] Available at: http://www.housing.org.uk/?page=52 [accessed 1 June 2011].

Norris, M., Paris, C., and Winston, N. 2010. Second homes within Irish housing booms and busts: North-South comparisons, contrasts and debates. *Environment and Planning C, Government and Policy,* 28, 666–80.

Norris, M. and Winston, N. 2010. Rising second home numbers in Ireland: distribution, drivers and implications. *European Planning Studies,* 17 (9), 1303–22.

O'Sullivan, P. 2007. *The Wealth of the Nation.* [Online] Available at: http://www.finfacts.ie/biz10/WealthNationReportJuly07.pdf [accessed 1 May 2009].

O'Toole, R. and Callen, G. 2008. *The Emerald Isle, The Wealth of Modern Ireland.* [Online] Available at: www.nationalirishbank.ie [accessed 12 February 2009].

Paris, C. 2006. Housing markets and cross-border integration, in *The Dublin-Belfast Development Corridor: Ireland's Mega-City Region?* edited by J. Yarwood. Aldershot: Ashgate, 205–30.

Paris, C. 2008. *Second homes in Northern Ireland: growth, impact and policy implications, final report.* [Online: Northern Ireland Housing Executive]. Available at: http://www.nihe.gov.uk.

Paris, C. 2011. *Affluence, Mobility and Second Home Ownership.* Oxford: Routledge.

Paris, C. 2012. Multiple homes', in *The International Encyclopedia of Housing and Home,* edited by S. Smith. Oxford: Elsevier.

Partridge, C. 2006. Going large in London. *The Times* [Online, 29 September]. Available at: http://property.timesonline.co.uk/tol/life_and_style/property/overseas/article2543181.ece [accessed 10 October 2008].

Pearman, H. 2007. Water margin. *Sunday Times* (Homes supplement). 6 May 2007, 41.

Phadnis, S. 2011. Desis home in on London in clusters. *The Times of India* [Online 18 June] Available at: http://articles.timesofindia.indiatimes.com/2011-06-18/india/29673706_1_cluster-savills-london [accessed 2 December 2011].

Sassen, S. 2006. *Cities in a World Economy.* 3rd edition. London: Sage.

Savills. 2007a. *Second homes abroad 2007.* [Online 19 March] Available at: http://www.savills.co.uk/research/Report.aspx?nodeID=8522 [accessed 25 July 2008].

Savills. 2007b. Second homes numbers continue to rise. [Online] Available at: http://www.savills.co.uk [accessed 2 January 2008].

Savills. 2008. *Second Homes Abroad 2008.* [Online] Available at: http://www.savills.co.uk/research/Report.aspx?nodeID=9899 [accessed January 2009].

Savills. 2011. *World Cities Review.* [Online] Available at: http://pdf.euro.savills.co.uk/uk/residential---other/insights---worldclasscities.pdf [accessed 2 December 2011].

Shah, O. 2011. £73m bonus for Knight Frank partners, *The Sunday Times* (Business Section), 23 October, 3.

Shelter. 2011. *Taking Stock.* [Online Policy Briefing] Available at: http://england.shelter.org.uk/__data/assets/pdf_file/0008/346796/Shelter_Policy_Briefing_-_Taking_Stock.pdf [accessed 20 July 2011].

Shearer, P. 2007. Luxury keeps its cool. *The Times*, [Online September 28] Available at: http://property.timesonline.co.uk/tol/life_and_style/property/overseas/article2543181.ece [accessed 10 October 2008].

Urry, J. 2007. *Mobilities*. Cambridge: Polity Press.

Woods R. 2007. Super-rich treble wealth in last 10 years. *The Sunday Times* [Online April 29] Available at: http://www.timesonline.co.uk/tol/news/uk/article1719880.ece [accessed 10 June 2008].

Statistical Sources

UK official housing: the Department of Communities and Local Government, www.communities.gov.uk/housing/housingresearch/housingstatistics

Irish official housing data: the Central Statistics Office and the Department of the Environment, Community and Local Government, http://www.cso.ie/px/Doehlg/Database/DoEHLG/Housing%20Statistics/Housing%20Statistics.asp

Chapter 2

Mortgaged Tourists: The Case of the South Coast of Alicante (Spain)

Tomás Mazón, Elena Delgado Laguna and José A. Hurtado

Introduction

This chapter focuses on the issues of residential tourism in the southern coastal area of the Spanish province of Alicante, which has been marked by the phenomenon of massive and anarchic growth of holiday homes. The lack of adequate urban and tourism planning resulted in an absolute hegemony of residential tourism combined with a paltry provision of hotel accommodation. As early as the 1990s, warnings were issued of the conflicts which this development model can generate owing to its structural deficiencies (Aledo and Mazón 1997). It was clear at that time that the tourism sector in particular, and the financial sector in general, faced serious future risk. Nevertheless, expansive urban development was adopted, creating developments and housing of dubious quality, without the necessary and appropriate infrastructures, provisions and services. New and overcrowded urban settings were created, away from traditional town centres, clearly lacking in quality and with an excessive consumption of land and severe impacts on the area. The situation has consequently become alarming.

The current situation in this area is illustrated by the opinions expressed by local representatives of social agents – the political elite, economic agents and experts – recorded in a field study of the municipalities of Santa Pola, Guardamar del Segura and Torrevieja. These municipalities have been characterized the overwhelming share (close to 80 per cent) of holiday homes in the total number of residences and of the residential tourism sector in this area's economic activity (Rodríguez 2010).

Methodology

The opinions were obtained through 45 in-depth interviews with various well informed social agents in this area, consisting of three groups according to their link with tourism:

i. those not directly involved in tourism: 4 bank managers, 1 trades union representative, 6 retailers and 1 representative from a retailers' association;

ii. those directly involved in the tourism industry: 5 hotel managers, 5 travel agency managers, 5 estate agency managers and 3 property developers; and

iii. people linked to local government: 5 councillors from the PP – Partido Popular (right-wing), 3 councillors from the PSOE – Partido Socialista Obrero Español (socialists), 2 councillors from the USP/UPSP – Unidos por Santa Pola (nationalists), 1 councillor from IU – Izquierda Unida (communist) and 4 local authority managers.

The interviews were conducted between March and October 2008 in the social agents' places of work during 30–60 minutes each. While relying on the findings from other research carried out in this same geographic area, the process of gathering and analyzing information for this study followed the stages suggested by Reissman (1993). On the basis of the transcribed recordings, the inductive and deductive discourse analyses were performed in accordance with Mantecón (2009) regarding essential thematic categories, such as tourism model, type of tourist, local economy, the natural environment, and management. As part of the discourse analysis, performed in accordance with the methodology set out by Ruiz (2009), textual analysis of the specific content of the interviews was carried out, followed by a contextual analysis in order to better understand the discourse.

This qualitative research was completed with a sociological interpretation of the discourse, considered to impart information about a social reality.

Findings

The title given to this chapter illustrates the dominant negative perception that social agents have of those who move to this strip of coastline of the Alicante province. They tend to be middle or lower class, affected by the financial crisis and, in many cases, are paying off the mortgage on their holiday home. Using a concept which will be expanded later in this chapter, a local politician stated that the status of tourists in this area does not go beyond that of the 'mortgaged worker'. They are people who feel forced to 'stay at home, unable to dine out', according to one hotel owner, convinced that many of the people who come here have trouble continuing to pay their mortgage.

Besides this strongly negative attitude towards the 'mortgaged tourists', the interviewees provided an indeed enriching discourse, offering information on a broad range of both positive and negative situations that should be taken into account when considering the future competitiveness of tourism in this area. Several key issues are highlighted in the social actors' perceptions, ranging from the established development policies when the tourism towns were first developed, to the projections for the future, foreseeing negative consequences if the necessary corrective measures are not urgently implemented. The tourist model, the types of tourists that come to the area, the complete hegemony of residential tourism compared with just a token number of hotel beds, the impact caused by the current financial crisis, the environmental problems associated with the development model and the high occupancy rates in the area, as well as the management of these issues by local councils, are but some of the issues that permeate the social actors' discourse.

The Development Model

Positive Views

There is unanimity of opinion among those interviewed with regard to the area's tourism origins. They state that the tourism development of their towns has led to great improvement at both social and economic level. A lot is owed to tourism. Many people earn a living from it, and they recognize that it is responsible for the significant boost to construction and commerce in this area, and the positive effects that these sectors have on employment and the economy as a whole. In the early 1970s, a few family hotels opened in response to growing domestic demand, particularly among people from Madrid, for summer beach holidays in these towns. According to the interviewees, it was the sun, sea and sand that attracted these people to this area as well as the quiet location. Another key factor is believed to be the low cost of housing in the area, particularly when compared with other places on the Alicante coastline, as well as the success of the area's development.

The interviewees rate the dramatic growth of recent years as positive, due to the arrival of foreigners, particularly pensioners from the EU, as confirmed by O,Reilly (2009), Gustafson (2009) and Krit (2011). These are people who are attracted by intangible factors such as a supposed better quality of life and the fact that 'they have realized that living in Spain is cheaper than in their countries of origin' (right-wing politician[1]). Although this development model shows a very clear seasonality as regards employment, (Mazón and Huete 2005), with most visitors arriving in summer, this is considered to have a positive side, as 'thanks to pensioners, Spaniards and foreigners alike, who spend long periods here over winter, there is work all year round, and trade and services benefit from this situation' (right-wing politician). The population increase resulting from residential tourism has also been beneficial for these towns; it has generated the highest number of services and infrastructures, as well as the growth of consumption, which benefits the whole of the town (Mazón 2001): 'the higher the number of inhabitants, the higher the budgetary allocation' (right-wing politician).

The population increase and greater share of the registered foreign than Spanish residents in the three towns of the study area is shown in Table 2.1.

---

1 Henceforth, a brief characterization of the interviewee who uttered the statement quoted will be inserted in brackets after each statement.

Table 2.1   **Share of foreign and Spanish holiday home residents in total population of the studied municipalities.**

Municipality	Population		Registered foreign holiday home residents (2010)		Registered Spanish holiday home residents (2010)	
	2002	2010	N°	%	N°	%
Santa Pola	20965	32507	8626	26.5	8483	26.1
Guardamar del Segura	10732	16423	6626	40.3	4764	29.0
Torrevieja	69763	102658	55863	54.4	13512	13.2

Source: Official Population Patterns of National Statistics Institute (INE), Madrid (2002–2010).

Negative Views

The interviewees feel that, from its origins, urban growth occurred without the necessary channels of organization and control. Their opinions are extremely negative as they perceive that their towns have been converted into areas that specialize in second (holiday) homes, and that the tourism industry aspect is insignificant. Another point of contention is the urban collapse that has occurred as a result of several decades of mass construction. Urban overcrowding is a reason for rejection, particularly when in certain areas property developers have used poor-quality materials in order to build houses at very low prices. The interviewees stated that the levels of overcrowding that have been reached are contrary to the interests of the area as they cause many people to leave these towns.

They blame those responsible for the management of town planning for having allowed a development model based, in hegemonic fashion, on the construction of holiday homes and for the way in which this has occurred through an infinite labyrinth of development projects and narrow streets (Figure 2.1).

One local authority manager interviewed stated that the development process has been marked by 'improvisation, unfettered building work and town planning atrocities, in a process in which the land has been lost, as regards its cultural and landscape value'. Figure 2.2 illustrates one such case.

The interviewees disagree with the way in which their towns have grown. This growth has occurred hastily and without sufficient time for it to be assimilated, as analyzed by Andreu et al. (2005). With the property turmoil, 'the town's identity, if it ever had one, has been lost' (hotel manager). Some negative

Figure 2.1   Adjoining properties on the coast (2009).

Source: T. Mazón.

Figure 2.2   Properties in Torrevieja developed with questionable quality. There are no pavements, proper roads or street lighting (2009).

Source: T. Mazón.

appraisals arise due to the fact that, as the towns' development was not properly planned, private initiatives have focused on the construction and sale of housing (López de Lera 1995), creating imbalances as a result of 'quick development in which some people have operated unscrupulously' (estate agent). Similarly, town councils have limited themselves to collecting the lucrative benefits that these developments provided for the constantly depleted municipal coffers, as confirmed by Mazón and Huete (2005). Consequently, there is now a lack of necessary infrastructures and services for the huge number of developments that are now a veritable burden, both for the users (who are the ones who lose out most) and for the local political scenario, who no longer have the economic means to rectify the situation.

The Tourism Model

Positive Views

The development model is thought to be satisfactory. The interviewed are aware that a model based on building holiday homes tends to attract a type of visitor that is rather different to what is generally understood by the term tourist. However, they consider it to be beneficial for the town as 'they buy and then they keep coming back, including at the weekends' (right-wing politician) or, because of the fact that they are owners, 'they come more, they take weekend and public holiday breaks' (travel agent), which is perceived to be beneficial as residential tourism helps to maintain economic activity and jobs throughout the year.

Some social agents go further when considering the tourism model. They state that residential tourism is beneficial for commerce, drives the economy significantly and creates a great number of jobs: 'residential tourism is much better, because people come back to their homes here and withstand times of crisis much better, which is the complete opposite to hotel-based tourism' (property developer). Furthermore, 'residential tourism increases the population with people who do not come to work, but rather to create work and demand services, so the increase is twofold' (travel agent).

The fact that holiday homes are only occupied a couple of months a year also has a positive side for some. This is because the owners pay their taxes all year round and only consume services for a very short period of time. The interviewed politicians and managers among the local authorities state that 'we have frequent and loyal clients' (manager) and 'holiday homes aren't so bad for us because in winter many people rent them out, and more people come here who also consume' (right-wing politician).

The overwhelmingly greater share of second (holiday) homes in relation to first homes, as well as of accommodation in second (holiday) homes compared to that offered in the small number of the hotels is shown in Table 2.2.

Table 2.2   Accommodation in hotels and holiday homes of the studied municipalities in 2010.

Municipality	Hotels	Beds in hotels	All homes	Permanent homes		Holiday homes		Beds in holiday homes (*)
	N°	N°	N°	N°	%	N°	%	N°
Santa Pola	3	670	33,753	7,183	21.3	26,570	78.7	120,000
Guardamar del Segura	10	1,600	15,706	3,560	22.7	12,146	77.3	55,000
Torrevieja	13	1,350	102,355	21,565	21.1	80,790	78.9	375,000

Source: Housing Census of National Statistics Institute (INE), Madrid (2001).

(*) Based on estimates of more or less 4.5 beds per holiday home.

Negative Views

Negative opinions about the tourism model go much further than positive ones. The interviewees not directly involved with tourism are critical of the model due to the striking lack of hotels and leisure services. They accuse the local authorities of thinking only about the profits that property development generates for their coffers and not about promoting a rational tourist development model. As earlier argued by Mazón (2001), they report that hotel tourism has greater purchasing power, greater expenditure capacity and generates more business to the local catering sector. Regarding visitors, participants state that because they are paying the mortgage on their holiday home they have less purchasing power, and as a result they come 'with their rucksacks full of tinned food' (retailer).

Participants criticize this model as visitors are concentrated at certain times of the year, which means that 'the normal development of any town is impeded when all services are made available for such a short space of time' (bank manager). In short, 'the model is for many people all year round but who are not registered as residents. Therefore, the services are not in keeping with such a population as fewer people contribute to the municipal coffers' (bank manager). They complain that 'the land on the coast has been consumed excessively, creating a large ghost town' (retailer) which only operates in the summer period. Similarly, and due to a lack of planning with crowded development, the town's charm is being lost. This situation leads to families being driven towards other tourist resorts, which they feel to be more attractive. Finally, the increase in delinquency has had negative effects on economic interests: 'tourists who come here to retire, peacefully and in the sunshine, find that they are being attacked and robbed, so they leave' (retailers' association representative).

The interviewed who are involved in tourism are also very critical of the lack of adequate hotel amenities. They feel strongly that the few hotels that do exist 'are not good quality, they're not in the right areas, they don't have sea views, they don't have the infrastructures that people expect in a hotel' (restaurant owner). They also feel that 'hotel tourism spends more and is more beneficial for local commerce as it leaves more money and real tourists come' (hotel owner).

The excessive prominence of holiday homes also comes under fire. The interviewees feel that this kind of residence has more negative effects than positive ones, lamenting the fact that instead of planning properly, construction has been allowed to go ahead in an abusive way, following 'the policy of building and leaving others to deal with the consequences' (property developer). The result is a 'lack of control over what has been built, overcrowding, too much cement and not enough greenery' (property developer). As amply argued by Ruiz and Guía (2004), they admit that it is very complicated to find a solution and they accuse the local authorities of backing development of this kind: 'For the councils' coffers, the benefits from property development projects are more immediate, it's more visible money, whilst in hotels, profits are only seen in the long term. However, a developer builds, sells and forgets about it, but this generates income for the council, and that's where the problem lies' (property developer).

Participants in the study state that there are property developments which, throughout the year, are completely uninhabited – 'the Gran Playa is dead in winter' (restaurant owner) –, which damages the economy and employment, added to the fact that summer holidaymakers 'lead the same kind of lifestyle as if they were at home' (retailers' representative). These are people with little purchasing power – 'many tourists arrive on Friday evening and bring everything with them and so they don't consume' – who come to have a good time spending as little as possible (restaurant owners' association) as well as 'people who come because they own a house' (hotel manager). The general feeling is that, with a few exceptions, 'this model leads to many tourists not spending any money outside their home' (restaurant owner), due to the fact that 'they have mortgages, they wanted to buy a house and until it's paid for they'll stay in and won't be able to go out for dinner' (hotel manager). Furthermore, 'several families live in these holiday homes in the summer' (estate agent), and do their shopping at large retail outlets, to the detriment of local trade.

Property developers have succeeded in profiting from 'people from Madrid who like to have their little villa with their bit of garden, which is just a plant pot. Apparently, it's the "in" thing to have a holiday home or a villa on the coast' (property developer), by building huge development complexes in which 'the streets can't be made any wider; they're just narrow passageways and the place looks like an anthill' (property developer), as shown in Figure 2.3. In short, and jokingly, the inhabitants of the area state that the tourism model, rather than corresponding to the three S's (sun, sea and sand), translates here into three P's: *playa*/beach, *paseo*/ gentle stroll and *pipas*/sunflower seeds to snack on (socialist politician).

Figure 2.3   Horizon of bricks in Santa Pola (2010).
Source: T. Mazón.

The politicians and local authority managers clearly do not agree with how things have been done. They state that urban growth has been unbridled. House construction and sale is the most prominent aspect of this growth, whereas there is an alarming lack of hotel beds, and what does exist is of medium- and low quality. With regard to the tourism model, one of the subjects interviewed stated that this model of tourism 'had been opted for, not as a political choice, but rather as an economic alternative, by property developers who built the houses' (socialist politician). The interviewees feel that planning for the town's future has been forsaken in favour of money-making. This type of urbaniation occurs beyond the control of official town planning; it relentlessly occupies the best areas and creates chaos in the development of urban structures. Only the financial crisis has been able to halt this incongruous growth. These interviewees claim that on some occasions local authorities have tolerated all kinds of aggressive urbanisation practices to the extent that the area has been plundered 'and I think that has been the key to how tourism has grown, the eagerness of politicians and developers to turn the coast into what we call a wall of concrete' (local authority manager), reaching a point at which 'we can no longer expand horizontally because practically all the land has been used up' (local authority manager).

The interviewees are aware that pernicious collateral effects arise with the growth model applied, as 'there has not been a corresponding rise in services' (local authority manager), due to a clear lack of municipal budget available for the increased provision of services and infrastructures. Similarly, the catering sector 'has grown with the same chaotic dynamics, with a lack of professional workers and a lot of underground labour' (communist councillor), which goes against the interests of the sector. There are permanent or semi-permanent residents who are not registered as such with the local authorities, so the area is not allocated resources from other authorities, which means it is impossible to foster the growth of services: 'there aren't enough services; that's what makes the situation here somewhat unbearable' (local authority manager).

The interviewed complain about overcrowding in summer: 'there isn't enough room for everyone on the beach, there aren't enough benches on the seafront, everywhere ends up being overcrowded; it's impractical for vehicles and even for walking in the summer' (communist councillor), as shown in Figure 2.4. They state that seasonality is very harmful because 'for a significant part of the year, entire neighbourhoods are empty of people' (local authority manager). A large town has been created all geared towards the coastline, but which has a seriously detrimental effect on the environment just for 'people who come to spend less than three months here, leaving it completely empty the rest of the year' (local authority manager). Participants in the study maintain that two completely different towns exist: one in the summer, when the town is overcrowded with tourists, and another in the winter, which is completely empty, with large areas taken up by villas which are closed up for ten or eleven months of the year. Finally, they think that a tourism model has been created whereby 'people who stay in their holiday homes spend much less money, because it's obvious that all they do is bring their normal home life along with them. They're different from the typical tourist who stays at a hotel and eats out, goes out for drinks and spends money shopping. These people come to their holiday homes and go about their normal lives' (right-wing politician). To sum up this point, 'no thought has been given to the future, as this area should really be promoted through hotel-based tourism' (local authority manager).

Types of Tourists

Positive Views

Concerning the opinions on the type of tourists who visit the area, the subjects interviewed gave very similar answers. They feel that there are two main types of tourists. In summer, the tourists who visit the area are Spaniards, particularly from Madrid. It is what they define as family-type tourism in an area where 'personal security issues are minimal' (right-wing politician), and fortunately the tourists who

Figure 2.4   Summer overcrowding. El Cura beach in Torrevieja (2009).
Source: T. Mazón.

come here do not cause any conflict: 'there haven't been any aggressive tourists, as has occurred in other towns' (travel agent), in a clear reference to Benidorm.

The second type is the foreign tourist, particularly retired couples who usually spend the winter months here to avoid the harsh winter of continental Europe. The interviewees state that these tourists 'are more polite and respectful of the environment' (retailer), and also 'foreign tourists are more grateful, value service and spend more than Spaniards' (restaurant owner). Spaniards are said to prefer living close to the town centre, whereas foreigners opt to live in housing developments further out of town. Finally, they are aware that 'if there are no tourists, there's no work' (right-wing politician), and as such they accept that it is almost obligatory to get on well with visitors: 'the relationship with tourists is good because they are seen as a source of income, so the people of Santa Pola have always been polite' (local authority manager).

Negative Opinions

All the interviewees have negative comments to make of the tourists who come to the area. They confess that tourists come who have 'very little to spend' and consume very little, and who 'try to spend as little as possible by eating meals at home' (local authority manager), with a predominance of 'tourists from Madrid, who have little purchasing power'. It is no surprise that the catering sector suffers with this, to the extent that 'even summertime can be difficult' (retailer). Because of tourists' low purchasing power these tourists are classed as 'mortgaged workers' (communist politician), a circumstance which is further aggravated by the fact that 'tourists from the middle or north of Spain expect a lot in exchange for the little that they contribute to the town', or that 'Spaniards expect more and are ruder' (local authority manager). There are even cases in which interviewees state that visitors 'are arrogant and think that as they have bought a house in this area, local people are able to put food on their tables' (estate agent). In short, 'today's tourist is the end result of what is on offer' (property developer), as the towns should no longer be promoted in the way they were years ago: 'you cannot promote this place by depicting it as a quiet town on the coast and so on...' (hotel owner).

The interviewees reveal a longing for past times, when the summer season lasted up to three months. Now, holidays are reduced to such an extent that they often barely last twenty days. This is clearly a damaging situation for business and for jobs. They stress visitors' low expenditure capacity. Some subjects interviewed provided worrying information: 'two or three families living in one apartment' (retailers' association), 'people have come here who could no longer make their mortgage repayments if the interest were to go up by €5 a month' (communist politician), or 'in Torrevieja there is no tourism, people have a house, they bring their potatoes from their town, do their shopping at a large retail outlet and don't go out to a bar for a beer or a glass of wine' (restaurant owner), or 'people from Madrid say they put food on our table, whilst local people say that the people from Madrid are just a nuisance' (local authority manager). There are those who cannot understand why tourists come, given that apart from an overcrowded beach there is little on offer for them. What is certain is that, according to one person interviewed, 'residential tourism encourages people to become mortgaged to the hilt, unable to make ends meet or to spend any money during their holidays. This kind of tourism is therefore of no interest to a tourist town' (hotel-owners' association). One retailer states that there is another type of visitor: 'old men who are separated and come here to meet Brazilian women'.

Participants complain that negative attitudes have begun to emerge towards the tourists who come to the area: 'when it became overcrowded people began to speak badly of tourists, who weren't precisely tourists but rather retired people who came to live here. No direct benefit is obtained from the type of people who do their shopping in large retail outlets. Then you go to the hospital and you have a queue with people who nobody can understand, or you need to report an incident at the police station and there's a terrific queue. This has made a lot of people upset and the friendliness towards foreigners is not the same as it used to be' (communist politician), or 'the townspeople aren't very friendly, particularly if they don't have anything to gain; people with a shop or a tourist business are very open and friendly, but other people just want them to leave as soon as possible, as all they do is take their parking spaces and places on the beach' (retailer).

Interviewees also have a negative outlook on how little foreigners become involved in the towns they live in: 'foreigners have closed circles, they have their bars, restaurants, shops, customs' (bank manager), as well as 'everything they need in their residential developments' (retailer). Thus, they feel that foreigners isolate themselves in their residential developments, 'they do not mix, they have their own bars and their own things' (restaurant owner), forming a kind of ghetto turning the town into 'the largest geriatric ward in Europe' (travel agent). They also complain about how little foreigners involve themselves in local culture: 'the English can't speak Spanish, and don't want to speak Spanish' (local authority manager), or 'there are British people who have lived here for ten years and they can't speak Spanish', and 'they don't adapt to the opening hours kept by businesses here' (bank manager). Finally, they claim that 'the Spaniards don't mix either, even those from Santa Pola don't get on well with those from Elche' (retailer), as they think that 'the people from Elche say that Santa Pola is an extension of their town with a beach' (retailer). Elche is a large industrial town 20 km from Santa Pola.

The Local Economy

Positive Views

The established development model is considered positive because it generates a considerable amount of jobs. Residential tourism has succeeded in making construction, tourism, commerce and catering decisive pillars in the growth and economic wellbeing of these towns. Interviewees are aware that the construction sector is the main driving force of the local economies, and that the arrival of new visitors is also advantageous for local trade. There is a consensus that a point has been reached in which there are many 'who earn a living indirectly from tourism and construction' (property developer). These sectors have helped families to increase their economic standing in the same way that, in the labour world, people have been able to develop professionally without having to resort to emigration, as has occurred in the past, as jobs can be found in the town.

Negative Views

The interviewed social agents are aware that, for some time now, there have been serious problems in their community that are attributed to its development model. With the onset of the current financial crisis, which has paralyzed the construction sector, alarm bells have begun to ring about the terrible consequences that this situation is having on the economy and employment. They state that 'a lot of for-sale signs and transfer of business signs can be seen, places closing down, in other words, it is a depressed area' (bank manager). Banking organizations state that 'late payments have increased' (bank director) and that visitors are spending less. Businesses are being closed down, 'the situation is going from bad to worse, businesses such as shoe shops, jewellery shops, travel agencies, clothes shops, even bars are closing down' (retailer). At the same time, the high season has become shorter and visitors are spending less, so the restaurant and catering sector is suffering as families 'cannot have an ice-cream every day because it's too expensive' (restaurant owner). In short, a situation has been reached in which 'people are disillusioned and it is very difficult for an entrepreneur to come and set up a business here' (local authority manager).

These opinions are confirmed by the data in Table 2.3, which illustrate the profound economic crisis affecting this area, with the unemployment at almost 40 per cent in two of the towns, and with a drop of more than 20 percentage points in the number of people registered in the social security system.

Table 2.3   Population, employment and social security status in 2002 and 2010.

Municipality	Population		Active population				Population registered in social security system				Unemployed population			
Years	2002	2010	2002		2010		2002		2010		2002		2010	
	N°	N°	N°	%	N°	%	N°	%	N°	%	N°	%	N°	%
Santa Pola	20,965	32,507	3,768	18.0	8,178	25.2	3,349	88.9	4,988	61.0	419	11.1	3,190	39.0
Guardamar del Segura	10,732	16,423	2,537	23.6	4,670	28.4	2,347	92.5	3,559	76.2	190	7.5	1,111	23.8
Torrevieja	69,763	102,658	12,909	18.5	26,588	25.9	11,210	86.8	16,542	62.2	1,699	13.2	10,046	37.8

Source: Occupation Observatory. Report on the Labour Market 2002 and 2010. Ministry of Labour and Current Affairs.

In the specific case of Santa Pola, its dependence on the neighbouring town of Elche is seen as a negative. Vicissitudes in the footwear sector, the main economic activity in Elche, have a direct effect on Santa Pola: 'if Elche has a cold, Santa Pola catches pneumonia' (nationalist politician). They are aware of the precarious economic situation that they have reached with regard to unemployment problems, and even though in summer 'services increase and staff numbers go up, it's not enough' (socialist politician), a situation which is made worse by the fact that 'wages in the catering sector are very low' (trade unionist).

It is widely thought that foreign visitors with less than healthy finances are arriving in increasing numbers: 'There are people in England who do not even have enough money to pay their heating bills, and you can't live there without heating; there are people who die from the cold. So they come here and live in a climate where you turn the heating on one week a year. What's more if your pension goes further in Spain, it's obvious what people are thinking: I've got sun and I'm saving money which is worth more here' (right-wing politician).

Finally, this transcript of a comment made by one of the social agents interviewed explains the current situation better, in terms of the type of 'tourism' that is dominant at the moment: 'There are families who bring their grandmother here and then expect Torrevieja to look after her; first they register her as a resident, with no income other than her widow's pension. Then she applies for all the council's welfare benefits, she applies for free water which is provided up to a certain number of cubic metres, she applies for an annual payment of €200 available from the Council, she applies for home help and she even applies for the 'meals on wheels' service, with meal vouchers that the council provides' (communist politician).

The Natural Environment

Positive Views

The interviewed subjects are very clear in their opinion that the land that is protected in their towns is so due to interventions carried out by the Valencian Regional Government in this regard. It is their understanding that if the decision to protect land with a certain environmental or landscape value were left to the local authorities, much less of it would be protected. This feeling arises from the different councils' proven voracity when urbanizing their towns. Participants also state that the environment is well cared for in the area as there is increasing environmental awareness. They are conscious that the conservation of ecosystems brings social and environmental benefits and also attracts quality tourism, as was argued by authors such as Wearing and Neil (2000); Martínez Jiménez (2003) and Hall (2005). Therefore, strengthening environmental protection provides a wealth of possibilities for these towns. A range of interesting steps are being taken, such as the creation of various walking routes, green corridors, and the prohibition of cars in protected places.

Negative Views

The interviewed feel that in environmental matters things should be looked after rather more: 'They're not protected enough, nor cared for or explored for tourism' (hotel manager), just as for some there is an 'excess of protected land, strangling the town's growth' (estate agent). Contrary to this, there are those who state that the area's tourism and town planning development models are steamrolling everything and 'property development should have been more controlled' (local authority manager), as it has been carried out with complete 'lack of environmental control' (socialist politician), with the assurance that 'the more property development, the greater the negative effects on the environment' (bank manager). There are those who are convinced that there is an over-saturation of construction, and a shortage of parks and green areas (Fayos 1989). In short, 'a lot has been built very quickly, which brings contamination along with it' (property developer). It is the interviewees from the local authorities who are most critical of the environmental situation. There are those who think that 'we are running out of areas of greenery because of the eagerness to build' (nationalist councillor), as 'not even the few forests, or land, or historic wooded areas have been respected' (local authority manager). For others, despite the existence of a few protected areas, these are not sufficiently cared for. Finally, there are those who go further and say that 'the sea and the mountains have been ruined' (communist councillor).

Management Issues

Negative Views

Nothing is offered in the interviewees' discourse that could be classified as positive. That things are being done reasonably well is the most repeated sentiment in this regard. However, complaints abound about the lack of good town planning and tourism policy. Those interviewed who are not involved in tourism complain about the general lack of infrastructures: 'There are areas that aren't even tarmacked' (bank manager). They blame the authorities for not having been able to rein in construction work, with the added problem that the necessary services that go hand in hand with urban growth have not been created. Retailers are more critical: 'There are shops that have closed down on the high streets' and 'the intention is to move the commercial focus to the suburbs, damaging traditional trade and impoverishing the town centre'. It is said that politicians 'don't have a clue, regardless of which party is in power; they live with their backs to their citizens' (retailer), that 'the eagerness shown by politicians has a negative effect on the town' (retailer) and that politicians 'don't have a clue about what they represent or the consequences of their actions' (bank manager). This point can be concluded with the comment made by one retailer: 'you rarely find a councillor in their office; they have no sense that this is their job and that they have to fight for the town'.

For those involved in tourism, the way things are managed can clearly be improved. They also blame the local authorities for the lack of infrastructure: 'it is substandard, for tourism and for any other activity' (travel agent), 'in summer there are power cuts on a daily basis' (bank manager), 'serious problems with parking' (hotel manager), 'water is going to be a very serious problem' (travel agent), and 'they are not creating an infrastructure that is in proportion with the population growth' (travel agent). Interviewees are also against overcrowding and the collapse that occurs in summer: 'Things that make it really awkward to live here due to the summer chaos, when there is a lack of everything' (estate agent).

With regard to local authority politicians and managers, there are those who cling to the recurrent discourse about quality, by stating that 'they should have stopped thinking about quantity and sought more quality' (socialist politician), or 'it should be a more manageable town, with specialist services and things to grab people's attention' (nationalist politician). There is an awareness that many apartments have been built in the area without the creation of green areas or leisure spaces, leaving the seafront as the only place where people can meet: 'you have to go there if you want to go for a walk' (local authority manager). There are complaints that a large town has been created from a small village, with a growth that has developed from speculation. As such, the result is none other than a hieroglyph of a town that means that 'there is less police pressure on delinquency and the characteristics of a town favour delinquency' (communist politician).

Discussion

The general assessment is that the tourism industry development has been positive, because economic growth and progress have been possible, with significant development in catering, trade and services. However, there are those who state that 'things couldn't be worse than they are now; things have to get better' (right-wing politician). They admit that with another kind of tourism 'things would have been better' (retailers' association representative) and that residential tourism should no longer be advocated as it has been so far.

Those not involved in tourism say that they do not feel that there are any future development policies and that 'problems are just being patched over' (retailer). There are housing developments with serious problems due to the fact that they have been built in such a way that 'the urban infrastructures are not equipped in any way, the streets are narrow, badly urbanized and with a lack of services' (bank manager). The village has been turned into 'a commuter town where young people have to go elsewhere for their daily activities – work, leisure, etc.' (bank manager) and 'a ghost town' (retailer). They state that 'there is time to change to a better kind of tourism' (bank manager) provided that a different type of development model is applied, creating 'more hotels, with leisure and cultural initiatives' (retailer), as well as more services in the town and 'infrastructures that are better adapted' (bank manager).

Those involved in tourism feel that 'the local authority does not see tourism as a priority' (bank manager) and that the time has come for appropriate tourism planning, as the only thing that has been done is 'build, build and build' (estate agent). They are also critical of residential tourism whose seasonality has been negative for the town, (hotel manager) as it is impossible to live all year round on merely on summer-based tourism.

Finally, local authority politicians and managers express opinions much more in agreement with the current situation: 'the positive aspects have greater weight' (socialist politician), considering that there should be 'an important commitment' to tourism (socialist politician) and that 'a planning policy that does not go against the interests' of these towns is needed (right-wing politician).

Conclusions

In general terms, the social agents interviewed for this study feel that the tourist development of the southern coastal area of the Alicante province has led to great social and economic improvements. The development of the construction sector has been responsible for the major economic boost of recent decades, attracting people who have moved here to enjoy the sun, sea, sand and low housing prices. They recognize that urban growth has occurred in a disorganized way and without suitable controls, leading to a tourism model that has focused on residential tourism, and with a glaring lack of hotels. Housing developments have been built that are completely uninhabited for much of the year, generating the presence of tourists with little spending power and causing major disruption in the summer period due to the overcrowding and agglomeration of services and infrastructures.

Two main types of tourists are identified: the families who come to the area in summer, and the retired foreigners who spend the winter months here. Both are described as tourists with low purchasing power, this being a cause for complaint among the local population, who do not look favourably upon the fact that foreigners tend to remain isolated in their own residential developments.

The current economic situation is weighed down by the depression suffered by the construction sector and the harsh effect that this has on employment. Consequently, businesses are closing down and debt is on the rise.

With regard to the environment, the subjects interviewed clearly feel that the local authorities have acted greedily by allowing an urbanization process that has steamrolled valuable areas of nature, and that the little land that has been protected has only been done so due to interventions by the Valencian Regional Government against alleged municipal interests. In terms of management, criticism is levelled at politicians, who are accused of having turned their backs on citizens and of being unaware of the often negative consequences brought about by their poor management.

Finally, looking to the future, the subjects interviewed strongly feel that municipal policies should take action which considers tourism as a priority, strengthening hotel-based tourism and halting the type of residential growth that has occurred so far.

Acknowledgements

This chapter is based on findings from a research project 'Social perception of tourism in the Region of Valencia (Spain). Sociological analysis and prospective study' (CSO2009–10293; sub-programme GEOG) funded by the Ministry of

Science and Innovation, Spanish Government, with Tomás Mazón as the main researcher.

References

Aledo, A. and Mazón, T. 1997. *El Bajo Segura. Análisis turístico.* Alicante: Diputación Provincial de Alicante.

Andreu, N., Galacho, F.B., García, M. and López, D. 2005. Técnicas e instrumentos para el análisis territorial, in *Planificación territorial del turismo,* edited by S. Antón and F. González. Barcelona: Editorial UOC, 61–142.

Bardin, L. 1986 (1977[1]). *Análisis de contenido.* Madrid: Editorial Akal.

Coller, X. 2000. *Estudios de casos.* Madrid: Centro de Investigaciones Sociológicas (CIS).

Fayos-Solà, E. 1989. Environment and Tourism. The Valencian case. *Papers de Turisme,* 1, 86–108.

Gustafson, P. 2009. Estrategias residenciales en la migración internacional de jubilados, in *Turismo, urbanización y estilos de vida. Las nuevas formas de movilidad residencial,* edited by T. Mazón, R. Huete and A. Mantecón. Barcelona: Editorial Icaria, 269–84.

Gutierrez Brito, J. (coord.). 2007. *La Investigación Social del Turismo. Perspectivas y aplicaciones.* Madrid: Thomson Editores.

Hall, M. 2005. *El turismo como ciencia social de la movilidad.* Madrid: Editorial Síntesis.

Ibañez, J. 1979. *Más allá de la sociología: el grupo de discusión, teoría y crítica.* Madrid: Editorial Siglo XXI.

Janoschka, M. 2011. Imaginarios del turismo residencial en Costa Rica. Negociaciones de pertenencia y apropiación simbólica de espacios y lugares: una relación conflictiva, in *Construir una nueva vida. Los espacios del turismo y la migración residencial,* edited by T. Mazón, R. Huete and A. Mantecón. Santander: Editorial Milrazones, 81–102.

Krit, A. 2011. El análisis del entorno construido como un medio para entender la interpretación que los migrantes residenciales hacen sus nuevas vidas: los británicos en España, in *Construir una nueva vida. Los espacios del turismo y la migración residencial,* edited by T. Mazón, R. Huete and A. Mantecón. Santander: Editorial Milrazones, 179–202.

López de Lera, D. 1995. La inmigración en España fines del siglo XX. Los que vienen a trabajar y los que vienen a descansar. *Reis,* 71–2, 225–45.

Martínez Jiménez, E. 2003. Los centros de turismo ante el reto de la gestión medioambiental. *Papers de Turisme,* 34, 6–37.

Mantecón, A. 2009. *La experiencia del turismo. Un estudio sociológico sobre el proceso turístico-residencial.* Barcelona: Editorial Icaria.

Mazón, T. 2001. *Sociología del Turismo.* Madrid: Editorial Fundación Ramón Areces.

Mazón, T. and Huete, R. 2005. Turismo residencial en el litoral alicantino, in *Turismo residencial y cambio social*, edited by T. Mazón and A. Aledo. Alicante: Editorial Aguaclara, 105–38.

Merton, R. and Kendal, P. 1946. The Focused Interview. *American Journal of Sociology*, 51(6), 541–57.

O'Reilly, K. 2009. Migración intra-europea y cohesión social: el grado y la naturaleza de la integración de los migrantes británicos en España, in *Construir una nueva vida. Los espacios del turismo y la migración residencial*, edited by T. Mazón, R. Huete and A. Mantecón. Santander: Editorial Milrazones.

Reissman, C.K. 1993. *Narrative Analysis*. London: Sage.

Rodríguez Gonzaléz, P. 2010. *La elaboración de estrategias empresariales en el sector turístico andaluz. Prácticas y discursos ante la reconversión turística*. Doctoral thesis, Universidad de La Laguna.

Ruiz García, E. and Guía Julve, J. 2004. Financiación del municipio turístico y competitividad: estudio de los municipios turísticos de Cataluña. *Papers de Turisme*, 35, 59–76.

Ruíz Ruíz, J. 2009. Análisis sociológico del discurso: métodos y lógicas. *Forum Qualitative Social Research*, 10(2), Art. 26 [Online]. Available at: http://csic.academia.edu/JorgeRuizRuiz/Papers/133931/Analisis_sociologico_del_discurso_metodos_y_logicas [accessed: February 2012].

Wearing, S. and Neil, J. 2000. *Ecoturismo. Impacto, tendencias y posibilidades*. Madrid: Editorial Síntesis.

Chapter 3
Residential Roots Tourism in Italy

Antonella Perri

Introducing the Emigration Phenomenon in Italy

Emigration is a phenomenon which has always existed and, though the motivations that have driven people to move may be different, the economic reasons for migrating have surely dominated at all times (Thomas 1942). Italy is definitely a country that was much affected by this phenomenon in the past, and still continues to be so, albeit in moderation. In fact, over the past 150 years it has seen important migratory movements, and the main reason for abandoning the native land has unquestionably been the poverty in which most of the Italians found themselves.

The Italian history is marked by two major migratory phases: the 'Great Migration' and 'European Emigration and Internal Migration'. The Great Migration began in the late nineteenth century and the target destinations of migratory flows were mainly overseas countries: Argentina, Brazil and the United States. Those who chose Europe in this period emigrated mainly to France and Germany. The exodus was extraordinary: in 1901 Italy had about 33m inhabitants; between 1861 and 1940 approximately 20m fled from Italy (Sori 1979: 19). What drove the Italians to emigrate was the sudden collapse of food prices with the consequent depression of the entire world market. Those who emigrated during this period were overwhelmingly male, fairly young and had no professional qualification, in fact, mostly labourers.

The first process of Italy's industrialization also influenced the migration process. In the period from the early twentieth century to the First World War, the industrial boom did not involve in a uniform manner the whole Italian territory and was not sufficiently intense as to absorb the surplus labour. Economic remittances sent by Italian emigrants were crucial for the State, in that they served to support development in the form of economic aid and public procurement.

After the Second World War, the job prospects were minimal. Widespread unemployment, which had weakened the whole country, induced the government to consider emigration the only way out, fostering the exodus. Indeed, Italy established unilateral pacts with many European countries, like France, Belgium, the Netherlands, Great Britain, Switzerland, among others, which had to deal with a strong lack of skilled and unskilled workers (Romero 2001).

This new phase of Italian emigration, called the 'European Emigration', also helped to lighten the demographic weight and increase incomes and consumption due to emigrants' economic remittances, indirectly stimulating the economic development of industrial areas (Pugliese 2006). It was numerically much smaller than the 'Great Migration'; moreover, those who emigrated to Europe did so temporarily, while those who had emigrated overseas enacted a clear break with the country of origin. In the second half of the 1950s and all through the 1960s, countries such as Germany and Switzerland became the main destinations as a result of growing demand for foreign labour to be employed in an increasingly growing industry.

The major emigration flows occurred in the poorest areas of the country, namely in the South. In these areas, emigration has caused important changes in the communities: the decline in population in the region of origin, and in particular of men, depleting the human capital; the disruption of social relationships within the community, with family breaks up, causing the breakdown of relations between neighbours and relatives; the flow of money that emigrants transferred to their families was used as a means of income for entire families, giving a huge boost to the economy of the areas of origin (Ascoli 1979).

In this migratory flows those who moved were young, usually single, men. The married men after some time reunited their family by having their wife and children join them. Their belief was, nonetheless, that one day, albeit not in the near future, they would return to Italy. And it is also and especially for this reason that their earnings were invested in the hometown for the purchase or renovation of the house and/or the purchase of land. The ability to purchase or renovate the house also represented, from the social point of view, a social redemption in the eyes of the community.

The Return of the Emigrant

The emigrant, generally, continues to feel part of the original social group and always has a strong desire to return to homeland, even if there are people who, before moving, decide never to return to their hometown. As a rule, the emigrant always looks for the contact with the *millieu* where he/she belongs through the constant demand for news and maintaining ties with the community (Giovane 1974). The return is seen as a basic need and a fundamental step in a previously undertaken journey. And, anyway, there is no single motivation that drives the emigrant to return, but rather the combination of several reasons: the emotional connection to the region of origin, the geographical proximity to the country of emigration, unemployment and/or reaching retirement age, family or individual vicissitudes, lack of and/or little integration and adaptation in the receiving society, changes in the economic and social situation of the recipient region, among others (Brenna 1928, Blumer 1970, Pugliese 2006,).

There are different types of return, depending on the criterion adopted. On the basis of a time criterion, there are two types: final and temporary return.

The final return is when the emigrant goes back and settles permanently in the country of origin. The reasons for the return may be manifold and the wish for the final return is inherent in most of the people who are forced to flee their homeland. Remittances are a tangible sign of their desire to return, as they not only serve as support to family members left behind, but also serve to win over the group who, when they return, will receive and welcome them (Giovane 1974).

The return is not always easy: re-entering the social system of the group left to depart and getting used to the economic and political context that is found in the place of return is not always as dreamed. Remittances are often used exclusively for the purchase of homes or consumer goods, thus not being used productively. Moreover, the reintegration into the economic context is not easy, in that many times in the place of origin there are no conditions to put into practice the knowledge and skills learned through the experience of emigration. Even more complex is the situation for the return of an entire household, where the social integration of children is very difficult (problems of education, language, etc.) (Cafiero 1964). In the case of returning after retirement, emigrants do not have to face major problems. This is due to the fact that, having economic means of subsistence, they are not confronted with the need to look for work and thus integrate the economic fabric. Their challenge will be the need to re-accustom themselves with the original environment, clashing with social patterns and a way of life now abandoned for years.

Temporary returns reflect migrant's objectives and conditions abroad and at the place of origin (e.g., geographical proximity may foster short returns to visit family, friends and relatives) and may indeed gain a tourism connotation.

From Emigrants to Tourists

While waiting to return, emigrants keep the relationship with the community of origin alive, and the purchase of a house or land on which to build a new one is a critical step for the return itself. Owning a home has always represented a symbol of economic and social affluence. For the emigrant, even more, it shows that his/her experience has been helpful and it reinforces the progress achieved. The house, therefore, is a symbol of social success, of the temporariness of emigration, of belonging to a group and of the ties to a place. The return has a sweet flavour: visits to relatives and friends, participation in traditional events, among others. Whenever emigrants have a chance they prefer the context of origin in order to practice these social and cultural models that they carry with them to share with the community that knows and recognizes them.

Once, several emigrants made an event coincide with their return. On this occasion the rites were celebrated in the home community (weddings, christenings, baptisms) and these actions were designed to maintain social relationships with the community, with the aim of preparing the long-awaited final return. Today, these assumptions of sharing and socializing are declining, and social relations as well as personal ties are slowly shrinking and becoming increasingly rare. Now the temporary return of emigrants is increasingly stronger, and the place where they live permanently is at this point their place of belonging, where they have created their own family and social relationships. The spirit in which they return has changed. The dream of returning to the homeland has become the desire to return to those places of their childhood to spend a holiday, surrounded by the affection of their relatives and/or friends.

The emigrant, therefore, no longer feels a citizen of the place of origin but rather a tourist in the place of origin. Returning, he/she undertakes a journey to discover their origins.

The Roots Tourism

The term roots tourism means 'the movement of people who spend leisure stays in the place in which they themselves, and/or their families, were born and where they lived before emigrating to places which, in time, have become the ones where they now live permanently' (Perri 2010: 147). This definition includes both the emigrants who were born and lived part of their lives in the place of origin, and their children, grandchildren, great grandchildren and family, who were born and raised elsewhere, but feel, nonetheless, a strong bond with that land where the roots of their family lie.

According to existing literature, this particular form of tourism is closely linked to residential tourism. The mode of accommodation of the roots tourists is, in most cases, a private residence, whether owned or granted for use by friends and/or relatives at no cost. The property house remains empty and uninhabited for most of the year and is considered a true vacation home.

The residential roots tourism is a widespread phenomenon and involves many countries around the world. In Portugal, for example, migrants of the 1960s, visit the places of origin during the summer holidays, and in some cases even on weekends. The link with their homeland is still very strong and many have revamped their old homes using them as holiday homes (Colás and Cabrerizo 2004). In Spain, there are many second homes that lodge retired tourists native of the place but who, for various reasons, have spent their lives in other countries (Chimoniti-Terroviti 2001). Greece has many unoccupied homes that are used as holiday accommodation by their owners, who are native of the place. And so also in Morocco second homes are used to accommodate the Moroccans working abroad, returning during the holidays to spend time with their relatives. In some areas of Italy, the development of residential tourism is mainly due to its roots tourism, to emigrants who come to their region of origin as tourists and use private homes to stay there.

Some countries, through the creation of *ad hoc* tourism products, have invested in this particular form of tourism. England, for example, is fostering its lineage tourism aimed at a niche of people and integrating it in the field of cultural tourism. An increasing trend in the U.S. is the study of the ramifications of family lines, which has led to the displacement of people in search of their roots (genealogic tourism).

Even Ireland has focused on roots tourism, establishing an organization called the Irish Genealogy Limited (IGL). In addition, there are several organizations working together to promote and increase roots tourism in this country.

In Italy this phenomenon is not recognized and it does not receive the same attention other countries give it, although there is a strong interest on the part of emigrants, even third or fourth generation, towards the homeland of their ancestors. Italy thus underestimates the positive impact that this phenomenon may have in some areas of the country, particularly in inland and rural areas that have been most affected by emigration and that are and/or may witness temporary returns.

A Case Study on Residential Roots Tourism

For some years, in Italy, studies on residential roots tourism have begun to emerge. Among these, a comprehensive field research was carried out in a hinterland community of southern Italy in cooperation with the University of Calabria in 2009. It was based on observation and semi-structured interviews with roots tourist and local residents.

The studied community has been subject to intense emigration processes in the past. It has now a few thousand permanent residents, living from agriculture and small business and it is not, definitely, a tourist destination. It was observed during the field research that in the summer thousands of people who are native to the community but migrated to other areas come to this place. They spend vacation in the place where they or other members of their family were born. Among the research issue was the dynamics that arise from the relationships between the roots tourists and the local population.

One of findings was that in the encounter between roots tourists and residents, whether it occurs in public or in private, the economic and social status achieved by emigrant becomes relevant. The tourist willingly seeks the encounter with the local population. The same is true also for the local population. In the meetings the conversation revolves mainly around having information about the social and economic situation of the tourist, who despite being a native of the place, is seen as a part-foreign subject. The local resident will seek to get information on the tourist's work and family, on any successes in the workplace, on the financial resources and possession of symbols of affluence. This creates an ambivalent mood in tourists: on the one hand, they are very annoyed at having to give an account of their private life to people they actually know very little; on the other hand, they know they cannot evade this social practice.

Conversely, the community, interacting with the roots tourist, often has to suffer much criticism of what in the country and in the place of origin does not work, reluctantly enduring comparisons with the territory where the emigrant lives permanently today, comparisons which often tend to emphasise only the conditions of backwardness of the place of origin.

Several residents have noticed how the emigrant tends to exhibit the symbols of economic well-being achieved which place him in a different social situation from that which had determined the need to emigrate, such as high powered cars, fashionable clothes, typical objects of the affluence society, lifestyles and ways of speaking different from those of origin, state-of-the-art tools and equipment, and so on. Even the manner in which tourists present themselves is different from what they were before emigrating: a new accent, the use of second person plural to emphasize the different customs and the affiliation with a new social group, to name but a few.

The behaviour and the attitude of roots tourists towards the local population can seem cold and aloof sometimes, typical of a tourist who has no personal relationship and no connection with the place where he is sojourning. Actually, this attitude is only a psychological defence and protection mechanism *vis-à-vis* the social status possessed, as well as a way to avoid having to admit publicly that the return to their birthplace is in most cases unlikely and untimely.

The impression is that roots tourists would like to be treated as one would a soldier who went to war, who when returning to their homeland is greeted as a hero even though the war was lost. The interviews revealed that what roots tourists really carry is the dream of returning as winners. In fact, however, even given the work and/or economic opportunity, this desire, as has been proven, is rarely able to satisfy them, whether because it is difficult to move the whole family or because there is a fear of reintegration into the society of origin.

In any case, home ownership in the place of origin plays an important role in maintaining ties with the place. In the interviews conducted, many roots tourists stated that even in cases where they do not use them, they have no intention of leasing, renting, let alone selling the house they own: it is a symbol of belonging to the place of origin. In the mountainous areas, usually the restored and modernized old village houses are converted into such homes, while in the seaside they tend to be newly built dwellings (Figures 3.1 and 3.2).

The detachment from their native region, already lived at the time of emigration, reappears and is re-experienced every time people return to the place of residence. To return permanently to the places they emigrated from means giving up what has been achieved elsewhere with sacrifice; not returning at all, however, involves feeling homesick for that place they left reluctantly. The best alternative is, therefore, to return to their places of origin for temporary stays which we call vacations today, in other words with the status of roots tourists.

Regarding the mode of stay, research shows that the roots tourists look for tranquillity and relaxation. They visit relatives and friends, walk and read, spend time in the village square or pubs to socialise with the residents, occasionally visit nearby towns enjoying natural amenities (sea, mountains, and others) and participate in local festivals and fairs.

The testimonies given by some interviewees (community leaders, professionals, among others) confirm the strong involvement and the importance that the community assigns the phenomenon of returning emigrants in the summer time.

Figure 3.1   Historic village of Pedace (Calabria, southern Italy), a residential roots tourism destination.

Figure 3.2   Seaside housing development in Cassano Allo Ionio (Calabria, southern Italy), a residential roots tourism destination.

These respondents are convinced that there are large differences between the emigrants of the past and those of today. The former left their homeland for economic reasons related to lack of work. These were people of humble origins with little or no education. The today's emigrants, conversely, abandon their region not only for lack of work but also for reasons that are not tied to personal or family economic survival. Today's emigrants are perceived as educated, skilled and competent people. However, according to the interviewed, the reasons for the return of emigrants are the strong bonds that connect them to their relatives in the places of origin and the fact that, for most of them, the emigrants who return temporarily to their country of origin should not be considered tourists. They say that, in order to be tourists, they need to set out on a journey not for sentimental reasons or for family ties, but on entertainment, recreation or cultural grounds. Moreover, in their opinion, tourists travel to new places, which they do not know, and behave differently from the resident of that place. Basically, the interviewees distinguish returning emigrants from tourists based on journey motivation and behaviour.

Nevertheless, the term commonly used by locals to indicate returning roots emigrants is 'tourist'. This makes us think that the term tourist is 'simply' a way to highlight the fact that their fellow emigrants are now 'other' subjects regarding the communities from which they emigrated in the past.

A Survey on Residential Tourist and Residential Roots Tourists

Introduction

The research that the Centre for Research and Studies on Tourism of the University of Calabria has carried out since the late 90s on the residential tourism, besides having led to the collection of information to help assess its impacts on environmental resources and on the economic and social development of territories, has also generated knowledge on the profile of residential tourists and on their relations with the host population.

In our studies on residential tourism, but more specifically in the context of the research on the profile of residential tourists, particular attention has been paid to roots tourists. In fact, the research carried out showed the presence of a very special relationship which, sometimes diffusely, binds residential tourists to their vacation destination. Below, we present some of the results of the research aiming at collecting information on the motivations and on the behaviour of roots tourists and residential tourists and also at comparing these two types of tourists.

The research was carried out in fifteen sea and mountain municipalities in Southern Italy. Data collection was obtained through the application of a questionnaire directed only at persons who said they were on vacation and staying in a private home for whatever reasons. During August 2007, a total of 577 interviews were conducted, of which 312 to residential roots tourists and 265 to residential tourists. The interviewees' profile is outlined in Box 3.1.

Box 3.1 **Characterization of the interviewed residential tourists and residential roots tourists in the municipalities of Acquappesa, Amantea, Belvedere Marittimo, Bonifati, CassanoalloIonio, Cetraro, FagnanoCastello, Longobucco, Pedace, San Giovanni in Fiore, Sangineto, Trebisacce, and Villapiana (percentages).**

Gender: male 47.7; female 52.3
Age distribution: 24 years and under – 7.3; 25–34 years – 25.8; 35–44 years – 29.9; 45–54 years – 21.7; 55 years and over – 15.3
Place of origin: South and Islands 37.5; Centre 17.0; North 38.1; Abroad 7.4
Marital status: single 23.9; married 62.7; engaged 5.4; divorced/separated 3.1; widowed 1.9; unmarried partnerships 3.0
Education: no formal education 0.5; basic education 3.8; middle education 14.7; secondary education 48.6; higher education 32.3
Occupation: entrepreneurs 21.1; industrial/agricultural workers 12.3; unemployed 3.3; housewives 5.8; retailer/self-employed 3.5; teachers/clerks 37.5; students 8.2; pensioners 8.2

Source: Centre for Studies and Research in Tourism – Department of Sociology and Political Science of the University of Calabria: 2007 Survey on residential tourists.

The following analysis takes into consideration only the data provided by the tourists we here call roots tourists. In fact more than half of the valid interviews concerned roots tourists, in other words, respondents who reported having relatives, or being born themselves, or someone in their family being born in the place where they were spending a holiday period (see Tables 3.1 and 3.2) and who reside in a municipality other than that where the interview was conducted

Table 3.1 **Answer to the question: 'Were you or was someone in your family born in this town?' (percentages)**

Answers:	Yes	No	Total
I was	70.5	29.5	100.0
My parents	61.2	38.8	100.0
My grandparents	43.3	56.7	100.0

Source: Centre for Studies and Research in Tourism – Department of Sociology and Political Science of the University of Calabria: 2007 Survey on residential tourists.

Table 3.2   Answer to the question: 'Do any of your relatives reside in this town?' (percentages)

Answers	Total
Yes	95.7
No	4.3
Total	100.0

Source: Centre for Studies and Research in Tourism – Department of Sociology and Political Science of the University of Calabria: 2007 Survey on residential tourists.

Results of the Research

Travel to affections

In the vast majority of cases, roots tourists decide to spend a holiday in the chosen location because relatives live there. The percentage of roots tourists who stated this is so high that that there is no possible misunderstanding or incorrect interpretation: the roots tourists feel the lure of the origins, of the village and of the affections (see Table 3.3).

Table 3.3   Answer to the question: 'Why are you here?' (percentages)

Answers	Total
- because my relatives live there	83.9
- because my friends come here	4.5
- because friends or relatives recommended it	1.0
- because it is customarily	2.3
- because the travel agent recommended it	0.0
- because it is close to where I live	2.3
- because it is rich in attractions	0.3
- for other reasons	5.7
Total	100.0

Source: Centre for Studies and Research in Tourism – Department of Sociology and Political Science of the University of Calabria: 2007 Survey on residential tourists.

Local rooting

In most cases roots tourists are regular visitors to the town; it is not the first time they are on holiday in the place where they currently are, thus demonstrating clear attachment to the place.

Emphasizing the affection for the place where they are spending their holidays there is another fact: three out of ten roots tourists say they come for a vacation, even a brief one, in other periods of the year, and in these cases private residence remains the type of lodging widely used.

Organizational mode

Our respondents were tourists who travel mainly with their families; in fact, this was stated by nearly 80 per cent of them. In the vast majority they arrived at their holiday destinations by car, the means of transport mostly used to travel but also to bring the many things which a family may need during a period away from their residence, which can last from a minimum of two weeks to a maximum of thirty days although a considerable number of them remain on holiday more than a month.

Housing situation

The house is the place where basic services such as food and accommodation are organised, and for that reason it becomes the main reference point for all the family members. First, it is worth noting that, in over half the cases, the house in which our respondents were staying is owned; also relevant is the percentage of people who enjoy the home where they spend vacation free of charge.

This could not be different, given that the vast majority of people who emigrated have built a house in the places of origin over time and with their remittances, almost always from the ground up, with the idea of using it during the holiday period and then after returning permanently, may be as pensioners.

The houses are commonly composed of 2–3 rooms besides sanitary facilities (bathroom and kitchen) and a prevailing number of 5–8 beds. They are not, therefore, minute homes, even if these vary greatly depending on whether the house is on the coast or in the mountains (usually the former are smaller than the latter).

It is clear that roots tourists have a tendency to spend their holidays in the company of other people, with whom they make living arrangements within a single housing unit. A possible explanation for this, besides sharing the holiday period with relatives, is also the opportunity of entertaining friends with whom they live different experiences from those they normally live.

Tourist behaviour

The choices of roots tourists are governed by criteria that favour self-determination, free time management, and enjoyment of natural and cultural amenities freely available in the region where they stay.

To confirm this, among the vast list of activities that respondents said they enjoy during the holiday season, those linked to the use of existing natural amenities, reading, walking, relaxing take the lead.

When asked about the places visited during the holiday period, the respondents indicated that the most popular ones are those where the local landmarks (meaning natural and cultural amenities, festivals, gastronomic events) can be found, in addition to places of public gathering and, to a smaller extent, places for shopping and town centres.

Satisfaction and perception of places

As highlighted above, the vast majority of interviewed tourists were very close to the place where they spend their vacations, visiting it for a long time and returning to it even for short periods of vacation whenever they can. This means they are familiar with the village, the surrounding region and population, therefore they seem to be able to make judgments and assessments regarding the place that are the result of reflections and observations accumulated over time.

In any case, at least six in ten respondents would urge a relative and/or friend to come on holiday to the town where they are. It is logical to assume that this is meant to express satisfaction with the place; otherwise, why would recommend it to others at all?

In this context the set of answers to questions which asked respondents to list the advantages and disadvantages of the place where they were staying takes on a particular meaning. In fact, most of our respondents proved to have very clear ideas about the subject. Among the qualities emerge natural amenities and the opportunity to relax; regarding the flaws, the lack of services and leisure activities, and the lack of care of the landscape emerge.

It is evident that various factors shape their opinion on the place. For this reason, respondents were asked to indicate their level of satisfaction with some aspects of the quality of life of the place as well as tourist services and general services available there. The responses indicate some dissatisfaction concerning various aspects of the place from the point of view of preserving and maintaining environmental resources, and from the point of view of infrastructures and services for tourism.

On the one hand, therefore, the majority of respondents claimed to be generally satisfied with the location in which to spend their holidays, both on the seashores and in the hinterland, in places that have indeed turned into the root tourism resorts (Figures 3.3 and 3.4); on the other hand, however, many disappointing situations are mentioned. It is therefore evident that those judged satisfactory, although fewer, weigh more, because it is in the eyes of residential roots tourists that they have a higher value.

Figure 3.3 **Popular residential roots tourism location on the seashores: Roseto Capo Spulico (Calabria, southern Italy).**

Figure 3.4 **Typical residential roots tourism location in the hinterland: the historic village of Pedace (Calabria, southern Italy).**

Concluding Remarks

The tourism industry is very active in the construction of tourist products, whose complexity increases with increasing market competition. The problems which this process gives rise to are largely unknown, or belong to a very limited extent to self-organized tourism. In other words, to those tourists who decide by their own volition in which places to sojourn, in which house to stay, which means of transport to use, and which road to travel to reach their goal, how often to go to the restaurant during the holidays, what kind of food and menu, who to visit during the holidays, how much and when to enjoy existing tourism resources, whether to participate in local tourist initiatives or not, and so on. In this situation, the places that are interested in securing some of these tourist flows have, first and foremost, the task of preparing welcoming communities and ensuring access to and use of environmental resources.

Towards residential roots tourists, host communities should have an obligation to reserve a privileged path of welcome, not only because they are offspring of those communities, but also because they demonstrate towards it particular attention. Roots tourists do not need incentives and/or particular enticements to holiday in the community in which they or any member of their family was born. They do this regularly for years because they feel the obligation to do so regardless of everything else. While at first there were also instrumental as well as affective reasons, today these have become less relevant.

Roots tourists are at once a cultural and an economic resource, but they are not given the attention they deserve. They still tend to be considered disadvantaged people, people who had to abandon their hometown because they did not find there the opportunity of a decent life in economic as well as in social terms. Yet, for some time now, there has been an ongoing change in attitude on part of the roots emigrant *vis-à-vis* the community of origin of their own family: namely a progressive loosening of emotional ties to it. This depends on the fact that they have established stable relations with the community in which they have normally lived for years. In their new community they feel citizens, whereas in their hometown they feel like tourists. When one is able to understand this one will be able to recover the useful relationship with the people who once inhabited the place of emigration.

The return of the emigrant has indeed changed, because personal relationships as well as friend and parental ties have decreased substantially. The return is experienced as a holiday to a place that is familiar and where one can meet friends and relatives. Emigrants who return are tourists who have roots in the place where they stay, where they were born or where they have relatives. The reasons for their choice are related to the presence of relatives and their primary activity is to visit them. They stay in their own home, now turned second home for vacation purposes. This feature of the second home indeed links roots tourism with residential tourism.

Residential roots tourists are tied to the traditions, to the customs of the place they come from and which they expect to find every time they return. The community sometimes acts in an incomprehensible way towards tourists, while other times it does not fully understand them. On the one hand, there are the emigrants, who return as tourists and would like to be welcomed triumphantly for what they have accomplished with their experience of emigration. On the other hand, there is the community, which sees emigrants return as tourists to the birthplace they left, that birthplace where part of the people remained, and who must demonstrate they were able to remain in a useful way. This triggers a sort of competition on the social level between the two subjects that weigh one another based on the ownership and the achievement of symbols of social and economic affluence.

In sum, since the return of emigrants may indeed represent development opportunity for the receiving inland communities, especially those that are still economically and socially lagging, tourism industry agents and local authorities should attribute greater attention and due importance to roots tourism.

Acknowledgments

The author would like to thank Dr. Lucia Groe and, especially, Prof. Isabel Canhoto, for their valuable assistance in the translation of the text into English.

References

Ascoli, U. 1979. *Movimenti migratori in Italia*. Bologna: Il Mulino.

Blumer, G. 1970. *L'emigrazione italiana in Europa*. Milano: Feltrinelli.

Brenna, G.P. 1928. *Storia dell'emigrazione italiana*. Roma: Libreria Editrice Mantegazza.

Cafiero, S. 1964. *Le migrazioni meridionali SVIMEZ*. Roma: Giuffrè.

Chimoniti-Terroviti, S. 2001. *Ampliamento ed evoluzione delle abitazioni nel corso degli ultimo decennia: osservazioni ed indicazioni sulle aree urbane.* Atene: Centro di Programmazione e Ricerche Economiche (CPRE).

Colás; J.L. &Cabrerizo, J.A. 2004. Vivenda secundaria y residencia múltiple en España: una aproximación sociodemográfica. [Online] *Geo Critica Nova, Revista Electrónica de Geografía y Ciencias Sociales*. 8: 178. Available at: http://www.ub.es/geocrit/sn/sn-178.htm [accessed January 2012].

Giovane, M. 1974. La nostalgia degli emigrati per il paese di origine. *Affari sociali internazionali*, II(5), 101–13.

Perri, A. 2010. Alcune riflessioni sul turismo residenziale delle radici, in *Il Turismo Residenziale. Nuovi stili di vita e di residenzialità, governance del territorio e sviluppo sostenibile del turismo in Europa*, edited by T. Romita. Milano: Franco Angeli.

Pugliese, E. 2006. *L'Italia tra migrazioni internazionali e migrazioni interne*. Bologna: Il Mulino.

Romero, F. 2001. L'emigrazione operaia in Europa 1948–1973, in *Storia dell'emigrazione italiana, I,Partenze*, edited by P. Bevilacqua, A. De Clementi and E. Franzina. Roma: Donzelli.

Sori, E. 1979. *L'emigrazione italiana dall'Unità alla seconda guerra mondiale*. Bologna: Il Mulino.

Thomas, W. 1942. *Population Problems*. New York: McGraw-Hill.

Chapter 4

Place Attachment among Second Home Owners: The Case of the Oeste Region, Portugal

Maria de Nazaré Oliveira Roca

Introduction

The level and sense of one's attachment to a place has evolved with changing mobility patterns. In traditional, low-mobility societies, place attachment and mobility were almost opposite concepts, as place attachment implied the existence of roots almost exclusively in a single place. In modern society, integrated in a globalized economy and culture, place attachment and mobility are not necessarily mutually exclusive phenomena anymore (Gustafson 2009). As Quinn (2004) points out, the increased mobility re-affirms place rootedness, allowing individuals to consolidate attachments with multiple places.

One distinct way of enhancing attachment to multiple places has been through owing a second home. However, there is no consensus about the kind and intensity of attachment that people can attribute to different places. On the one hand, Williams and Kaltenborn (1999: 227, in Williams and Patten 2006: 40) consider second homes a way of re-creating 'the segmented quality of modern identities in the form of separate places for organizing distinct aspects of a fragmented identity. It [second home ownership] narrows and thins out the meaning of each "home" by focusing the meaning of each on a particular segment of life (...).' McIntire et al. (2006: 127) counterpoint suggesting that it is also possible 'to envision an "integrated lifestyle" in which all sectors (home, work and leisure) are merged and the individual feels equally "at home" in each'. Similarly, Giddens (1991, in Frys and Nienaber 2007: 11) refers to the 'cosmopolitan man' that can feel at home in different contexts and places. Such an approach views a second home as an extension of modernity (Williams and Kaltenborn 1999: 227). As Perkins and Thorns (2006: 79) state, 'primary and second homes (...) represent a continuum of experience'.

Many authors consider features of place identity associated with the escape from modern life as the major factor that influences the intensity of bonds to the place of second home. For instance, in his fieldwork in Northern Wisconsin, Stedman (2006: 141) found that escape-related variables were the only significant predictor of place attachment among second home owners. Wolfe

(1952 in Hall and Muller 2004: 12) in his research on holiday cottage use in Canada, and Chaplin (1999) in his study of British second home owners in France, have also found that removal and inversion from everyday urban life are the main attraction of second homes (Hall and Muller 2004: 12). Still, in her field research on second homes in Ireland, Quinn (2004: 119) concludes that 'the escape in question is really an attempt to re-visit and rediscover experiences, times and places that create a sense of connectedness', given that 'memories of places associated with childhood, with family connections (...) create a bank of memories that influence subsequent mobility patterns' (Quinn 2004: 119). Thus, although it is less evoked in second home research, previous ties to places seem be important in explaining place attachment among second home owners too.

Territorial identity features, other than escape-related ones, can also become factors of place attachment. According to Roca and Roca (2007), territorial identity embodies sets of spatial fixes that constitute landscapes, and sets of spatial flows that determine the specific lifestyles of a place, or region; moreover, 'the search for empirical evidence of changing landscape- and lifestyle-related identity features can reveal the sense of belonging to a territory of residence, work and/or leisure, which reflects the levels of satisfaction with the environmental, social, economic, cultural and other conditions provided in that territory' (Roca and Roca 2007: 441). Representations of territorial identity, and hence the type and strength of the sense of place and place attachment, are highly subjective and normally vary between permanent residents and temporary ones, such as second home owners. One of the rare examples of significant relationship between place attachment and territorial identity features, other than the escape-related motivation, can be found in research by McIntire and Pavlovich (2006) in a coastal community in New Zealand, where social and community relations (in fact, a type of spatial flows) are considered a significant variable of place attachment among seasonal residents.

Assessing place attachment among second home owners can also be made from the developmental perspective by perceiving and/or recognizing them as local development stakeholders. However, such an approach is still rare. An important exception is Ankre's (2007) discussion of her research findings on coastal zone planning in the Lulea archipelago in Sweden, where the empirically supported arguments are put forward that land use and development planners and managers could indeed benefit from second home owners' place attachment to a specific territory. Furthermore, Overvag (2009) found that second home owners in Eastern Norway are relevant actors in rural development processes and in the implementation of related public policies, also arguing that their interests, opinions and political rights should be taken into account. Likewise, Muller (2011: 137) claims that second home owners 'form an important aspect of rural communities and in many countries they may be prominent agents of rural change'.

In this chapter, the above mentioned main factors that influence place attachment – previous ties and place identity, particularly escape-related features – are examined from a developmental perspective in the context of the

Oeste Region, an important area of second home expansion in the perimeter of influence of the Lisbon Metropolitan Area (LMA).[1] To this end, findings from the field research survey carried out as a part of the SEGREX project[2] on the second homes phenomenon are brought forward and discussed.

Methodological Aspects

Conceptual and Methodological Framework

Inspired by Gustafson's (2006: 19) comprehensive definition of place attachment as 'bonds between people and place based on affection (emotion, feeling), cognition (thought, knowledge, belief) and practice (action, behaviour)', the conceptual departure point in this study of place attachment among second home owners are the following two complementary notions: 'topophilia', coined by Tuan (1990:2), who defines it as 'the affective bond between people and place, or setting', and 'terraphilia', proposed and defined by Oliveira et al. (2010) as the experience-based affection between people and a territory (specific locality, place, or region) that encourage them to engage in local development intervention. In sum, the concept of place attachment, as it is understood and analysed here, encapsulates affection and action as its two fundamental and determining dimensions.

Since it is assumed that different purposes for the use and the activities (leisure, work, or combinations thereof) performed in or around second homes affect place attachment (Stedman 2006, McIntire et al. 2006), the following broad definition of second home was chosen: 'dwelling for different uses where no member of the household lives permanently'.[3]

The field research survey carried out as part of the SEGREX project addressed second home owners in the Oeste Region regarding, *inter alia*, different aspects

---

1   The Oeste Region is a NUTS III about 100 km NW of Lisbon. It is characterized by a dynamic and diversified economy with numerous small and medium-sized firms, especially in the agro-business and ceramics industry, a competitive market-oriented agriculture and a rising tourism sector. It is known for its rich cultural heritage, both material and immaterial, as well as attractive natural and cultural landscapes – from long beaches with cliffs and the Nature 2000 protected area of the Montejunto Mountain, to bucolic rural parts and charming small towns. Accessibilities are excellent, marked by very good, recently modernized and expanded roads.

2   The research project "SEGREX – Second home expansion and spatial development planning in Portugal", funded by the Portuguese Foundation for Science and Technology, was conducted in the period 2009-2012 jointly by TERCUD - Territory, Culture and Development Research Centre, Universidade Lusófona, and e-GEO Research Centre for Geography and Regional Planning, Universidade Nova de Lisboa, Lisbon.

3   This definition is the same as the one adopted in the SEGREX project, and is actually used in the Portuguese Population and Housing Censuses by the National Institute of Statistics (INE).

of their place attachment. The Oeste Region was chosen as the case study because of the strong presence of second homes (in 2011, 23.9 per cent of total housing stock) and high growth rates of their expansion since the 1990s (45 per cent and 26.6 per cent in 1991–2001 and 2001–2011, respectively).[4] In the first phase of the field research, a questionnaire was applied to the representatives of almost all (i.e., 112 out of 121) of the Parish Councils of the Oeste Region.[5] As the best informed and most directly involved agents of local governance, they were asked to identify localities and assess the main features and trends of the recent (since the 1990s) second home expansion in their parishes. It was revealed that the majority of second home owners are permanent residents in LMA, followed by Portuguese emigrants and foreign citizens. Predictably, these three groups differ in the frequency of use of second homes: the first group spends there mostly weekends; the second group summer vacations; while the latter group prefers long stays. Furthermore, both LMA residents and emigrants are mostly economically active couples with children, while foreigners are mostly retirees. Consequently, the former two groups are also similar in terms of the lower propensity to change second into first home while this tendency is significantly higher among foreign second home owners. Also, the degree of expansion of second homes owned by LMA residents and foreigners is much higher than in the case of emigrants. These findings were used in the sample design for the survey of second homes owners.

Sample Design

The sample of second home owners as potential respondents to the survey questionnaire was designed through a four-fold step-wise method. First, based on the latest available data from the 2001 Population Census, parishes in which the share in total housing stock or growth rate of second homes was above the average for the Oeste Region in the period 1991–2001 were identified. Second, based on the replies obtained from the Parish Councils representatives to the question 'How do you evaluate the degree of expansion of second homes in the 2000s in your parish?', parishes with 'strong or very strong' growth of second homes were defined. Third, parishes that fit both lists of the above mentioned sources of information were singled out. Finally, within these parishes, concrete locations that Parish Council representatives highlighted as those with 'strongest growth' of second homes were further identified. This procedure yielded 24 locations in eight parishes selected for the survey. According to the classification provided by INE, three of these parishes are considered 'rural', three 'rurban' and two 'urban'. Five of them are on the coast and three in the interior (Figure 4.1). All selected locations in coastal parishes are within 2 km from the ocean shorelines.

A total of 163 person-to-person interviews with second home owners were held in all these 24 locations.

---

4   Source: 2011 Population and Housing Censuses.
5   Parish is the smallest territorial administrative unit in Portugal.

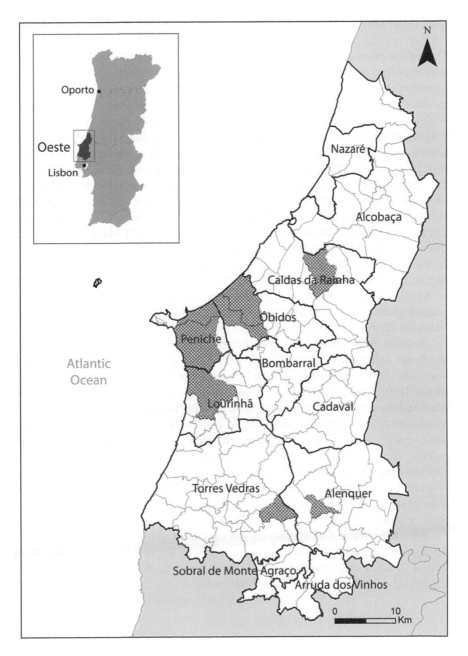

Figure 4.1   Parishes of the Oeste Region covered by the survey.

Questionnaire

One of the main objectives of the questionnaire applied to second home owners was to record the nature of their place attachment. To this end, they were asked to define their relationship with the local institutions and social groups, as well as about their affective bonds with these places (topophilia), their familiarity with local development problems and the ways and means they have contributed and/or intend (or not) to contribute towards solving them (terraphilia). As in several other place attachment studies reported in the literature, the analytical processing of the obtained replies was controlled by variables that could influence the degree of place attachment, such as: (i) socio-demographic characteristics of second home owners, i.e. gender (Jansson and Müller 2004, Stedman 2006, Bjerke et al. 2006), age (Quinn 2004, Stedman 2006, Bjerke et al. 2006, Cottin 2011), education (Stedman 2006), occupation (Cottin 2011), place of birth, place of first residence (Cottin 2011), and type of owner's family; (ii) variables related to the second homes themselves, i.e., characteristics of their location (rural, rurban or urban parish), frequency of use, purpose of use (i.e., leisure, work, or both), and duration of residency (Cottin 2011); (iii) motives for choosing second home location (Quinn 2004, Jansson and Müller 2004, Bjerke et al. 2006, Stedman 2006, Vágner and Fialová 2011).

In order to determine whether the relations of second home owners with the local community can influence their place attachment, they were asked about the degree of relationship with (i) institutions such as Parish council, Town Hall, Bank, Post Office, sport club or association, cultural club or association, civic association, social care institutions, local media, farmers' cooperative, entrepreneurs' association, church; and (ii) local individuals and social groups, such as activists, artisans, artists, retailers, entrepreneurs, local priests, representatives of the parish council, farmers and food producers. Furthermore, some questions were included which addressed second home owners' participation in local cultural events (celebrations, fairs, processions, pilgrimages), local protests and volunteering activities, as well as the nature of their relations with neighbours.

With the intention of identifying the existence and nature of topophilia and terraphilia, second home owners were asked to express their (dis)agreement with a series of statements (Box 4.1), as well as give reasons for and examples (open end questions) corroborating their reply.

Box 4.1   Sample of questions about topophilia and terraphilia in the questionnaire

---

Topophilia:
- Sometimes I feel like I really belong to this place because...
- I like/dislike to spend time here than anywhere else, because...
- Things that I miss when I am away from this place are...

Terraphilia:
- I know enough about the problems facing this place, such as...
- I identify with the goals and values of the local community, such as...
- I am willing to invest, or have already invested my time and effort for the benefit of this place in order to...
- I am willing to spend, or have already spent, my financial resources for the benefit of this place, in order to...
- My contribution to improve this place was recognized, or not, by local authorities, in relation to...
- My contribution to improve this place was recognized, or not, by the local community, in relation to...

Findings

Basic Profile of the Interviewed Second Home Owners

Most respondents (58.9 per cent) are male, almost half are middle aged and more than a third are 65 years or older. The share of those with college education (30.7 per cent) is almost equal to that of those with completed four years of schooling only (31.9 per cent). Along with pensioners (30.1 per cent) and professionals (27 per cent), white and blue collar workers (21.5 per cent) are also significant. More than two thirds of respondents were born and live in a parish other than the one where the second home is located. The primary residence of most respondents is in LMA (52.7 per cent), abroad (20.8 per cent), or in the Oeste Region (18.4 per cent). Accordingly, almost half of the interviewees uses the second home every, or almost every, weekend, while less than a third (30.7 per cent) use it during the summer vacations (30.7 per cent). Most (77.9 per cent) do not intend to change the established frequency of use. A great majority spends time in the second home with their children and/or grandchildren. Almost half started using their second home less than 20 years ago. While second homes are predominantly located in urban and rurban parishes within the perimeter of a town or a village (76 per cent), an overwhelming share of the respondents spend their time almost exclusively in leisure activities (88.3 per cent), mostly walking (81 per cent), hosting friends (66.9 per cent), reading (53.4 per cent), watching TV or a DVD (49.7 per cent), listening to music (40.5 per cent). The respondents spend much less time on the internet (19.6 per cent), on the beach (17.8 per cent) or practicing sports (17.1 per cent).

Reasons for Choosing the Place of Second Home

Among the reasons for choosing the second home location, a wide range of previous bonds to the place prevails (50.9 per cent), specifically 'the place of birth', 'the place of former permanent residence', 'inherited house', 'family ties to the place', 'friends living there'. The share of place identity features related to escape such as 'beautiful nature', 'peace and calm', 'to get away from city life' represents about a quarter of all replies.

Similar findings were reported by researchers in other, quite different European settings. For example, in a field research carried out in an Irish coastal area Quinn obtained fairly similar results: 'for 58% of the sample, strong personal connections with the area had influenced their decision to buy a second home in Wexford (…) thus the decision was clearly not founded simply on general "placeless" factors such as amenity value and a desire to relax' (Quinn 2004: 125). In their field research in an area with a high concentration of cottages in Bohemia in the Czech Republic, Vágner and Fialová (2011) concluded that 30 per cent of second home users had ties with the place from childhood and 40 per cent had friends there. González (2009), in his research on the inner province of Albacete in Spain, and Perri (2010), in her field work on coastal and mountainous places in Southern Italy, showed the importance of previous bonds to the place, particularly through family ties of second home owners who are emigrants or out-migrants. Also, Duval (2004) focused on the importance of transnationalism in second home tourism in the Caribbean Region.

On the other hand, Bjerke et al. (2006), who studied second home owners' motivations in quite a different environment (i.e., mountain areas of Norway), found that escape-related variables prevail among reasons for buying a cabin. Furthermore, Jansson and Müller (2004) showed that escape variables were predominant reasons for choosing a second home in Kvarken, a transfrontier region in northern Finland and Sweden. Also, Williams et al. (2004) highlighted the importance of escape as the main reason for lifestyle migration of British second home owners in southern Europe.

Thus, it could be argued that in territories with considerable emigration or out-migration, previous bonds to the place is a common key factor in explaining the reasons for second home ownership, while escape-related qualities of the place predominate in other geographical contexts.

Major Groups of Second Home Owners

Bivariate analyses of the aforementioned control variables revealed two major groups of second home owners. The much larger group consists of permanent residents in the LMA, born in a parish other than the parish of the second home, who are more educated, work as executives and professionals, and prefer spending weekends in their second home, which is mostly located in an urban or rurban parish. The other group, less frequently represented, is composed of

Portuguese emigrants who were born in the parish of the second home, are less educated, are blue-collar or white-collar workers who commonly spend the summer vacations in the second home, located in a rurban or rural parish.

In both these groups, employed middle aged individuals dominate even though a significant share of second home owners is formed by retirees. Also, neither group intends to change the frequency of use, including converting the second home into primary home. Although among both Portuguese emigrants and LMA permanent residents the 'previous bonds to the place' prevail as the reason for choosing the location of the second home, 'place identity features' and 'location factors' are strongly present in the latter group.

As mentioned earlier, these two groups were also singled out by the representatives of Parish Councils of the Oeste Region as the main groups of second home owners.[6] Unlike the case of North-Western and Northern Europe, the considerable presence of emigrants among second home owners is common in Portugal and also in other Southern European countries such as Spain (González 2009), Italy (Perri 2010) and Greece (Karayiannis et al. 2010). While intense emigration flows marked these countries in the 1960s and 1970s, first and second generations of emigrants still maintain strong links to their place of origin by keeping (and often renewing) their former primary home dwellings, or building new ones.

The findings on how the profile and motivation of the two main groups of second home owners, i.e., LMA permanent residents and Portuguese emigrants, influence topophilia and terraphilia are presented and discussed in the following sections.

Relations between Second Home Owners' and Local Institutions and People, and Participation in Local Activities

The strengths of relationship of second home owners with local institutions do not seem relevant in explaining their attachment to the place. Indeed, a great majority never or very rarely contact with most of the local institutions (Table 4.1 below). It is likely that such weak relationship with local institutions has to do with the fact that most owners spend time in their second home when these institutions are actually closed (weekends) or function in shorter periods for public attendance (summer vacations), as well as because they do spend time mostly in leisure activities.

---

6   The interviews of foreign citizens, the third most numerous group of second home owners, will be carried out in a follow-up phase of the SEGREX project.

Table 4.1   Degree of intensity of relations of second home owners with local institutions.

Local institutions	Intensity of the relations (%) N = 163						
	Always	Frequently	Occasionally	Rarely	Never	Do not know/ No reply	Total
School	0.6	0.6	0.0	4.3	92.0	2.5	100.0
Business association	0.6	0.6	1.2	1.2	93.9	2.5	100.0
Social care institutions	0.6	0.6	1.8	6.7	87.7	2.5	100.0
Media	0.6	0.6	3.1	2.5	90.8	2.5	100.0
Farmers' cooperative	0.6	1.2	2.5	1.2	92.0	2.5	100.0
Civic action club/ association	1.8	1.2	2.5	6.7	84.7	3.1	100.0
Cultural club/ association	5.5	4.9	6.7	11.7	67.5	3.7	100.0
Town Hall	1.2	3.7	17.2	30.7	46.0	1.2	100.0
Sports club/ association	4.9	5.5	10.4	15.3	61.3	2.5	100.0
Post office	1.2	4.9	21.5	16.6	52.7	3.1	100.0
Parish Council	1.2	9.2	17.8	31.3	38.0	2.5	100.0
Church	5.5	5.5	18.4	17.2	49.7	3.7	100.0
Banks	1.2	16.6	17.8	21.5	41.7	1.2	100.0

Source: Questionnaire applied to second home owners, SEGREX project, 2011–2012.

The intensity of the relationship of second home owners with local social groups is also very weak (Table 4.2). They mostly replied that they do not contact members of such groups or did not reply. The exception is contacts with retailers with whom the relationship is considered even friendly, close or cordial by 47.9 per cent of respondents. Most owners (72.4 per cent) maintain close, cordial relations, or established friendships with their neighbours.

Table 4.2   Degree of intensity of relations of second home owners with local social groups.

Local social group	Intensity of the relations (%) N = 163							
	Friendly	Close	Cordial	Formal	Formal and distant	No relation	Do not know/No reply	Total
Activists	1.2	1.2	1.2	0.6	3.1	77.9	14.7	100.0
Artisans	1.2	0.6	3.1	1.8	1.2	77.3	14.7	100.0
Artists	1.2	0.0	3.1	1.8	1.2	79.1	13.5	100.0
Retailers	12.9	17.8	17.2	8.6	16.0	20.9	6.7	100.0
Businessmen	4.9	6.1	9.8	5.5	11.7	52.1	9.8	100.0
Priest	0.6	3.7	8.0	6.7	23.9	49.7	7.4	100.0
Politicians	6.7	11.0	7.4	6.1	11.7	49.1	8.0	100.0
Farmers	6.7	9.8	11.0	9.8	5.5	47.2	9.8	100.0
Food producers	6.7	7.4	11.7	6.7	4.9	52.1	10.4	100.0
Neighbours	40.5	20.2	11.7	5.5	11.0	7.4	3.7	100.0

Source: Questionnaire applied to second home owners, SEGREX project, 2011–2012.

Concerning the participation in local activities, second home owners never or rarely participate in local volunteering activities or in local protest actions. On the other hand, 57.7 per cent of them occasionally, frequently or always take part in local cultural events (Table 4.3).

Table 4.3   Degree of participation of second home owners in local events.

Local events	Degree of participation (%) N = 163						
	Always	Frequently	Occasionally	Rarely	Never	Do not Know/ No reply	Total
Cultural events	9.2	16.0	32.5	16.6	23.3	2.5	100.0
Protest actions	1.2	0.6	4.9	4.3	85.9	3.1	100.0
Volunteering activities	1.2	0.6	1.2	3.1	90.8	3.1	100.0

Source: Questionnaire applied to second home owners, SEGREX project, 2011–2012.

Compared to LMA permanent residents, emigrants as second home owners have stronger relations with neighbours and local groups such as local politicians, farmers, retailers, the local priest and neighbours. They also participate more intensely in cultural events. It is safe to say that previous bonds to the place, which are stronger among emigrants, play an important role in these relations.

It seems that, in generalized terms, the intensity of the relations between the second home owners and the local community is not important in determining their level of place attachment. In fact, only relations with neighbours and participation in cultural events, which are also related to topophilia, proved to be intense, while relations related to terraphilia are insignificant. Stedman (2006: 141–2) also found from his field work in Northern Wisconsin that most social relations and participation in local activities are not significantly correlated to place attachment among second home owners. On the other hand, McHugh and Mings (1996, in Stedman 2006: 133) registered a quite strong participation of second home owners in the USA 'Sunbelt metropolis' in volunteerism, while Green et al. (1996, in Stedman 2006: 133) evaluated social relations with the rural communities, also in the USA, as 'inclusive'.

Topophilia among Second Home Owners

The sense of belonging to the place of the second home is quite strong among all interviewees: 78.2 per cent confirmed that 'Sometimes I feel like I really belong to this place because...'. Furthermore, it was found that the motives that prevail correspond to spatial flow-related dimensions of territorial identity, such as social environment (38.6 per cent), as illustrated by the following replies: 'I have already spent quite a lot of time in this place', 'I like the people', 'I feel well here', 'I am happy here', 'I like the milieu'; 'I feel integrated', 'I am always welcomed here'. The previous bonds to the place are also often mentioned (25.8 per cent): 'I was born in this place/parish'; 'I have family here'. Reasons related to spatial fixes, such as nature's place amenities (climate, landscape, natural environment), are referred by only 7.3 per cent of respondents.

While, as expected, previous bonds to the place predominate among emigrants, the social environment was chosen as the main reason by the majority of LMA permanent residents. Also, the more recent the year respondents started using the second home, the stronger the feeling of not belonging to the place. Such feeling is less intense in urban than in rurban or rural parishes.

Among the respondents who said that they enjoyed spending time in the place of second home, which is nearly all (90.8 per cent), most (51.5 per cent) corroborated this with motives linked to territorial identity features, particularly the escape-related ones ('because it is peaceful and quiet', 'because of the direct contact with nature'). More than a quarter of them (27 per cent) emphasised the previous bonds to the place ('because I go back to my childhood', 'because I was born here', 'because my family's roots are here'). And, again, the two main groups reacted quite differently: emigrants pointed to the previous bonds to the place as the main motive to enjoy spending time in the place of second home; LMA permanent residents considered escape-related place identity features to be the most important reason for enjoying spending time there. (Figure 4.2 below).

When asked if they miss the place of second home when they are away from it, the overwhelming majority (86.5 per cent) replied in the affirmative. Among the explanations for such replies, place identity features prevail as the key motives, particularly those associated with escape, such as 'calmness, tranquillity, quietness' (34.4 per cent), as well as natural features, such as landscape, beach, sea, countryside, pure air (20.2 per cent), and social relations (18.4 per cent), most often embodied in the reply 'spending time with friends and family'. Emigrants most frequently referred to social relations, while LMA permanent residents evoked escape features as the most important reason for missing the place of second residence when away (Figure 4.3 below).

As persistently reported in the literature on second homes, factors related to escape features of place identity (Coppock 1977, Chaplin 1999, Williams and Kaltenborn 1999, Williams et al. 2004, Stedman 2006, McIntire et al. 2006) and previous bonds to the place (Quinn 2004, Muller 2004, Perri 2010, González 2009) also predominate in this survey's records on topophilia among the owners of second homes in the Oeste Region. The main type of second home owner, LMA permanent resident, uses the second home as an escape from the stressful urban life during weekends, while the second main type, Portuguese emigrants, looks forward to spending the summer vacation there, resuming and/ or strengthening ties with the family and friends who remained in their place of origin, as well as participating in cultural events.

'Previous bonds with the place' has not been used so frequently as a place attachment variable in the literature probably because most of the research on place attachment has been developed in countries of Scandinavia and Anglo-America where emigration and out-migration is not a contemporary socio-spatial phenomenon. On the other hand, in countries of Southern Europe, where these forms of human mobility and their effects on settlement-related and other place identity features are still significant, research on place attachment among second home owners has been almost absent for a long time.

It is also worth mentioning here that, besides escape-related features, other spatial flow dimensions of place identity, particularly changing dynamics of social relations, are also important in determining place attachment both among second home owners who are emigrants and LMA permanent residents. As regards spatial fixes, although the Oeste Region has been known for its rich and diverse cultural and natural amenities, in this survey these elements of place identity showed lower levels of significance. In sum, these findings reinforce the importance of affective bonds in explaining place attachment.

Figure 4.2 *'Refúgio da Filipa'* ('Filipa's escape'), a second home owned by a resident from Lisbon Metropolitan Area (Óbidos County, Oeste Region).

Figure 4.3 Second homes on the Atlantic shorelines (Óbidos County, Oeste Region), mostly owned by residents from Lisbon Metropolitan Area.

Terraphilia among Second Home Owners

Terraphilia is much less intense than topophilia among second home owners in the Oeste Region. Indeed, more than two thirds (67.5 per cent) of the respondents acknowledged that they do not know enough about the problems facing the place of their second home. Among the minority that confirmed their familiarity with the local situation, socio-economic problems surpass the environmental ones. Although LMA permanent residents and emigrants do not differ in terms of their knowledge of such problems, the former group is more concerned with environmental issues while the latter referred more to socio-economic concerns. Only in rural parishes did most owners (61.5 per cent) reply that they are aware of the local problems, particularly the environmental ones.

When subsequently questioned about terraphilia, a sizable minority of owners replied 'yes'. Indeed, 41.1 per cent replied positively when asked if they identify with the goals and values of the local community. Spatial flows of place identity such as 'friendship ties' (12.3 per cent) and 'social features of the local community' (11.1 per cent) were the most frequently evoked values.

LMA residents and emigrants differ considerably in their attitudes and opinions. Positive references in the reactions prevailed among the latter, particularly regarding friendship ties, while most of LMA permanent residents gave no reply, or claimed that they do not know, or just did not identify themselves with the goals and values of the local community. Those few LMA residents who replied positively highlighted social features of the local community as the most important value. And, here again, in the rural parishes the majority of second home owners replied positively, most frequently mentioning friendship ties.

A considerable minority (44.8 per cent) of interviewed second home owners have already invested, or is willing to invest their time and effort for the benefit of the place of the second home, particularly by engaging in social activities (20 per cent) such as 'helping to organize events', 'helping people', 'maintaining or enhancing the work of associations'. Accordingly, almost the same number of respondents (44.2 per cent) confirmed that they had already spent financial resources, or declared willingness to do so, mainly for the support of social activities (18.4 per cent) and also, to a much lesser extent, for helping to maintain or preserve the built heritage and the environment (6.7 per cent). The willingness to spend time and money for the benefit of the place is more present among emigrants than among LMA permanent residents. Also, among second home owners in rural parishes such willingness is stronger than among those in rurban or urban environments. Yet, only about a fifth of the respondents considered that their contribution for the improvement of the place was recognised by the local people and local authorities.

Thus, it could be argued that previous bonds to the place influence terraphilia, that is, the need for intervention that could encourage local development and that is much stronger among Portuguese emigrants than among LMA permanent residents. Also, regardless of the origin of the second home owner, terraphilia seems to be much more pronounced in settings of rural than rurban and urban parishes.

In Lieu of Conclusion

The phenomenon of second home expansion has induced significant changes in the environmental, social, economic and cultural identity features of many places and regions in Portugal, and has indeed become an important challenge to the sustainability of spatial organization and to development planning and management at all levels, from local to national. However, its comprehensive scientific interpretations have been manifestly lacking (Roca et al. 2010). The research findings presented in this chapter are an initial, indeed pioneering contribution to a better understanding of this phenomenon at regional level, and thus a tribute to the search for answers to the famous Coppock's dilemma of second homes as 'curse or blessing' (1977).

It is worth reporting here that in the initial stages of this field research in the Oeste Region, the representatives of the Parish Councils transmitted a great deal of optimism about actual and potential positive economic, social, cultural and environmental impacts of second home expansion, particularly in the rural areas, ranging from the introduction of cosmopolitan spirit and creation of employment, to modernisation of communal infrastructure and services, and socio-demographic renewal (Roca et al. 2011). This, however, does not match the findings on second home owners' terraphilia, or keenness to contribute to local development. In fact, second home owners – though they do have a strong sense of topophilia – have a very weak sense of terraphilia. Also, their relations with local institutions and local groups have little, if any, intensity. Thus, how the 'potentially positive' impacts of second home expansion could become a reality, and second home owners grow to be local development stakeholders, has yet to be studied, especially in the rural areas where their sense of terraphilia is stronger.

The sense of topophilia among LMA permanent residents as the major group of all second home owners in the Oeste Region is mostly explained with spatial flows of place identity, namely social environment and social relations. However, most of them see the second home location as a place to escape, and actually do not identify themselves with the values and goals of the local community. Therefore, local authorities that are keen on having positive impacts should foster development actions that could engage second home owners by making them feel part of the local community, thus strengthening terraphilia among them.

On the other hand, Portuguese emigrants, the second but smaller group of second home owners, have, not only a stronger sense of topophilia and stronger relations with the local community, but also a firmer sense of terraphilia, most probably because of their previous bonds to the place of the second home. However, as amply evidenced in other parts of Portugal as well as in other Southern European countries, the probability that the emigrants would return to the place of second home is low, and the probability of topophilia weakening among their children and grandchildren is high.

To reverse such trends, local authorities should find ways and means of making the already important engagement and financial contribution of emigrants to social activities be more acknowledged by the local community.

Although none of the two major groups of second home owners intends to become permanent residents, the chances that their stays become longer when they retire are high. Such trend could not only enhance their role as development stakeholders but also contribute to the alleviation of negative demographic trends, while adding to the local economy as consumers of goods and services, both communal and personal.

Strengthening terraphilia among second home owners is a *sine qua non* condition for their joining the ranks of active development stakeholders. The firmer sense of belonging to the place of their second home, the higher the chances for owners' active participation in the implementation of development strategies aimed at enhancing the value of local and regional material and immaterial identity features by, for example, promoting environmental and socio-cultural awareness in the local community, encouraging the protection of natural and cultural heritage, fostering the efficiency and effectiveness of local institutions, etc.

To this end further research is needed about how the existing and potential second home owners (i) assess the 'attractiveness' of places and regions they wish to escape to, (ii) identify elements of territorial attractiveness that are vanishing, evaluate their relevance and promote their revalorisation, (iii) introduce new elements of territorial identity, and (iv) contribute to the implementation of public policies and investment projects that correspond to both their own and the local community's environmental, economic and cultural specificities and needs. Such new insights could lay forth firmer grounds for a comprehensive interpretation of the above mentioned Coppock's dilemma and for the search for appropriate local development solutions.

References

Ankre, R. 2007. *Understanding the visitor: A prerequisite for coastal zone planning.* [Online: Licentiate Dissertation Series No. 2007:09, Blekinge Institute of Technology, School of Technoculture, Humanities and Planning]. Available at: http://www.bth.se/fou/forskinfo.nsf/0/c7219801e0082a2dc125737800311bdd/$F ILE/Ankre_lic.pdf [accessed: 27 May 2012].

Bjerke, T., Kaltenborn, B.P., Vitterso, J. 2006. Cabin life: Restorative and affective aspects, in *Multiple Dwelling and Tourism: Negotiating Place, Home and Identity*, edited by N. McIntyre, D. Williams and K. McHugh. Cambridge: CABI, 87–102.

Chaplin, D. 1999. Consuming work/productive leisure: The consumption patterns of second home environments. *Leisure Studies*, 18, 41–55.

Coppock, J.T. (ed.). 1977. *Second homes, Curse or Blessing?* Oxford: Pergamon Press.

Cottin, I. 2011. The spatial and socio cultural impacts of second home development: A case study on Franschhoek South. [Online: Master's Thesis 2011, Ultrecht: International Development Studies, Utrecht University] Available at: http:// igitur-archive.library.uu.nl/student-theses/2011-0831-200740/Masterthesis%20 Ine%20Cottyn.pdf [accessed: 27 May 2012].

Duval, D.T. 2004. Mobile migrants: Travel to second homes, in *Tourism, mobility and second homes: between elite landscape and common ground*, edited by C.M. Hall and D.K. Müller. Clevedon: Channel View Publications, 87–96.

Frys, W. and Nienaber, B. (eds.). 2007. Developing Europe's rural regions in the era of globalization: An interpretative model for better anticipating and responding to challenges for regional development in an evolving international context. [Online: DERREG: WP2: International mobility and migration of rural population, Deliverable D 2.4: Work Package Summary Report]. Available at: http://www. derreg.eu/system/files/derreg_bull2_FINAL.pdf [accessed 27 May 2012].

Giddens, A. 1991. *Modernity and self-identity: Self and society in the late modern age*. Stanford: Stanford University Press.

González, J.A.G. 2009. El turismo de retorno: modalidad oculta del turismo residencial, in *Turismo, urbanización y estilos de vida: las nuevas formas de movilidad residencial*, edited by T. Mazón, R. Huerte and A. Mantecón, Barcelona: Icaria, 351–65.

Green, G.P., Marcoullier, D., Deller, S., Erkkila, D. and Sumathi, N.R. 1996. Local dependency, land use attitudes and economic development: Comparisons between seasonal and permanent residents. *Rural Sociology*, 61, 427–45.

Gustafson, P. 2006. Place attachment and mobility, in *Multiple Dwelling and Tourism: Negotiating Place, Home and Identity*, edited by N. McIntyre, D. Williams and K. McHugh. Cambridge: CABI, 17–31.

Gustafson, P. 2009. Mobility and territorial belonging. *Environment and Behavior*, 41, 490–508.

Hall, C.M. and Müller, D.K. 2004. Introduction: Second homes, curse or blessing? Revisited, in *Tourism, mobility and second homes: between elite landscape and common ground*, edited by C.M. Hall and D.K. Müller. Clevedon: Channel View Publications, 3–14.

Jansson, B. and Müller, D.K. 2004. Second home plans among second home owners in Northern Europe's periphery, in *Tourism, mobility and second homes: between elite landscape and common ground*, edited by C.M. Hall and D.K. Müller. Clevedon: Channel View Publications, 261–72.

Karayiannis, O., Iakovidou, O. and Tsartas, P. 2010. Il fenomeno dell'abitazione secondaria in Grecia e suoi rapporti con il turismo, in *Il Turismo Residenziale: nuovi stili di vita e di rezidenzialità, governance del teritorio e sviluppo sostenibile del turismo*, edited by T. Romita. Milano: Editore Franco Angeli, 94–113.

McHugh, K.E. and Mings, R. 1996. On the road again: seasonal migrants to a Sunbelt metropolis. *Urban Geography*, 12, 1–8.

McIntire, N. and Pavlovich, K. 2006. Changing places: Amenity coastal communities in transition, in *Tourism, mobility and second homes: between elite landscape and common ground*, edited by C.M. Hall and D.K. Müller. Clevedon: Channel View Publications, 239–61.

McIntire, N., Roggenbuck, J.W. and Williams, D.R. 2006. Home and away: revisiting 'escape' in the context of second homes, in *Multiple Dwelling and Tourism:*

Negotiating Place, Home and Identity, edited by N. McIntyre, D. Williams and K. McHugh. Cambridge: CABI, 114–28.

Müller, D.K. 2011. Second homes in rural areas: Reflections on a troubled history, *Norsk Geografisk Tidsskrift – Norwegian Journal of Geography*, 65, 137–43.

Oliveira, J., Roca, Z. and Leitão, N. 2010. Territorial identity and development: From topophilia to terraphilia, *Land Use Policy*, 27, 801–14.

Overvag, K. 2009. *Second Homes in Eastern Norway: From marginal land to commodity*. [Online: Thesis for the degree of Philosophiae Doctor, Trondheim: Norwegian University of Science and Technology, Faculty of Social Sciences and Technology Management, Department of Geography]. Avaiable at: http://ntnu.diva-portal.org/smash/get/../FULLTEXT02 [accessed: 27 May 2012].

Perkins, H.C. and Thorns, D.C. 2006. Home away from home: the primary/second home relationship, in *Multiple Dwelling and Tourism: Negotiating Place, Home and Identity*, edited by N. McIntyre, D. Williams and K. McHugh. Cambridge: CABI, 67–81.

Perri, A. 2010. Alcune riflessioni sul turismo residenziale delle radici in, *Il Turismo Residenziale: nuovi stili di vita e di rezidenzialità, governance del teritorio e sviluppo sostenibile del turismo*, edited by T. Romita. Milano: Editore Franco Angeli, 145–56.

Quinn, B. 2004. Dwelling through multiple places: a case study of second home ownership in Ireland, in *Tourism, mobility and second homes: between elite landscape and common ground*, edited by C.M. Hall and D.K. Müller. Clevedon: Channel View Publications, 113–30.

Roca, M.N., Oliveira, J.A. and Roca, Z. 2010. Seconda casa i turismo della seconda casa, in *Il Turismo Residenziale: nuovi stili di vita e di rezidenzialità, governance del teritorio e sviluppo sostenibile del turismo*, edited by T. Romita. Milano: Editore Franco Angeli, 111–30.

Roca, M.N., Roca, Z. and Oliveira, J. 2011. Features and Impacts of Second Homes Expansion: The Case of the Oeste Region, Portugal. *Geographical Bulletin*, 73, 111–28.

Roca, Z. and Roca, M.N.O. 2007. Affirmation of territorial identity: a development policy issue. *Land Use Policy*, 24, 434–442.

Stedman, R.C. 2006. Places of Escape: Second-home Meanings in Northern Wisconsin, USA, in *Multiple Dwelling and Tourism: Negotiating Place, Home and Identity*, edited by N. McIntyre, D. Williams and K. McHugh. Cambridge: CABI, 129–44.

Tuan, Y.F. 1990. *Topophilia: A study of environmental perception attitudes and values*. New York: Columbia University Press/Morningside Edition.

Vágner, J. and Fialóva, D. 2011. Impacts of second home tourism on shaping regional identity in the regions with significant recreational function. [Online: *Book of proceedings vol. I – International Conference on Tourism and Management Studies* – Algarve 2011, 285–94]. Available at: http://tmstudies.net/index.php/ectms/article/../250 [accessed: 27 May 2012].

Williams D.R. and Kaltenborn, B.P. 1999. Leisure places and modernity: the use and meaning of recreational cottages in Norway and the U.S.A, in *Leisure/Tourism Geographies: Practices and Geographical Knowledge*, edited by D. Crouch. London: Routledge, 214–30.

Williams D.R. and Patten Van R. 2006. Home and away? Creating identities and sustaining places in a multi-centred world, in *Multiple Dwelling and Tourism: Negotiating Place, Home and Identity*, edited by N. McIntyre, D. Williams and K. McHugh. Cambridge: CABI, 32–50.

Wolfe, R.I. 1952. Wasaga Beach: The divorce from the geographic environment. *Canadian Geographer*, 2, 57–65.

PART II
Back to Nature: Between Urban Sprawl and Countryside Idyll

PART II

Chapter 5

The Multiplicity of Second Home Development in the Russian Federation: A Case of 'Seasonal Suburbanization'?

Tatyana Nefedova and Judith Pallot

Visitors to the Russian Federation cannot but be struck by the mass of small wooden shed-like houses standing cheek-by-jowl, each on its own plot of land, that encircle that country's towns. On summer weekend evenings when their owners can be seen hard at work weeding, tending fruit bushes or relaxing playing cards on the shed-house's porch, the scene might be mistaken for the spontaneous suburbs of the Global South. But in Russia, rather than being the destination of recent in-migrants to the city, these makeshift suburbs are evidence of a reverse movement, from the town into the countryside; they are a very twenty-first century version of that traditional institution, the 'dacha', which stands alongside sputnik, vodka and the 'babushka' in the popular iconography of Russia. The dacha is a manifestation of an historical relationship between town and countryside stretching back to the eighteenth century when the Tsar Peter-the-Great rewarded the service classes with small estates – the noun dacha derives from the verb *davat'*, 'to give'. Nobles and boyars would retreat in the summer months to their new country patrimonies to supervise their land and serfs thus beginning the tradition of the annual exodus from the town, which continues to the present day. As has been documented by historians, the use to which the dacha is put - its physical form and the meanings vested in it - have changed over time (Lovell 2003, Khauke 1960).

In this chapter we discuss the dacha as a possible agent of sub-urbanisation in contemporary Russia. It will become clear that the dacha does not fit neatly into the categories used to analyse settlement formation in the modern world, even though it shares many of the features of post-industrial second-home development in Western Europe and North America. Following the path first marked out in 1965 by Pahl, with his conceptualisation of urban to rural reverse migration as creating 'urbs' in 'rure', recent second homes development in Russia has been variously theorised by Golubchikov and Phelps (2009) and Lovell (2003) as 'seasonal suburbanisation', 'quasi-suburbanisation' or 'ex-urbanisation'. Stryuk and Angelica (2007: 234), in contrast, using data from a survey of seven cities, are prepared to view it as suburbanisation proper; they argue that the temporary or seasonal occupation of rural houses by urbanites is the first stage in a process leading to permanent settlement and, therefore, to the creation of suburbia.

Similarly, Mason and Nigmatullina (2011), drawing attention to how a relaxation of planning controls has allowed the city to spill over into the green belt, have identified the conversion of second-homes into full-time residences as an example of delayed suburbanisation, alongside the proliferation of satellite cities on the urban fringe and emergence of élite housing estates, such as on the Rublevo-Uspenskoe highway near Moscow, to which we might also add the re-drawing of the city boundary to absorb extra-housing estates and dacha clusters. In West European countries the purchase of the second-home has been conceptualised as an 'ideal-type life course' that essentially replicates the process of gentrification in the city; hence, the identification of the second-home development as rural gentrification or the 'class colonisation' of the countryside (Paris 2009: 297). The stages of second-home development – starting with the pioneer renovators, continuing through the revitalisation of areas and purchase of existing homes from former lower-income residents, through to fully commercial involvement – is certainly a process that can be observed in the outskirts of Moscow. Furthermore, it is extending wave-like outwards ever further into the rural periphery stretching to 300 kilometres from the city.

But there is another way of viewing the proliferation of seasonally-occupied houses in the ownership of urban dwellers in rural Russia which, inverting Pahl's formulation, sees it less as a process of the urban colonisation of the countryside than as evidence of the 'peasantisation', 'ruralisation' or 'provincialisation' of city life (Lovell 2003). As will be shown below, much of the movement into the countryside has been associated with the use of rural resources for household reproduction in the form of food production. In the crisis years immediately after the collapse of the Soviet Union, when there was a systemic breakdown in the labour market and welfare systems, urban dwellers used plots of land on the edge of cities to grow food for subsistence and this was a driving force behind the proliferation of small allotments around Russia's towns and cities in the 1990s. Twenty years on, when the need for an allotment to grow food is less urgent, the weekend retreat to the countryside in the summer remains an essential feature of the Russian urban way-of-life.

One of the problems with interrogating these various positions on the dacha is the paucity of relevant statistical data. Local authorities in rural Russia record the number of houses belonging to *dachniki* – dacha dwellers – that are of urban registered dwellers who own or rent a house in their authority, and this can be compared with the houses occupied by rural registered households. However, these figures are not collated at national or regional level and they do not, in any event, record whether these houses are occupied permanently or temporarily. Many urban residents who might spend the whole of the year in their rural house retain an apartment in town and they remain classified, therefore, as urban residents. Paris's (2009: 294) observation in relation to second homes in Western Europe that the idea of a first or second 'home' is problematic – since households do not necessarily identify a single place as their primarily home, and patterns of occupation of dwellings can change during a life course – is valid for the Russian case as well.

National censuses that use terms like 'normal place of residence' or that, as in Russia, register residency in one place only, are out-of-kilter with the way that living patterns have developed in the post-industrial world. For these reasons, the most reliable approach to analyzing second home development is by way of detailed case studies that draw on local statistics and household surveys. This chapter uses data collected in Moscow and adjacent regions over a period of two decades in a series of studies (Ioffe and Nefedova 2000, Nefedova 2003, Pallot and Nefedova 2007, Ioffe et al. 2006, Nefedova 2008, 2011a-b, Makhrova et al. 2008).

Urbanisation and Rural Depopulation in Twentieth and Twenty-first Century Russia

Before embarking upon a discussion of the role of dachas in the changing settlement geography in the Russian Federation, we need to recall the context within which the shift in the balance between town and countryside has been taking place in twenty-first-century Russia. Whereas in most post-industrial countries, the dominant movement has been settlement 'down-sizing' with the net migration balance shifting in favour of rural areas, in Russia second home development has been taking place against a backdrop of a rural-urban shift that has continued unabated since the third decade of the twentieth century and that has depopulated vast swathes of the rural periphery. Rural outmigration was, in part, a normal consequence of the economic development of a country which in 1917 was still predominantly peasant, but the intensity with which it took place was the product of the communist state's policy to transfer agriculture onto an industrialised footing through forced collectivisation. Peasants left the countryside destined for factories in the city in their droves where their labour was used to drive through the USSR's industrial revolution. In the post Second World War period the failure of the state to bridge the gap between urban and rural living standards meant that the rural out-migration did not abate. Rural out-migration at this time selected the young and energetic, and was particularly marked in central and northern regions, where every year saw a rise in the number of deserted villages.

There have been two notable reversals of the rural to urban stream of migrants coinciding with periods of extreme urban crisis; the first was during the Civil War in 1917–1918, when people left cities for the greater safety of the countryside, and the second was more recently, immediately after the USSR's collapse, when high inflation, unemployment and delays in paying wages drove people out of the cities to secure subsistence by reverting to self-provisioning on household plots. Also at various times in the twentieth century Soviet leaders, concerned with the rural exodus, promoted 'return migration' policies that had moderate and localised success. In the last Soviet decade, for example, incentives were offered to urban dwellers to take up farming on surplus land, whilst policies somewhat similar to the 'key settlement' approach in the UK attempted to stem rural out-migration by concentrating rural dwellers in settlements provided with superior services

(Pallot 1979, 1990). Stabilising the rural population was also implicit in the early agricultural reforms of Boris Yeltsin's first post-Soviet government and in policies to direct 'return migrants' from the post-Soviet republics to sparsely settled rural regions. In the past decade, such rural re-settlement schemes, in combination with the re-classification of so-called urban-type rural settlements (*poselki gorodskogo tipa*) as rural villages (*sel'skie sela*), has resulted in an absolute growth in the rural population, even though natural increase rates remain negative in most rural regions outside the South. Notwithstanding this modest rural population growth, the post-Soviet period has seen the resumption of urbanisation now under the influence of market forces.

Moscow, and the Moscow region, has been a particular magnet for in-migration in the 2000s, as its employment opportunities and wage levels have widened the gap between it and crisis-stricken provincial small towns and rural areas. The capital region has also seen an expansion of commuting not just involving the dormitory settlements encircling the city but also reflecting expanded car ownership now, too, from more distant places, such as Smolensk and Vladimir which neighbour Moscow oblast, and further afield, such as Kostroma and Vologda to the north, as well as the Volga oblasts to the east and south-east. Half the positive migration balance between Russia's regions is accounted for by Moscow and Moscow oblast. The pull of Moscow ripples outwards into the oblast and beyond in step-wise fashion. Thus, as inhabitants of the oblast migrate into the city, their place is occupied by people from further afield, who position themselves for a future onward move to the city proper. There are limitations on how far this process can progress at the present time, since the city tries to restrict its physical growth by Soviet-era instruments, such as the living permit, *propiska*. Rising rents and high property values place a move into the city out of the reach of many aspiring in-migrants, with the result that much potential city growth is taking place beyond the metropolitan boundaries in the peri- or ex-urban zone. Whereas in the 1990s the net migration balance in Moscow city exceeded by three times the oblast's, in the 2000s the pattern has reversed with net migration in the oblast (not including Moscow city) now exceeding Moscow's (in 2010, the oblast's migration balance was 108/1000, compared with 75/1000 for the city [*Regiony Rossii* 2011]). As it pulls ahead of other regions in Russia, the attraction of Moscow city and oblast is unlikely to abate in the future. The existence of people who are bucking the trend by moving out of the city may indicate the early stages of counter-urbanisation processes. Whatever the longer term prospects, current urban to rural flows are embedded within a much more widespread and culturally-mediated process which, characteristically, combines an apartment in the city with ownership of a dacha in rural areas lying beyond its boundaries, involving an array of different ways of combining work, leisure and home life.

The Dacha's Varied Form in Contemporary Russian Countryside

The 'Classic Dacha'

The 'classic dacha' is the oldest form of Russian seasonally-occupied second home. From its aristocratic roots in the eighteenth century, the dacha had been democratised by the beginning of the nineteenth century, becoming the universal aspiration for well-to-do sections of Petersburg society (Lovell 2003:58–65). The dacha was a place to which the middle classes repaired in the summer months to drink tea from a samovar, engage in languorous conversations and entertain neighbours. *Dachniki* or summerfolk thus came to be associated with a distinctive lifestyle that was centred on leisure and domesticity (Swift 2004: 527). By the century's end the myth of the dacha was firmly established and the genteel way of life with which it was associated was quintessentially middle class. In 1917 there were approximately 20,000 dachas in the environs of Moscow (Khauke 1960: 15) The dacha myth survived the 1917 Revolution and resurfaced when the communist state made available new plots of land outside Moscow to leading organisations and institutions, such as the Communist Party, Academy of Sciences and Union of Writers, in order to construct dacha settlements for members of their elite workforce. Later in the twentieth century these were transferred into the personal ownership of the users and, thereafter, they were passed down to successive generations (Figure 5.1 below). In 1989, prohibitions on the purchase and sale of second homes were raised (even though prior to this there had been a thriving shadow market in dachas). Under Soviet law all land was owned by the state and this included the plots on which dachas stood. In 1993, the post-Soviet government changed the law to allow private ownership of land plots as well as the buildings on them (although the right to buy and sell agricultural land remained prohibited until 2003).

During Soviet times, the classic dacha was most commonly found in the immediate environs of the capitals, Moscow and St Petersburg, where they remain a feature of the rural landscape to the present day. Soviet-era dachas are modest detached one- or two-storied wooden buildings with two to three rooms. Normally, they have only the most basic services – for example, few have running water – and they are suitable for use only in the summer months when they are used for leisure and relaxation. A common pattern is for pensioners to spend all the summer months at the dacha looking after young grandchildren with the working parents visiting for longer and shorter periods. There were more elaborate dachas built for top Party officials and other members of the apex of the Soviet elite that were sufficiently well serviced to allow winter living, but they were in the minority. In the latter decades of Soviet power, demand for second homes outstripped supply which was met by the emergence of a vigorous rental market in ordinary rural houses, especially in places nearest to the cities. This marked the first incursion of *dachniki* into villages proper and it resulted in the temporary intensification of rural densities in the environs of the cities in the summer months.

Figure 5.1   **A classic dacha from the Soviet period.**

Source: T. Nefedova.

Reflecting their primarily recreational purpose, classic dachas do not have a large amount of land attached; normally, the house and garden is less than 12–50 *sotok* (where one *sotka* is equivalent to 100 square metres, or to 1/100 ha).This reflects the fact that they were never intended for growing food, in which respect they are different from other types of dacha described below. Most commonly, land attached to dachas is left to wildflowers, bushes and trees, so that together they create a picturesque landscape, representing the rural idyll in urban residents' imaginary.

Garden Settlements

Garden settlements are the most widespread form of agro-recreational land in Russia. Like the classic dacha, they are not a new phenomenon but they appeared later, in the second half of the twentieth century. They first appeared in 1949 when a law was promulgated allowing for the formation of so-called 'collective gardens and allotments'. Under this law's provision, small plots of land equivalent to 600–800 m² were carved out of land in the use of urban or rural soviets (councils) or industrial enterprises and the state land fund to be re-allocated to rank-and-file workers for domestic food production. Some of this land was within urban boundaries but most was extra-urban located just beyond the city limits and within a suburban train ride. Initially in the 1950s, members of allotment and garden collectives were not permitted to use the sheds they built on their land for overnight accommodation but this changed in 1967–69 when the Soviet Union introduced a two-day weekend; in response to popular pressure garden and plot holders the prohibition on transforming garden sheds into accommodation was relaxed. However, building specifications put in place were strict aimed at discouraging long stays and they remained tight until 1990, when there was a further relaxation that led to a veritable flowering of new construction on allotments. The result has been the creation of what one Russian geographer refers to as 'slum cities' (*trushshobnye goroda*) (Rodoman 1993) that, lacking regulation, are notable for the absence of any amenity, sanitation and fire protection and the manifestation of various social and ecological problems.

Garden settlements with their small but ever more elaborately decorated house-sheds are a pale reflection of the classic dacha but numerically they far outweigh them (Figure 5.2 below). Already in 1950 there were 40,000 members of garden and allotment collectives in the USSR and the number grew to three million by 1970. In 1990, just before the USSR's collapse approximately 8.5 million households were members of garden and allotment collectives and an equivalent number were on waiting lists to join. In the first years of the 1990s, garden and allotment collectives grew again as urban workers sought to cushion the effects of economic crisis by growing their own food. By 2000 the number of collective members had nearly doubled to 14 million, which was equivalent to the total number of rural households.

Figure 5.2   A dacha in a garden collective from the Soviet period.
Source: T. Nefedova.

The land allocated to garden and allotment collectives is generally of poor quality for growing, and an enormous effort and some capital outlay is needed to improve its fertility. Garden collective members spend long hours cultivating the soil and, once the growing season begins, nurturing young plants to produce a good harvest of vegetables, potatoes, berries and fruit. In the early years of the 1990s this effort may genuinely had been a necessary survival strategy for some households (Pine and Bridger 1998) but studies after this have shown that when all the costs involved in producing food on such plots are added up (including the cost of travel), in many cases it would be more cost effective for the owners to purchase the equivalent produce (Clarke 2002). In that they are today as much about a way of life as survival and provided a summer retreat, the house-shed of collective gardens can be considered a type of dacha.

Vegetable Allotments

Vegetable allotments are more tenuously linked to the classic dacha than collective gardens. Officially, they differ from the latter in that it is absolutely prohibited to build a dwelling on them and they cannot be privately owned. Russian vegetable allotments are analogous to the sort of local-authority owned allotments that exist in rural and urban areas in the UK allocated to residents on a first-come-first served basis. They would not normally qualify for inclusion in a study of second homes were it not for the fact that in Russia there is often a gap between what is permitted by law and what happens in reality. In post-Soviet Russia vegetable allotments have turned out to be one of the multiple sites of summer house development, with the authorities unable or unwilling to enforce the no staying overnight rule (Figure 5.3). Like garden settlements, allotments in the 1990s were sometimes life-savers for the inhabitants of small and medium sized towns whose industrial base was destroyed by the transition to the market economy, but in 2000s their recreational function has surfaced so that, *de facto*, they little different from garden settlements.

Figure 5.3   Allotments with shed-houses outside the town of Pushkino, Moscow oblast.

Source: T. Nefedova.

The territorial blocks occupied by collective gardens and allotments add another element to the distinctive dacha landscape on the outskirts of Russian cities. Today, seventeen million households have access either to a dacha garden or allotment settlement (Rossiia v tsifrakh 2011); taking account of average family size, this means that half of the total population of Russia is connected with the 'dacha movement'. But the character of this movement is changing; the relative importance of the allotment has declined compared with the other forms, whilst changes in the law have transferred many plots in dacha settlements and garden collectives into private ownership (although there are still 80,000 non-commercial second-home collectives [Ovchintseva 2011]). In Moscow and St Petersburg oblasts, the classic dacha in its grove of trees remains the second-home of choice but, here again, changes in the last decade are noticeable as traditional wooden structures have been replaced by two-storey brick houses. The area of dacha development now extends for as much as 100–150 kilometres from the boundaries of the capitals, and the number which is occupied year round appears to be increasing. The greatest changes are taking place in the oldest dacha settlements closest to the cities that are slowly but surely being transformed into dormitory settlements.

Rural Houses

The forth type of dacha is a converted or newly built house within the boundaries of existing rural settlements. In Soviet times it was not permitted for an urban registered residents to buy a house in a functioning village – second-home development was confined in purpose-built dacha and collective garden settlements. Despite this restriction, by the end of the Soviet period there was a thriving shadow market in rural houses whereby local residents were the legal house owners but the occupiers were, in fact, urban residents. Wooden rural houses could be bought for three to four months wages by Moscow's better off strata. In 1989, this shadow housing market was legalized and urban registered households were given permission to buy up or build properties in existing villages. However, at this stage another layer of shadows was added as the local state had to set normative prices for rural housing plots that took account of differences in the location and quality of land. In theory, rural house plots should be bought and sold at the normative prices but, in reality, the money that exchanges hands when a rural house is bought can exceed the normative price by as much as ten times. The lack of correspondence between the official, normative, price and the market price creates scope for corrupt officials to line their own pockets, thus distorting the second home market and generally pushing prices upwards.

Apart from the possibility of financial advantage from the turnover of rural houses, local authorities have overcome their initial hostility to *dachniki* by recognizing that there are advantages associated with extending a welcome to the seasonal influx of urbanites; house repairs, land clearance and well repair create a demand for labour, and summer folk are a welcome market for the produce of household plots. As access to a rural house in existing villages has become more popular, children of rural residents who left for the town have re-engaged with the village, and are now willing to take on their parents' house and convert it into second home use; a decade ago these would have just been left to decay on the death of the original inhabitants. Such is the demand for rural houses in villages surrounding the large cities that the stock of vacated properties available for purchase is now exhausted and urban residents who want to buy a house in an authentic Russian village have to look much further afield. The annual influx of urban residents can swell village populations in the summer months in districts adjacent to the cities. In a study in Pereslavl' *raion*, Yaroslavl' oblast, it was found that the 27,000 rural households increased by over one-third to 37,000 families in the summer months, these in addition to a further 13,000 owners of dachas in purpose-built settlements (Nefedova 2003: 31–2).

Rural housing plots are normally quite large, 1,500–5,000 m², so there is ample space to use them for cultivation. However, normally it is only the children of rural residents who use their parents' house as a second home in the summer whilst they are still alive who practise domestic production. As the older generation of rural dwellers dies out and their houses pass on to the next generation, domestic food production also dies. New purchasers of rural houses rarely use the land for food production, although this does not mean that they are not interested in the land; a recent development associated with rural house purchase is the acquisition of additional land in, for example, woodlands and meadows of former collective farms. This additional land is known in Russia by the pre-revolutionary word for the nobles' estates, the '*pomest'e*' and it represents a form of land grab that can deprive rural residents of access to resources that they have long relied upon for domestic food production.

'Cottages' and 'Mansions'

The fifth type of dwelling that provides a seasonal home for Russia's urban population is newly built large brick detached house, called variously cottages (*kottedzhi*) or mansions (*osobnyaki*), neither bearing much resemblance to their English namesakes. Some of these new builds are extremely elaborate structures on which their owners, Russia's *nouveau riches*, have allowed their imaginations full sway so that wherever there are clusters of them they present a truly post-modern mixture of styles. They can be adorned with turrets and towers that reference medieval fortresses and surrounded by high walls, fences and secured gates and protected by armed guards reinforcing the message that the post-Soviet Russian's house is his castle (Figure 5.4 below). The architectural flights of fancy are most in evidence where a stand-alone house has been built; new housing estates have a greater uniformity of style but both are a very visible feature of the environs of the large cities. The estates consist of detached brick or breeze-block houses each with its own garden attached on plot of 1,000–1,500 m² in size that imitate the low density middle-class suburbs found in most European cities, only they can incongruously stand isolated on a field edge, accessed by a single narrow road, and separated from other signs of life by tracts of agricultural land and forest. The greatest difference of these estates from their West European equivalents is less visual than functional; they are inhabited by people who most likely have retained an apartment in the central city in which they stay during weekdays and in the winter, this despite the fact that the house may be in commuting distance of work. Thus, even these purpose built housing estates carry forward the dacha idea.

Figure 5.4   An elaborate dacha or 'mansion' in Chulkova, Moscow oblast.
Source: T. Nefedova.

Ruralization or Post-industrialisation?

The desire to acquire a house and a plot in the countryside is not the product of a particular Russian 'mentalité'. The Russian extra-urban second home is not a unique phenomenon, but rather should be understood as the manifestation of a universal desire in industrial society to combine the advantages of rural and urban living. It is true, however, that the realization of this desire in Russia has been mediated by historical and geographical circumstances that have given the resultant settlement forms features that distinguish them from sub- and counter-urbanization in other countries. Two features, in particular, are worth noting: first, their secondary or seasonal character and, secondly, their continuing, albeit declining, role in food production.

Principal among the reasons for dachas remaining the secondary home is the registration system – heir to the institution of the *propiska*, or living permit, which was used to control city size during the Soviet period in cities like Moscow. Individuals have one official place of residence that is shown in their internal passport. Fear of the loss of a Moscow registration, which can affect the prospects of securing employment in the city, is a strong deterrent to changing official place of residence to another town, since once it is surrendered it is difficult to get it back, not just for oneself but for other family members. The existence of a flourishing shadow market in Moscow oblast for registration documents to fictitious apartments is testimony to the continuing power of the registration system to affect people's decision about where 'home' is. The reluctance to sever ties with Moscow is also explained by rising property prices that encourage Muscovites to hang onto city apartments as a form of saving if they can.

The weak development of service and welfare infrastructure in rural Russia is another deterrent to quitting the city permanently; urban residents are not prepared to forego the schools, shops, health and leisure facilities available in the city by making the permanent move to the countryside, and poor rural roads make commuting to access these facilities arduous especially during the winter when roads can be clogged with snow. Russia's continental climate is the source of other problems, sub-zero temperatures increasing outlays on heating, assuming houses are equipped with central heating and running water, which the majority are not. Local authorities are provided with little incentive in the current tax system to invest in services for *dachniki* and so there is little prospect for improvement in the short term. Given the deep cultural predisposition towards second home ownership in Russia, these obstacles are more than sufficient excuses for retaining a foothold in the city.

In the Russian context, therefore, it is unusual for urban households to seek a permanent home in the extra-urban suburb but some households do, transforming the dacha into their exclusive or 'first' home. As distinct from Western European countries where counter-urbanisation has been a predominantly middle class movement, in Russia it is associated with the two polar social groups of the super-rich, who making a life-style choice move to fully equipped houses in purpose-built out-of-town elite settlements, or the poor, who taking advantage of high rents lease out their flats in the city and make what they can of over-wintering at their dacha. But these permanent counter-migrants are no more than a small drop in the vast ocean of households for whom the dacha is a seasonal second-home. A final, even smaller, group of urban to rural migrants consists of people who, rejecting consumerism, quit their jobs in the city to found 'eco-settlements' in the rural hinterland. These settlements, latter day communes, have attracted much media attention in recent years and they are, perhaps, the most extreme example of settlement downsizing in contemporary Russia.

The second distinctive feature of the Russian second home – its association with domestic food production – has carried forward to the present day as a result of successive economic crises brought about by the collapse of the command administrative system, global financial meltdown and the neoliberal transition. That the vulnerable urban household should respond by self-provisioning is understandable, but its status as the principal leisure pursuit among urban Russians speaks to a deep cultural embeddedness. Put this cultural attachment to the land together with the stresses of urban life (pollution, overcrowding, and rising crime levels, which have become a feature of Russian cities and towns in the last decades), and the movement to the countryside to grow food can be understood as a response to the stresses of post-1991 urban life. Registered urban households produce less on their small allotments than their rural counterparts but they are three times more numerous; two-thirds of all urban households have plots of one sort or another and they produce 12 per cent (in value terms) of Russian agriculture's vegetables, 21 per cent of it root crops and 37 per cent of its soft fruits. As regards dacha food production (Ovchintseva 2011), there are geographical differences between the large metropolitan and more peripheral regions, as Table 5.1 shows: whereas in Moscow oblast dacha plots are used for flower gardens, in Vladimir and Lipets oblasts, respectively, they are used for vegetables and perennials (such as asparagus).

Table 5.1   Land use in different types of dacha settlements in three oblasts.

	Buildings	**Lawns and flowers**	**Food plants**	**Unused**
Moscow Oblast				
Classic dachas	19.0	66.0	14.2	0.9
Gardens	16.3	49.8	34.7	0.5
Allotments	10.0	38.1	51.5	0.5
Vladimir Oblast				
Gardens	11.5	23.3	55.3	10.0
Allotments	1.7	3.5	59.3	35.5
Lipets Oblast				
Gardens	5.9	3.8	89.1	1.7
Allotments	0	1.2	97.0	1.8

Source: Ovchintseva 2011 using the agricultural census of 2006.

The New Geography of the Dacha

In the rest of this chapter the new geography of second home development in Russia is discussed by pointing to the transformations that have been taking place in successive zones around the country's 'two capitals', as shown in the shaded area in Figure 5.5.

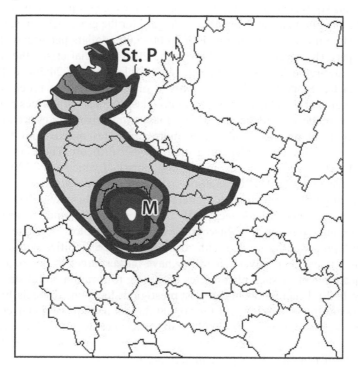

Figure 5.5   The zones of dacha expansion in the Moscow and St Petersburg regions. The intensity of shading indicates the near, middle and outer zones.

The Near Zone

The near zone of dacha development may be defined as the administrative districts located immediately outside metropolitan boundaries or within the boundaries of larger agglomerations. This zone occupies more or less the whole of Moscow oblast, an area larger than Switzerland, and it is characterised by a high density of dacha development, serving both as weekend retreat and as a base for commuting to work for all or part of the year. In the mid-2000s Muscovites owned three-quarters of the all the classic dachas in the near zone, 60 per cent of the collective gardens and allotments and one quarter of the newly built houses. This accounted for 36 per cent of the land under private housing in the oblast outside the city boundaries; the remainder is shared between inhabitants of the oblast's other eighty towns and other settlements (Makhrova, Nefedova and Treivish 2008). Muscovites are displacing the rural population as home owners in this zone and transforming the rural way of life.

Estates of two- and three-storey brick houses, some forming closed settlements equivalent to urban gated communities, begin just a few kilometres from the city boundary with apparent disregard for the green belt. In 2010 prices were high in this zone, in the range of 3,000–5,000 dollars per square metre of living space (Kuznetsov 2011). Prior to the 2008 financial crisis, the prices in the most desirable locations achieved remarkable highs of 24,500 dollars per square meter. Since the crisis, the top end of the market has fallen, creating space for greater development of 'economy-class' homes with prices per square meter of a 'modest' 1,500 dollars. Today, these extra-urban estates are in various stages of development – some little more than sites marked out for future housing – which makes calculating their number difficult. There are between 670 and 1,100 'cottage estates' in Moscow oblast of which only 320 are fully settled (Kuznetsov 2011). In addition to special closed settlements, there are free standing urban-style houses scattered throughout Moscow oblast. Aspiring second home owners do not necessarily buy a ready built home; in recent years it has become common for Muscovites to buy vacant plots in estates laid out for development and provided with roads and service facilities with a view to building a house themselves when they have the resources to do so. The price of a square meter of such plots varies according to location and distance from the city. In 2011, land plots in the most prestigious locations (especially west of Moscow city for example) sold for 50,000 dollars per square meter compared with the average of between 2–20,000 dollars (Kuznetsov 2011). Moscow oblast has the highest average addition to the living space per head of population in the whole of Russia – 1.1 square meters per head per year compared with the national average of 0.4 square meters and 15.0 for the city itself (Rossiia v tsifrakh 2011). This figure includes new floor space added in rural areas and in suburban towns, where Muscovites also are engaged actively in purchasing apartments and new houses at prices that compare favourably with those in the city (Golubchikov, Phelps and Makhrova 2010).

Despite the expansion and visual dominance of the new estates encircling Moscow, the more modest summerhouses in dacha settlements, garden collectives and on allotments are numerically dominant. For example, the large territory to the north of Moscow is covered by 'pseudo-towns' consisting of two-storied wooden houses covering an area exceeding that occupied by the three nearby large towns of Pushkino, Shchelkovo and Korolevo, shown in Figure 5.7. The same can be observed to the east of the city along the main highways. But to the west they are fewer and smaller, nestling in the forests or attached to existing settlements. The more pleasing landscape west of the city is reflected in higher land prices. The population of rural districts in Moscow oblast more than doubles as a result of the

Figure 5.6   An 'economy' estate of *kottedzhi* in the near zone outside Moscow.

Source: T. Nefedova.

influx of these small-scale dacha owners each summer (in 2010 from 1.4 million in winter to at least 3–4 million in summer).

The high demand for second homes has negative environmental and social consequences. Litter and waste left in the wake of the influx of summer folk and a consequence of the low level of servicing of new settlements is a major problem. There are only 47 waste dumps in Moscow oblast, two-thirds of which are used for waste from the city itself. It is not unusual to encounter household waste and rubbish piled at roadsides, not unusually next to local authority 'no dumping' signs. The expansion of dacha settlements has also led to forest clearance and threats to agricultural land from developers who in various (dis) guises buy up farm land from large enterprises and the rural population. This land grab is aided by corrupt officials who help developers exploit loopholes in the law to secure changes in land use. The result is that perfectly viable and profit-making agricultural land is lost; in 2010 the sown area in Moscow oblast was 550,000 hectares, a 2.5 reduction on the 1990 figure (Regiony Rossii 2011). Despite this contraction in agricultural land, large scale farming remains a feature of the landscape in the near zone which is the location of some of Russia's most profitable farms. Good infrastructure, abundant labour and the proximity of Moscow has stimulated agriculture at the city edge competing with second homes for land (Nefedova 2003: 305–45). Remarkably, two thirds of Moscow oblast's

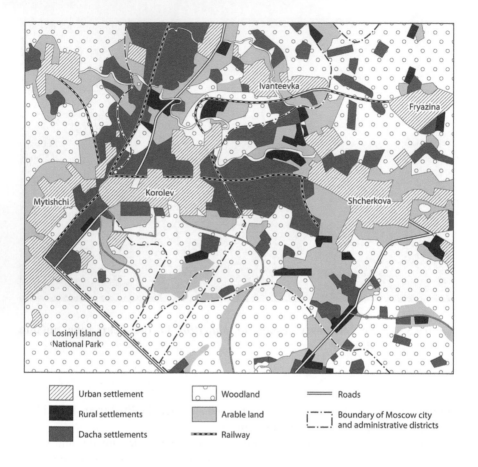

Figure 5.7   Settlement land use in the Pushkino-Shschelkovo-Korolevo triangle, 20 km from Moscow city.

livestock products that are destined for large food processing plants supplying the whole country originate in precisely those rural districts that are undergoing most intensive housing development near to the city.

The environmental disadvantages of the near zone are beginning to make themselves felt as formerly desirable rural locations are acquiring the character of 'edgelands'; *dachniki* are finding their rural idyll invaded by food processing factories, intensive livestock farms, out-of-town stores, markets, petrol stations, and traffic. Socially, there are also negative consequences of such an intensity of dacha development. The rural suburbs are becoming places of pronounced socio-spatial segregation and potential conflict. The newcomers have brought urban architecture into rural Russia, replaced meadows by grass lawns and potato patches by flower beds and put boundaries were there were none before that are resented by local populations. Whilst *dachniki* can bring benefits to local suppliers of building materials and provide a market for the produce from vegetable allotments stimulating the local economy, other benefits such as the demand for labour have not always materialized when, for example, incomers employ guest-workers from the former Soviet republics or from Moscow to build and repair their houses in preference to local workers. But the potential conflicts are not just with local people; different generations of *dachniki* can be at odds with one another – as for example when new developments cut off access to favourite picnic spots or lakes – and can precipitate conflicts between local residents who might have different attitudes towards selling land for dacha development.

In 2011 the decision was taken to extend Moscow city limits to the south west of the city taking it to the border of Kaluga oblast. This will incorporate a vast number of *kottedzhi*, town houses, dachas and garden collectives into the city. The development of 'New Moscow', as this extension of the city is to be called, is currently out to tender. When completed it will represent the first Western-type suburb within the boundary of Moscow city offering a sharp contrast to the high density high rise micro-regions of the socialist city (French and Hamilton 1979).

Middle Distance Dachas: Muscovites in the Capital's Neighbouring Oblasts

All Russian regions, but especially those in the non-black earth zone, have sharp centre-periphery population density, infrastructure and wealth gradients (Nefedova 2003, Ioffe et al. 2006). The peripheries of oblasts, therefore, provide a different social and economic environment for second home development compared with districts close to cities. The labour pool is smaller, roads poorer and services more limited, but, against these disadvantages, land is cheaper and more rural houses are on the market. In the case of Moscow oblast, this middle zone of dacha development, in fact, begins at the very boundary of the oblast at a radius of 100 km from the city and spills over into its neighbours – Vladimir, Yaroslavl', Tver, Smolensk, Kaluga, Tula, and Ryazan' oblasts – extending to about 300 km from the city centre. In the districts near the inner perimeter of this zone purpose-built dacha settlements as well as garden and allotment collectives and converted rural houses are all to be found, but towards the outer perimeter the latter of these become dominant. Petushinskii *raion* (district) in Vladimir oblast bordering Moscow to the north-east and equidistant (65 km) from Moscow and the regional capital is an example of the inner perimeter.(Makhrova et al. 2008: 276–7). In the 2000s, the local district authority agreed with Moscow to allocate housing plots to Muscovites for *dacha* development, in return for which it was to receive deliveries of dust-trucks, buses and ambulances. To date, land has been allocated for 139 garden settlements under this exchange agreement, with 30,000 plots reserved for inhabitants of Moscow oblast towns. Apart from this planned development, residents of Moscow city and oblast have bought rural houses in Petushinskii *raion*. In twelve villages, the native population has been completely replaced by *dachniki*. Land prices in the district are two to five times lower than in Moscow oblast, i.e., plots can be bought for a few thousand dollars a *sotka* (100 m²).

Whilst there are obvious financial advantages associated with purchasing a dacha in the middle zone, a weakly developed service sector creates familiar difficulties; if local government in the near zone has only latterly begun to understand the need to respond to *dachniki*'s demand for adequate services, in the middle zone local officials invariably insist that dachas are a drain on local services and impact negatively on local budgets. They are supported in this attitude by the complaints of local residents that the influx of 'rich Muscovites' pushes up prices in local shops and their waste pollutes the landscape. Therefore, even though agriculture and the rural economy are in a parlous state in the rural peripheries, local authorities restrict the housing plots they let onto the market annually. They are motivated by the hope that by hanging onto land that might otherwise be used for dacha development they might attract more lucrative industrial purchasers unable to find suitable vacant plots in the zone near the city.

Hostility towards dacha development is a relatively new phenomenon and appears to be a function of numbers. *Dachniki* appeared in the middle zone in the 1970s and 80s but they were not numerous and in the next two decades they transmitted the traditional dacha 'aesthetics' to new arrivals, who often came as part of a chain migration. *Dachniki* knew one another and, as a general rule, got on well with local populations. Initially, they were welcomed by local authorities because they relieved them of responsibility for unemployed villagers and invested in village restoration. The *dachniki* came in and repaired houses, contributed towards road and well maintenance, and their purchasing power helped to revive village shops.

The breakdown of the former good relationship between local people and *dachniki* in the twenty-first century is notable in conflicts over boundaries, land prices and other disagreements. In other words, the middle zone is beginning to witness a re-run of the sort of problems that characterised the near zone as dacha numbers began to rise there, although there are special features. Low population densities in the middle zone, for example, mean that *dachniki* have no alternative but to bring in labour from outside; often there simply is no-one left in middle zone village who can be trusted to do odd jobs. Meanwhile, old patterns of barter whereby *dachniki* exchanged deficit products from city for locally produced potatoes and milk have broken down and been replaced by monetary exchange. Instead of the former symbiotic relationship between *dachniki* and local inhabitants, today there are two separate communities both convinced the other is living parasitically on it.

Distant Dachas or the Outer Zone

The dacha zones of Moscow and St Petersburg meet and overlap in the southern part of Pskov and Novgorod oblasts: for example, in the Valdai district of Novgorod oblast 400 kilometres from Moscow, where population trebles in the summer, Muscovites occupy the south-east shore of Lake Valdai, and St Petersburgers the north-west. In the dying villages of the distant zone it is not uncommon to find whole streets that have been rebuilt by *dachniki* from the two capitals in search of the rural good-life. *Dachniki* appeared in the outer zone in the Soviet period, the forerunners of the wave that arrived after 1991, who took advantage of low house prices and surplus land as agriculture here collapsed. Twenty years on it is still possible to buy a good quality wooden house with 5,000 m² of land for 3–6,000 dollars.

The periphery of Kostroma oblast, which has been subject to analysis over a period of ten years from 1990–2000, illustrates very well the character of dacha development in what is known in Russia as the *glubinka* – a remote rural periphery (Nefedova 2008, 2011b). Manturovskii *raion* (district) is 600 km from Moscow and 250 km from the nearest large towns (Kostroma, Kirov, Nizhnii Novgorod) and is situated on the river Unzha, a tributary of the Volga, between the two *raion* centres of Manturovo and Makar'ev. It occupies a typical 'spatial hollow' in the coniferous forest that stretches across the north of European Russia and east into Siberia. In common with all such northern regions this district has experienced out-migration during the twentieth century so that today the population is only 14 per cent of its total in 1926. The *raion*'s economy, traditionally based on a combination of farming and forestry, has been in a perpetual state of crisis since the collapse of USSR – a tenfold contraction in arable land and the re-invasion of pastures by forest and shrubs – and the only employment today is in the heavily subsidised public sector. Although there is land for all in the district, nobody wants to take it on, unless, it is vacant housing plots in one of its shrinking villages. In nearly all 'living' villages in Manturovskii *raion* some houses are occupied by *dachniki* and the smaller and less populated the village, the relatively more numerous they are. In one cluster of villages that are home to a permanent population of 220, located on the upper reaches of the most picturesque parts of the Unzha river, 30 per cent of all the privately owned or rented houses are in the hands of *dachniki*. In smaller villages with fewer than 30 or 40 residents, the proportion is closer to 40 per cent and in the smallest one, with less than ten residents, the share of *dachniki* is 70–90 per cent (see Table 5.2 below). But in the distant zone, completely deserted villages are eschewed by *dachniki* – because they are easy prey for gangs of house burglars – second homes are not sustainable without a minimal local population to oversee them in the long winter months. In some villages where just a few pensioners remain *dachniki* combine resources to set up a family, for example of recent return migrants to Russia, to guard their property.

The dacha owners in the outer zone typically belong to older age groups (fifty plus), are middle income earners and from the professional classes, or they may be from the artistic and academic intelligentsia who have flexible work regimes allowing them to make the eight to nine hours journey by car (or overnight by train) to their distant house, once or twice a year. They stay for anything from two weeks to several months. With echoes, therefore, of communities like Peredelkino and Abramstevo, the legendary rural haunts of writers and artists of the Soviet era just outside Moscow, distant villages in the outer zone become home in the summer to clusters of journalists, teachers, artists and musicians. Most of these summer folk are Muscovites – 85 per cent of the total – with the remainder from

St Petersburg and the regional capital, Kostroma. In the latter case the *dachniki* are mainly people who have inherited a rural house. Many of the residents of the district centre, Manturovo, who are in the middle and younger working age groups, have parents in the surrounding villages and they engage in the time-honoured reciprocal arrangement of leaving their children with the grandparents and helping to cultivate vegetables and soft fruits at the weekends.

Table 5.2   The number of permanent residents and *dachniki* in the villages of Ugorskii *raion* in 2007.

	The total population	Population Dynamic, 2007 in % to 1926	Land plots belonging to local residents	Land plots belonging to *dachniki*	Share of *dachniki* in %
Ugory	227	34	99	46	32
Davidovo	40	10	16	14	47
Medvedevo	10	5	4	15	79
Khlyabishino	59	14	31	20	39
Dmitryevo	10	4	1	12	92
Zashil'skoe	6	5	5	12	71
Bazhino	0	0	0	7	100
Poloma	10	9	4	12	75
Stupino	2	2	1	10	91
Total	386	14	175	161	48

Source: Local administration data (Nefedova 2008).

The outer zone is not the site of new house building. The attraction of such a distance is precisely the opportunity they afford for newcomers to buy a house in the vernacular style to restore. Rural houses are large here, covering a ground area of 100–150 square metres, consisting of a dwelling space of 50–65 square metres connected to the livestock barn and a hayloft running the length of the building above. The outward appearance of these houses is indistinguishable from those of local residents', but the substitution of flowers or a grass lawn for potatoes on the land attached is a sure sign that the house is a dacha. Rather than till, weed and harvest, the Muscovite incomers spend their time reading, tending their flower gardens, making forays into the forest to collect berries and mushrooms, restoring the house to its original condition and visiting like-minded people in neighbouring villages.

Figure 5.8   Muscovites arriving at a dacha in Maturinskii *raion*, the outer zone.

Source: T. Nefedova.

A survey of 110 households (80 local residents – 20 per cent of the total – and 30 *dachniki*) in two villages in Maturinskii *raion* in August 2008, revealed that a proportion of *dachniki* used to own a second home closer to Moscow but were 'forced' to look further afield for the tranquillity and classic dacha landscape they yearn. One fifth still own a dacha in Moscow oblast for weekend use (Nefedova 2008).

Relationships between newcomers and existing residents in the outer zone are mostly cordial; local people recognize the role the former play in saving otherwise dying villages and clearing them of encroaching scrub to improve fire safety. One-quarter of all respondents to a questionnaire survey conducted among permanent residents in 2008 stated that they had a positive attitude towards *dachniki* describing them as 'pleasant' and 'sociable'. In the most distant villages of the outer zone the traditional symbiotic relationship between the newcomers and local people has established itself with village 'grannies' dependent upon owners to bring in provisions or take them to the district centre, and the *dachniki* dependent on these same pensioners to keep an eye on their house during the winter.

As for the *dachniki* themselves, they feel part of a diaspora community. Even though dispersed across separate villages they make the effort to visit one another and to socialize. In this way they re-enact the way of life associated with the classic dacha that harks back to an age before over-development of second homes in the near- and middle-distance zones. But their social life is self-contained; they are not integrated into the cultural and social life of the local community revolving as it does around the administrative and economic centres of the *raion* that were bolstered by the Soviet equivalent of key settlement policy in the 1970s. One such centre in Maturinskii *raion* is the settlement of Ugory which was developed in the late 1970s and 1980s as a result of the merger of two collective farms. At that time it acquired new housing and cultural and social facilities which attracted households in from surrounding villages accelerating the process of depopulation and the decline of the latter that began after the Second World War. Today, Ugory remains the centre of a network of disappearing villages and its share of population is increasing, but its social and cultural role is much diminished as the successor large farm to the collective has struggle to survive. The settlement is the home to the rural administration, a post office, school, shop and medical 'point' that employs a feldsher (health care professional), but all rely on subsidies to continue functioning. The focus of economic and social activity has, in fact, shifted to the new community that now provides the stimulus for the genesis of new businesses servicing its needs.

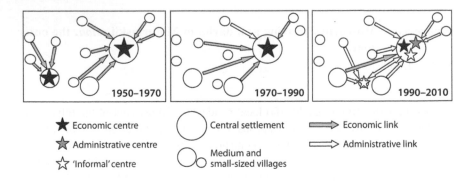

Figure 5.9   Changes in the production and social networks in a rural district as a result of the in-migration of *dachniki* (based on Kostroma region).

Despite the economic interdependence of permanent residents and the new population of second-home owners in the outer zone, there is a clear symbolic boundary between the two communities. The *dachniki* maintain a discourse of self-reliance that emphasises the inability of local people to meet their demands for repairs or for fresh food products. A commonly heard refrain is that 'we have to organize everything for ourselves' and criticism of local government for its failure to provide even the most basic level of servicing in more remote villages is vocal. They complain, with justification, about impassable roads, poorly maintained wells, the absence of running water, gas and a waste collection service. Local residents have long since become accustomed to such deprivations, a remarkable 40 per cent claiming to be satisfied with service provision.

There are also significant differences in the attitude of newcomers and residents towards the conservation of local culture and landscape. Whereas the local population might deny that there are any traditions to preserve or profess difficulty understanding the question, incomers are concerned about preserving vernacular architecture and institutions such as 'the bathhouse' and handicraft industries. *Dachniki* collect wooden implements and crafts of the region and display them in private museums. They may take up handicraft production (carving wooden dolls and basket weaving) or try to live an 'authentic' rural lifestyle cooking on the traditional stove and drawing water from the well. These efforts to preserve the local material culture and way of life are not necessarily valued by permanent residents who, rather, insist that the very presence of the urban intelligentsia is destroying the traditional rural way of life. They draw a contrast between their exoticised imaginings and the real lived experience of trying to secure subsistence in the hostile physical and employment environment of the region.

It is important not to underestimate the challenges presented by seasonal migrants to the outer zone of dacha settlements. Were this Western Europe or North America, the households that have restored old houses and learned craft skills could reasonably be theorised as pioneers of a more sustained colonisation as the zone of second home development pushes ever further outwards from its metropolitan hearth. Indeed, with its with its pristine environment and mature in-migrants the villages of Maturinskii *raion* and its neighbours would be candidates for retirement settlements, as on the south coast of England or the forests of New Hampshire in New England, USA. But in Russia there is no certainty that the second home colonisation of the further away rural peripheries will not simply fizzle out. The people who have bought dachas in the most distant places were young at a time of rural revivalism in the 1970s and the dawn of an environmental movement in the USSR and they have not fully come to terms with the rampant consumerism of the post-Soviet era, even though they have benefited from it. This is a generation that is most likely to feel nostalgic for classic dacha and most comfortable occupying the threshold space between town and country. It is far from certain that they will be followed by another generation in search of this life. Although nearly three-quarters of second home owners in Ugory said that they had had visits from their adult children, only seven per cent predicted that these visists would become regular with twenty per cent uncertain that they would do so. Where the *dachniki*'s own plans are concerned, staying on into old age is not an option at the present time given the paucity of infrastructural development and sheer difficulty of living all year round in such places. The stay of execution that they have brought to the most distant villages may be only temporary, therefore.

The tragedy of the situation is that local authorities in Russia's remote rural places still operate under the misconception that the central state will resume the role is played in the Soviet era and re-inject life into the local economy by investing in business and industry and by restoring the agrarian economy. This fixation on the state as the solution to the problem of the declining rural economy and depopulating villages is blinding local and regional government to the economic and social advantages to be gained from investing in the infrastructure needed to help the second home phenomenon develop. Local government in the rural peripheries is still stuck in a 'productivist' time-warp that equates rural development with large agricultural and forestry enterprises and supplementary industries (Shubin 2006) though the post-industrial path of recreation, conservation, eco-tourism and retirement may well offer better prospects.

Conclusions

The Russian dacha is a highly varied phenomenon but one in which all strata of the population are involved, from the poor to the richest. The former have their small houses, little more than garden sheds, on their allotments to which they repair at weekends for recreation and food production. The super-rich, for their part, go to their 'cottages' and 'castles' to 'perform' the country life behind high fences and protected by private security guards. The feature they have in common is that both are heirs to the tradition of the summer retreat from the city, each pursuing their own version of the rural idyll. The extraordinary expansion in the number of people who take part in the weekend and summer exodus to the country since the collapse of communist regime two decades ago has fuelled an enlargement and simultaneous fragmentation of the spaces of 'dacha colonization'. Negative social and environmental externalities associated with the pressure of numbers in the most accessible destinations have added impetus to the outward spread, and, at the same time, have been a factor in the failure of 'dacha settlements' in their various manifestations to transform themselves into suburbs proper. Migration between 'first' and 'second' or even 'third' homes remains the dominant characteristic of outward urban spread in the Russian Federation today. The sharp urban-rural divide observed by Tony French and Ian Hamilton in their seminal *The Socialist City* – the consequence of high rise blocks extending right to the outer boundary of the city – is still a feature of most large cities in Russia, though Moscow New City will mark a departure from this pattern. Beyond the city's edge, a complicated and fragmented landscape of vegetable allotments, makeshift semi-houses, gated estates of new brick houses, and faux manors and castles, interspersed with large industrialised farms and remnants of forest and green belt has developed. Economic, demographic, infrastructural and cultural intensity declines with distance from the city edge, but second homes remain a feature of the landscape. Even several hundred kilometres from the city the dacha is to be encountered, evident in a carefully restored traditional wooden house or a cluster of newly built brick and breeze block houses standing at a field or village boundary.

The social geography of the extra-urban dacha landscape has been slowly coming into clearer focus in the past two decades. In the immediate environs of the city, *dachniki* are drawn from a broad social spectrum but they manifest a high degree of socio-spatial fragmentation. In this zone of intensive dacha development the seeds are being sown for the emergence of a post-industrial suburban landscape of service villages, albeit interspersed with large agricultural and industrial enterprises. In contrast, dacha development in the middle zone is taking place against a backdrop of a rural economy ravaged by farm bankruptcies and in villages that are ill-prepared to cope with the ever-growing seasonal influx of second home owners. With time, dachas may stimulate the service sector, but until this happens it is unlikely that there will be changes in the pattern of use of second homes in this zone. In the outer zone, the initial stages of dacha colonization of the Russian countryside are being replayed, albeit in a twenty-first century version. In their quest to rediscover the rural idyll their forebears enjoyed, today's *dachniki* are combining the pursuit of leisure with the task of saving Russia's disappearing villages. But, as observed above, there is no certainty that this movement is sustainable once the current generation of both older residents and newcomers dies out. In all three zones, much depends upon the attitudes of local administrations. Whilst local officials have generally welcomed the opportunity to profit from land and house sales, this has not been followed up with policies designed to meet the needs of the seasonal influx or to encourage changing patterns of occupancy, including extending the period that newcomers stay. There is evidence of tension in the relationship between older resident households and newcomers but, as the experience of other countries has shown, this is not a necessary accompaniment to second home development (Cloke and Thrift 1987, Smith and Krannich 2000). Sensitive handling and identification of issues by community leaders can defuse potential conflict. Local administrators in Russia are still hampered by a mindset inherited from the Soviet era that prioritizes large scale agro-industrial development and fails to recognize the alternative ruralities associated with a globalized and post-industrial world. It is obvious that rising incomes combined with a strong cultural predisposition to own a place in countryside, the second home phenomenon is likely to grow in Russia in the twenty-first century. If this phenomenon is not to be associated with negative externalities that hurt both rural residents and urban second home owners, a change in the understanding of what constitutes rural development – to include the recreation, leisure and conservation – is urgently needed.

References

Clarke, S. 2002. *Making Ends Meet in Contemporary Russia; secondary employment, subsidiary agriculture, and social networks*. Edward Elgar.

Cloke, P. and Thrift, N. 1987. Intra-class conflict in rural areas. *Journal of Rural Studies*, 3(4), 321–3.

Ioffe G. and Nefedova T. 1997. *Continuity & Change in Rural Russia. A geographical perspective*. Westview Press.

Ioffe, G. and Nefedova, T. 2000. *The Environs of Russian Cities*. Edwin Mellin Press.

Ioffe, G., Nefedova, T. and Zaslavski, I. 2006. *The End of Peasantry? The Disintegration of Rural Russia*. University of Pittsburg Press.

Golubchikov, O. and Phelps, N.A. 2009. Post-Socialist Post-Suburbia? Growth machine and the emergence of 'edge city' in the metropolitan context of Moscow, Paper presented at the 3rd International Workshop of Post-Communist Urban Geographies, University of Tartu, Estonia.

Golubchikov, O., Phelps N.A., Makhrova A. 2010. Post-Socialist Post-Suburbia: Growth Machine and the Emergence of 'Edge City' in the Metropolitan Context of Moscow. *Geography, Environment, Sustainability* 1(3), 44–55.

Khauke, M.O. 1960. *Prigorodnaya zona* (The suburban zone). Mysl: Moskva.

Kuznetsov, I. 2011. *Moskovski region.Analiticheckaya spravka*. (Moscow region: an analytical digest). [Online]. Available at: http://www.cottage.ru/articles/analytics/209366.html [accessed: January 2012].

Lovell, S. 20003. *Summerfolk; A History of the Dacha 1710–2000*. Ithaca: Cornell University Press.

Makhrova, A., Nefedova, T. and Treivish A. 2008. *Moskovskaya oblast' segodnya i zavtra* (Moscow Oblast today and tomorrow). Moscow: NovyiKhronograf.

Mason, R.J and Nigmatullina, L. 2011. Suburbanization and sustainability in Metropolitcan Moscow. *The Geographical Review*, 101(3), 316–33.

Nefedova, T.G. 2003. *Sel'skaya Rossiya na pereput'e. Geographicheskie ocherki* (Rural Russia at the Crossroads.Geographical essays). Moscow: Novoe Izdatel'stvo.

Nefedova, T.G. 2008. Rossiyskaya glubinka glazami ee obitatelei (Russian remote places through the eyes of their inhabitants). Ugorsky Project: Environment and People in the Middle North, edited by N.E. Pokrovski. Moscow: Community of Professional Sociologists, 98–120. See: www.ugory.ru Ugorsky project, publications.

Nefedova, T. 2011a. Rossiiskie dachi kak sotsialnyi phenomen (Russian dachas as social phenomena). SPERO (Social Policy: Expertise. Recommendations. Overviews). 15, 161–73.

Nefedova, T. 2011b. Russian Rural Nechernozemye: Collapse or New Ways of Development? *Geography. Environment. Sustainability*, 4, 10–24.

Ovchintseva, L.A. 2011. Economicheskoe znachenie I socialnaya rol sadovyhtovarishestv (The economic significance and social role of garden collectives). *Razvitie APK*, 3, 50–55.

Pallot, J. 1979. Rural settlement planning in the USSR. *Soviet Studies*, 31(2), 214–30.

Pallot, J. 1990. Rural depopulation and the restoration of the Russian Village under Gorbachev. *Soviet Studies*, 42(4), 655–74.

Pallot, J. and Nefedova, T. 2007. *Russia's Unknown Agriculture. Household Production in Post-Socialist Rural Russia*. Oxford: Oxford University Press.

Paris, C. 2009. Re-positioning second homes within housing studies: household investment, gentrification, multiple residence, mobility and hyper-consumption. *Housing Theory and Society*, 26(4), 292–310.

Pine, F. and Bridger, J. 1998. *Surviving Post-Socialism: Local strategies and regional responses in Eastern Europe and the Former Soviet Union*. London: Routledge.

Regiony Rossii. 2006 (Regions of Russia). Moscow, GoskomstatRossii.

Rodoman, B. 1993. Problemy sokhraneniya ecologicheskikh funktsii prigorodnoi zony Moskvy. *Problemy zemlepolzovaniya v svyazi s razvitiem maloetaznogo stroitelstva*. Moscow.

Rossiia v tsifrakh. 2011. (Russia in figures), Moscow, Goskomstat Rossii.

Shubin, S. 2006. The changing nature of rurality and the nature of rural studies in Russia. *Journal of Rural Studies*, 22(4), 422–40.

Smith, M.D. and Krannich, R.S. 2000. "Cultural Clash" revisited: newcomer and longer term residents' attitudes toward land use, development and environmental issues in the Rocky Mountain West. *Rural Sociology*, 65(3), 396–421.

Swift, A. 2004. Review of Summerfolk: A History of the Dacha 1710–2000. *Journal of Social History*, 38(2), 526–8.

Stryuk, R.J. and Angelica, K. 2007. The Russian Dacha Phenomenon, *Housing Studies*, 11(2), 233–50.

Chapter 6

Second Homes and Outdoor Recreation: A Swedish Perspective on Second Home Use and Complementary Spaces

Dieter K. Müller

Introduction

Second home tourism has recently experienced a rejuvenation, which is also manifested in a growing number of scientific publications (e.g. Hall and Müller 2004, Gallent et al. 2005, McIntyre et al. 2006). The renewed interest can be explained by a number of changes in current western societies (Müller 2002a). Accordingly, demographically ageing socities, economic restructuring, technical changes, as well as modernity and globalization all contribute to provide greater opportunities for using second homes at home and abroad by entailing greater leisure time and new mobility patterns and forms, to name a few. Hence academic focus has been on related processes such as retirement homes (King et al. 2000, McHugh 2006), second home induced displacment of permanent residents (Gallent et al. 2003, Marjavaara 2007a, 2007b) and the internationalization of second home mobility (Buller and Hoggart 1994, Müller 1999). The greatest attention, however, has been given to issues related to the meaning of second homes. Pitkänen (2008), for example, discusses the different meanings of Finnish landscape from a second second home perspective, whereas Tuulentie (2007) scrutinizes second homes as a part of their owners' life stories. In this context second homes are often depicted as places for escape from modernity (Kaltenborn 1998, McIntyre et al. 2006, Stedman 2006, Van Patten and Williams 2008) offering not least contrasting experiences from everyday life (Jaakson 1986, Chaplin 1999), which is also seen as the major motivation for second home use (Perkins and Thorns 2006).

Nonetheless, little attention has been given to how this meaning is manifested in action. Haldrup (2004) provides an exception by discussing the process of inhabiting a second home, and Kaltenborn (1997a, 1997b) assesses place attachment scrutinizing activities in the surrounding area of second homes. Similarly Sievänen et al. (2007) have mapped the outdoor recreation activities of different second home user groups in Finland and found that active users seem to be more nature-oriented than ordinary and infrequent users, who are more oriented towards sports. Non participants in second home tourism were generally less active regarding outdoor recreation.

This chapter tests these findings for a Swedish context. It aims at revealing to what extent second homes facilitate greater engagement in outdoor recreation for second home users. This is achieved by a quantitative analysis of data collected in a national survey on outdoor recreation (Friluftsliv 2007). First, however, the relationship of second home access and outdoor recreation is discussed, drawing on previous research in the field. Then methodology and geographical setting of the study are introduced. The chapter continues with a presentation of result and a concluding discussion.

Second Homes as Complementary Spaces

Motivations for second home use are plentiful. Jaakson (1986) mentions the duality of routine and novelty as an important explanation for second home use, but also inversion of everyday life. Moreover, back-to-nature, identity, surety, continuity, creative work are listed as reasons for purchasing a second home. Jaakson (1986) also recognized elitism, sometimes related to aspirations for the area, as an important motivation. Müller (2007) points at the role of tradition and heritage in explaining second home tourism. Similar accounts were given by Williams and Kaltenborn (1999) who, however, interpret and summarize many of Jaakson's categories as escape from modernity. Kaltenborn (1998) even asks whether second homes are not truly first homes since they sometimes are inherited and passed on to the next generation. At least, he concludes, second homes are alternate homes.

This notion implies a contrast between life at the cottage and life in the often urban everyday environment. The argument indicates that being at the cottage is being at home – being at the cottage becomes a 'natural' state, and consequently second home activities become desirable. Meanwhile, urban life and mobility can be perceived as alienated and at least not as embraced as cottage life. Even if this may be true for certain second home owners, there is evidence that second homes can be the opposite, a commodity (Müller 2002b, 2004), or in itself an expression of modernity and globalization (Müller 1999). This argument is related to the notion of changing mobilites (Williams and Hall 2002, Hall and Müller 2004), multiple dwelling and place attachment (McIntyre, Williams and McHugh 2006) implying that mobile lifestyles are a common feature of current western societies owing to the societal changes discussed above.

Of course, being at home is a phenomenological experience and hence basically individual. Nevertheless in terms of activities it is possible to ask whether life at the second home differs from life at other homes, and whether second home owners engage in different acitivities than non-second home owners as indicated by Sievänen, Pouta and Neuvonen (2007). This is particularly true for a Nordic context with great availability of second homes. Evidence from Sweden indicates, for example, that contrast between permanent and second home is relative since the majority of second homes are just about 30km away from the owners' permanent residences (Müller 2006).

Obviously second homes are mainly located in rural areas (Gallent et al. 2003, Hall and Müller 2004). Many of their owners have on the other hand their permanet residence in urban areas. Hence, the trip to the second home is not only a switch of homes, it is also a shift in environments. Nevertheless, there are different experiences to what extent this mobility is also followed by a shift in lifestyle. Already 50 years ago Wolfe (1952) claimed that second home areas in Canada meant mainly a seasonal relocation of urban life into the countryside, a notion that is confirmed by Halseth (1998) who argues that second home areas function mainly as separated communities within communities. In contrast Müller (1999) states, with regards to German second home onwers in Sweden, that these are guided by a rural idyll and thus also adapt their activities to the rural surrounding. This is also confirmed by Kaltenborn (1997a, 1997b) who sees specific activities at the cottage as a way of attaching to place.

The Nordic experience points at second homes as complementary spaces offering qualities not available at the permanent home. Access to natural areas, outdoor recreation activities, relaxation and quietness are often mentioned as major assets available at second homes (Kaltenborn 1997a, 1997b, Müller 1999, Jansson and Müller 2003). It is likely that similar qualities also are present elsewhere; however second homes may be experienced as suitable spaces for regression from modern life (Jaakson 1986) or refuges of modernity (Kaltenborn 1998). In any case, second home owners obviously expect them to fulfill their desires and needs.

Second Homes and Outdoor Recreation

Historically not all second homes orignate from the same time period, and thus they have been established under different conditions. Hence Müller (2004, 2006) states that second homes in Sweden can be divided into two major groups, i.e., purpose-built homes often in the recreational hinterlands of urban areas and amenity-rich destinations, and converted homes that previously functioned as permanent housing. The former are spatially concentrated, while the latter are dispersed in the countryside, offering therefore another contact to other rural dwellers and probably rural lifestyles. Different regimes of land ownership entail however contrasting outcomes (Gallent et al. 2003, Paris 2010), and hence the Swedish situation with abundant access to land and the right of public access is radically differing from a UK-situation with a scarcity of rural land and very limited access rights. However, the latter has to be seen as the exception rather than the norm since 'Planning regimes in most countries do not share the British obsession with preserving attractive locales only for the rich...' (Paris 2010: 183).

Independent of these differences it can be assumed that the rural arena offers a different opportunity spectrum compared to urban areas that also can be used more intensively owing to the fact that second home owners visit the countryside mainly during their leisure time. Differences are however also partly related to the absence of urban entertainment supplies, partly to the presence of not least natural areas suitable for outdoor recreation activities. This is particularly true for the Nordic countries, where the public right to roam allows accessing the rural landscape almost without any constraints (Kaltenborn et al. 2001). And obviously, second home owners engage in outdoor activities to a remarkable degree which Kaltenborn (1998) interprets in the context of place attachment arguing that engaging in outdoor activities in fact creates place. Accordingly 85 per cent of the Norwegian second home owners engaged in skitrips, 76 per cent in nature walks and 53 per cent in fishing (Kaltenborn 1998). Similarly a Finnish study revealed that walking for pleasure, berry picking, boating and fishing were the most popular activites among second home users (Sievänen et al. 2007). Moreover, second homes in amenity-rich areas as mountains and seaside resorts lure particularly with specific activities such as alpine skiing and sun-bathing, respectively (Lundmark and Marjavaara 2005, Müller and Marjavaara 2007). It is also claimed that second home ownership has a positive impact on the second home owners' health status and retirement age (Hartig and Fransson 2009).

Access to natural areas varies geographically. Hall (2007) claims that an area's naturalness increases with distance. Still, in a Nordic context access to nature is sometimes portrayed as ubiquitarious. Hence, Hörnsten and Fredman (2000) reveal that more than 60 per cent of Swedish households have access to a recreational forest within 1 km from their permanent homes. Nevertheless, many would prefer an even shorter distance even among those living in close proximity. Thus it can be concluded that households usually have everyday access to recreational forests, meanwhile more grand and comprehensive outdoor experiences are to be found in countryside and wilderness locations.

In summary, second homes allow for regular access to these natural areas. Figure 6.1 illustrates a modern second home near Stockholm. Of course access is also given to housholds not owning a second home, but it can be argued that property ownership does imply strong ties to place facilitating frequent visits and motivations that at least partly can be related to outdoor activities. Hence second home owners are expected to have different patterns of outdoor recreation than households lacking similar access to the countryside indicating that second homes form complementary spaces offering activities beyond everyday patterns.

Second Homes in Sweden

Second home tourism in Sweden has a long tradition and thus in fact about 45 per cent of all Swedes have access to a second home (Müller 2004, 2007). The second home stock comprehends two types of second homes, i.e. previously converted permanet rural homes and purpose-built second homes, both with distinct geographies. The latter dominate in number and typical locations can be found in amenity-rich locations along coastlines and in mountain resorts, and in the vicinity of larger settlements. Converted homes dominate in peripheral regions and rural areas lacking specific amenities. Since these second homes

Figure 6.1   Modern second home in the Stockholm archipelago.

are a result of rural decline, Müller (2004) depicts areas where this type is dominating as disappearing areas. Nevertheless, second home use is to a high degree a regional phenomenon. Most second home owners have their leisure property close by. Hence, 25 per cent of all second homes are within a distance of 8 km from their owners' permanent residences, 50 per cent within 32 km and 75 per cent within 93 km (Müller 2006). Hence, it can be argued that most second homes can be characterized as weekend homes (Müller 2002a). This implies no radical change in environment, but still the distance may make an important difference to their owners opening up a new opportunity spectrum apart from the often urban everyday environment. The average distance between primary residence and second home is 87 km indicating that there are numerous exceptions, and thus there is also a stock of vacation homes not least in more peripheral locations such as the northern mountain range (Lundmark and Marjavaara 2005, Müller 2006).

A previous study on second home ownership in northern Sweden revealed a wide array of motivations (Jansson and Müller 2003); relaxation, outdoor recreation and the ability to spend time in an attractive environment were the top three motivations given. Nevertheless, these motivations are not necessarily mirrored in the activities pursued at the second homes. Here reproductive activities particularly related to the maintanance of the second home dominated.

Still mushroom and berry picking as well as recreational walks were also mentioned as important activities. The data provides however only a snapshot of second home activities in northern Sweden. Comprehensive data on second home use in Sweden are absent.

Data and Methodology

The study is based on a national survey of outdoor recreation in Sweden 'Friluftsliv 2007', the first comprehensive investigation of Swedish outdoor recreation patterns ever. The survey targeted a representative sample of 4,700 inhabitants of Sweden aged between 18 and 75 years. Altogether 40 per cent of the addressed persons answered 55 questions regarding their outdoor recreation activities and opinions towards various aspects of outdoor recreation (Fredman et al. 2008). A thourough follow up study of non-responses was conducted to reveal whether there was a systematic bias caused by non-responses. Nevertheless, the responding population proved to be representative showing similar age structures and geographical residence patterns as the total population of Sweden.

One aspect covered in the survey was access to and use of second homes. In this context the respondents were asked regarding the frequency of use and the location of the second homes. Other questions addressed outdoor recreation habits during youth, current outdoor recreation activities and habits, differences between everyday, weekend and holiday activity patterns, future expectations regarding outdoor recreation, spending patterns and general attitudes towards outdoor recreation and outdoor recreation management. Only relevant parts of the survey results are used for this study.

The study results were also checked for their representativness regarding second home access in Sweden. In fact the result for access to second homes differs only marginally from previous national surveys asking the same question (Fredman et al. 2008).

For this chapter data was retrieved from the 'Friluftsliv 2007' database and analyzed. Respondents were given the opportunity to state whether they own a second home, have access to a second home in other ways, regularily rent a second home, or do not have access to a second home. This classification was subsequently used to control for differences between the groups regarding their activities. This was conducted using statistics comparing means and OLS regression in SPSS 18.

Results

Altogehter 52 per cent of the respondents do not have access to a second home. The remaining group with access to a second home is distributed into three subgroups; 23 per cent own a second home, 21 per cent have access to a second home, for example, the parents' property, and 4 per cent stated that they regularly rent a second home. Visitation patterns differ mainly between second home owners and groups having access through rental or without ownership (Table 6.1). The latter groups have only few one-day visits to second home and a median amount of 10 overnight stays in a second home. In contrast, half of second home owners stay at least 35 nights in the second home. The geographical pattern of second homes accessed by the study population largely mirrors the global distribution of second homes in Sweden as described above.

Table 6.1   Access and visits to second homes.

Access to second home		Median	Mean	Std. dev.
Access through rental	One-day visits	0	5.5	18.4
	Overnight stays	10	21.0	28.9
Access without ownership	One-day visits	2	10.3	32.8
	Overnight stays	10	22.1	38.2
Access through ownership	One-day visits	5	24.3	56.0
	Overnight stays	35	56.2	65.9

The Second Home Users

Second home ownership is overrepresented in high income groups. Accordingly, more than 43 per cent of all households with a monthly income over 5,000 € own a second home. Nevertheless, second home access is also evident in households of lower income levels (Table 6.2 below). Access to these second homes is, however, granted to other households, presumely the owners' children and their families, since this access category does not show any pattern contingent of income. People with access to second homes are on average older than those lacking access. This picture is also supported by an analysis of educational patterns among the different access classes. People with higher education, and hence presumably higher income, are significantly more likely to have access to a second home. Still, differences between different access groups are limited. For example, employment patterns of people owning a second home and those who lack access do not differ significantly.

Similarly, results also indicate that housholds with children having access to second homes are more likely to rent the second home or have access to other owners' second homes. In contrast, second home ownership is more common for households without children. Access patterns among men and women show no significant differences. Even the environment where respondents grew up does not influence differences between different access categories significantly.

Table 6.2   The study population.

	No access	Access through rental	Access without ownership	Access through ownership
Number (%)	856 (52%)	69 (4.2%)	342 (20.8%)	378 (23.0%)
Mean age	39	40	44	45
Std.dev.	15.7	13.4	15.1	12.6
Mean monthly income	2 700 €	3 600 €	3 200 €	3 900 €
Std. Dev.	1 900 €	1 900 €	2 000 €	2 300 €
Secondary school	178 (61.8%)	9 (3.1%)	39 (13.5%)	62 (21.5%)
High school	385 (53.8%)	32 (4.5%)	152 (21.2%)	147 (20.5%)
Higher Education	285 (45.2%)	28 (4.4%)	150 (23.8%)	167 (26.5%)

Table 6.3   Access to second homes and home environment.

Home environment/ Access to second homes		Countryside with only few other houses in sight	Small rural settlement, villa dwellings only	Small rural settlement, mixed dwellings	Urban area, villa dwellings only	Urban area, mixed dwellings	Urban area, rental housing only
No access	Count	120	103	102	175	210	136
	%	60.9%	59.2%	49.5%	47.7%	51.9%	48.7%
Access through rental	Count	10	5	11	21	13	9
	%	5.1%	2.9%	5.3%	5.7%	3.2%	3.2%
Access without ownership	Count	32	32	49	68	96	63
	%	16.2%	18.4%	23.8%	18.5%	23.7%	22.6%
Access through ownership	Count	35	34	44	103	86	71
	%	17.8%	19.5%	21.4%	28.1%	21.2%	25.4%
Total	Count	197	174	206	367	405	279
	%	100.0%	100.0%	100.0%	100.0%	100.0%	100.0%

$p < 0.05$

Anyway, second home access is not regularily distributed among the Swedish population. No access is particularly common among rural residents (Table 6.3). However the urban respondents showing the greatest access to second homes are in fact residents of urban areas with villa type or mixed dwellings. This indicates that second home access is also a contingent of economic assests and not merely a question of contrast to the everyday environment, though the figures also support the idea of second homes as complementary space.

Second Home Access and Visits to Nature

The locational patterns of second homes induce that access to second homes usually implies access to nature, too. In this context distinctions can be made for everyday visits, visits during the weekend, and visits during longer leaves, for example, vacations. Table 6.4 (below) reveals diverging patterns for visits to nature among the different access groups. Obviously, people lacking access to second homes do not visit natural areas particularly during weekends and holidays as often as people with access to second homes. This differs dramatically particularly from the answers given by people owning a second home. The latter group claims to visit natural areas very often during everyday life, weekends and holidays. The group with regular access without ownership shows a similar pattern for everyday life and weekends with very frequent visits to natural areas. However, holidays are used to engage in other activities, where second homes and visits to natural areas obviously only play a lesser but still important role. Figure 6.2 (below) depicts a second home with beach house for sauna.

Hence. It can be argued that place attachment to the second home is not as developed as for owners, which is also supported by the fact that second home owners to a far greater extent rate gardening and other activities related to the second home as their most important activities during longer leaves. For the group gaining access through regular rental agreements visits to natural areas takes place particularly during weekends and holidays indicating that the second home at least in some cases is rented to facilitate visits to natural areas.

It is, however, not only visits to natural areas that distinguishes different access groups. Even the choice of activities varies between the groups (Table 6.5). Particularly recreational walks are significantly more frequent among second home owners and other households with regular access. Even biking on roads and various other activities are more often pursued by groups with access to second homes. Particularly regarding winter activities it is obvious that people with access to second homes are more engaged than those lacking access. Other activities such as walking with dog, mountain-biking, diving, hunting, canoeing and other water activities do not show any significant differences between the groups. Although Table 6.5 does not reveal whether second homes are a precondition for becoming involved in these activities, this appears to be likely, since second homes obviously offer environments suited for outdoor recreation. This is not least true for outdoor activities as alpine skiing which only can be pursued in relative few and remote

Table 6.4   Access to second homes and frequency of visits to natural areas.

Visits to natural areas		Never	Seldom	Often	Very often	Total
No access	Every day	7.4%	40.5%	33.5%	18.6%	100%
	Weekends	2.4%	25.6%	44.3%	27.7%	100%
	Holidays	0.6%	14.0%	40.8%	44.6%	100%
Access through rental	Every day	2.9%	44.9%	31.9%	20.3%	100%
	Weekends	0.0%	13.0%	49.3%	37.7%	100%
	Holidays	0.0%	5.8%	24.6%	69.6%	100%
Access without ownership	Every day	5.1%	40.4%	39.2%	15.4%	100%
	Weekends	0.6%	16.2%	49.8%	33.3%	100%
	Holidays	0.6%	6.4%	33.3%	59.7%	100%
Access through ownership	Every day	5.9%	32.4%	40.2%	21.4%	100%
	Weekends	0.0%	11.4%	45.4%	43.2%	100%
	Holidays	0.3%	4.1%	31.0%	64.7%	100%

$p < 0.05$

Figure 6.2   Vernacular second home with several buildings including a beach house for sauna, Stockholm archipelago.

locations in the Swedish mountains, requiring accommodation outside the everyday environment. Figures 6.3 and 6.4 depict traditional sports cabins and modern cabins in the mountains for the practice of sports.

Figure 6.3   Traditional sport cabins in the Hemavan mountain resort.

Figure 6.4   Modern sport cabins in Borgafjäll in the Swedish mountain range.

Table 6.5   Selected activities and access to second homes (grey shading marks the dominating category).

Activity	Access to second homes	Never	1–5 times	6–20 times	21–60 times	> 60 times
Recreational nature walks	No access	12.3%	32.2%	25.6%	16.8%	13.1%
	Access through rental	3.0%	31.3%	22.4%	22.4%	20.9%
	Access without ownership	6.2%	24.9%	31.8%	22.0%	15.1%
	Access through ownership	5.1%	18.0%	27.7%	26.9%	22.3%
Other recreational walks	No access	6.5%	14.2%	22.7%	28.3%	28.3%
	Access through rental	1.5%	9.1%	18.2%	37.9%	33.3%
	Access without ownership	3.3%	10.4%	19.3%	28.5%	38.6%
	Access through ownership	3.8%	11.7%	21.5%	25.9%	37.1%
Mountain hiking	No access	88.7%	9.5%	1.7%	.1%	.0%
	Access through rental	75.0%	19.1%	5.9%	.0%	.0%
	Access without ownership	78.6%	16.0%	3.9%	.6%	.9%
	Access through ownership	79.2%	15.0%	4.9%	.8%	.0%
Nordic walking	No access	76.5%	8.9%	6.5%	5.5%	2.6%
	Access through rental	69.1%	4.4%	8.8%	10.3%	7.4%
	Access without ownership	77.2%	7.7%	7.4%	2.7%	5.0%
	Access through ownership	68.6%	10.4%	8.7%	6.6%	5.7%
Biking on roads	No access	28.7%	17.3%	20.4%	14.1%	19.5%
	Access through rental	16.2%	17.6%	23.5%	25.0%	17.6%
	Access without ownership	19.6%	18.4%	20.5%	13.1%	28.5%
	Access through ownership	22.4%	18.9%	22.6%	17.0%	19.1%

Activity	Access to second homes	Never	1–5 times	6–20 times	21–60 times	> 60 times
Outdoor swimming	No access	29.7%	31.9%	23.7%	9.8%	4.9%
(lake, sea)	Access through rental	13.2%	26.5%	27.9%	22.1%	10.3%
	Access without ownership	15.8%	31.3%	33.0%	14.6%	5.4%
	Access through ownership	19.8%	28.7%	29.8%	13.0%	8.7%
Recreational fishing	No access	68.5%	21.2%	6.3%	3.0%	1.0%
	Access through rental	49.3%	34.3%	14.9%	1.5%	.0%
	Access without ownership	51.8%	29.2%	10.7%	5.1%	3.3%
	Access through ownership	49.2%	31.1%	14.9%	3.8%	1.1%
Picknick/outdoor barbecue	No access	24.6%	43.1%	23.3%	6.3%	2.6%
	Access through rental	10.3%	39.7%	42.6%	5.9%	1.5%
	Access without ownership	11.0%	40.5%	32.1%	13.1%	3.3%
	Access through ownership	17.8%	37.7%	34.8%	7.0%	2.7%
Cross-country skiing	No access	82.6%	10.6%	5.1%	1.5%	.1%
	Access through rental	67.2%	13.4%	14.9%	3.0%	1.5%
	Access without ownership	68.6%	17.4%	9.3%	3.6%	1.2%
	Access through ownership	60.1%	19.6%	14.1%	4.3%	1.9%
Alpine skiing	No access	83.2%	11.5%	3.9%	1.0%	.5%
	Access through rental	56.7%	14.9%	20.9%	6.0%	1.5%
	Access without ownership	70.8%	16.7%	9.8%	1.8%	.9%
	Access through ownership	72.0%	15.5%	9.5%	2.2%	.8%

$p < 0.05$

A difference between the groups can also be noted regarding the most common activities during everyday life, weekends and holidays, respectively. Though recreational walks are the most common activity for all groups in everyday life, second home renters and owners tend to prefer nature walks during weekends, and weekends and holidays, respectively (Table 6.6). Alpine skiing seems to be an important activity for second home renters, too.

Table 6.6　Most important activities during everyday life, weekends and holidays (only one choice).

Access to second homes		Top 3 activities		
No access	Everyday life	Recreational walks (32.1%)	Gardening (11.2%)	Road cycling (10.4%)
	Weekends	Recreational walks (25.2%)	Nature walks (14.9%)	Gardening (10.2%)
	Holidays	Recreational walks (14.9%)	Nature walks (10.5%)	Sunbathing (12.4%)
Access through rental	Everyday life	Recreational walks (42.4%)	Road cycling (10.6%)	Jogging (9.1%)
	Weekends	Nature walks (29.2%)	Recreational walks (20.0%)	Gardening (9.2%)
	Holidays	Recreational walks (15.2%)	Alpine skiing (12.1%)	Sunbathing (12.1%)
Access without ownership	Everyday life	Recreational walks (36.1%)	Gardening (9.5%)	Jogging (8.9%)
	Weekends	Recreational walks (21.3%)	Nature walks (16.5%)	Gardening (8.7%)
	Holidays	Recreational walks (13.6%)	Nature walks (11.4%)	Sunbathing (9.9%)
Access through ownership	Everyday life	Recreational walks (31.7%)	Gardening (9.8%)	Nature walks (9.5%)
	Weekends	Nature walks (21.5%)	Recreational walks (16.3%)	Gardening (15.7%)
	Holidays	Nature walks (17.1%)	Recreational walks (13.5%)	Gardening (13.8%)

The fact that second home access facilitates outdoor recreation can also be proved by using analytical statistics. For that purpose, an index was created mirroring the outdoor recreation habits of the respondents. Accordingly, the responses to the questions regarding the frequency of being out in nature in everyday life, at weekends and on holidays were recoded in the following way: never = 0; seldom

= 1; often = 2; and very often = 3. The resulting index, summing up the values for the time categories, thus resulted for each respondent in a value between 0 and 9 with an average of 6.2, a median of 6 and a standard deviation of 2.0. A stepwise linear regression analysis identified the variable 'No access to second home' as the single most powerful in explaining variations in the index values ($r^2 = 0.026$). Accordingly, respondents with no access to second homes had an index value that was 0.63 lower than the one for respondents with some kind of access.

After excluding the 'No access' variable, the other categories of second home access were included in the model. For the analysis, variables representing sex, age, income, education, occupational status, everyday home environment, and access to second home categories were used. The final model, which has a limited explanatory value for the variations in the index ($r^2 = 0.129$) included a total of 11 significant variables (Table 6.7).

Table 6.7   The impact of second home access on being out in nature.

	B	Std. error	t	Sig
(Constant)	5,622	,121	46,511	,000
Sex (Female = 1)	,414	,096	4,317	,000
Retired	,694	,134	5,189	,000
Single	-,342	,126	-2,702	,007
Household with children	-,237	,107	-2,209	,027
Highest education secondary school	-,531	,134	-3,952	,000
Countryside with only few other houses in sight	,999	,150	6,676	,000
Small rural settlement, villa dwellings only	,804	,160	5,033	,000
Urban area, rental housing only	-,450	,133	-3,385	,001
Access through rental	,735	,234	3,136	,002
Access without ownership	,499	,124	4,013	,000
Access through ownership	,747	,121	6,177	,000

$r^2 = 0.129$

The model indicates that a rural home environment has a strong positive influence on respondents' being out in nature. Besides, being retired and female are other factors positively influencing the value for being out in nature, while children, a single status and a basic school education only have negative impacts.

Regarding the access to second home categories, it becomes clear that access through ownership and rental have almost the same positive impact on being out in nature. However, even access without ownership, for instance to the parents' second home, has a positive impact on the frequency of nature visits. Access through ownership and rental have higher impacts mirroring the respondents' active decisions for a second home.

Conclusions

This chapter departed from the notion that second home owners' motive for having a second home was the desire for being out in nature. Hartig and Fransson (2009) even argued that second home owners enjoy better health than people lacking access to second homes. Analyses have, however, often been limited to second home owners, while other groups with access to second homes have been neglegcted. The results of the 'Friluftsliv 2007' survey show that second home owners and households renting second homes on a regular basis are more often in nature in everyday life, at weekends and on holidays than people without access to second homes and people with access through others. Many second home owners have already in their everyday life easy access to nature and thus second homes are no necessary means for gaining access. Hence second homes cannot be said to be complementary spaces for all. Rather, they are to be seen as investments for households with an interest in outdoor recreation enabling easy access to nature and nature-based activities. Thus, the assumption that access to second homes is beneficial for good health should also be seen in the context of outdoor recreation habits. The results of this study indicate that outdoor recreation at home and at the second home go hand in hand.

It is therefore not possible to conclude that improved access to second homes, for example facilitated by investment programs, would improve people's engagement in outdoor recreation (Figure 6.5). Second homes are no cure for a potentially decreasing level of interest in these activities. They are rather a mirror of current interest, which may be the result of childhood experiences of outdoor recreation and second homes.

This study, however, did not ask whether people with and without access to second homes, respectively, have similar or different motives and interests for outdoor recreation. It can thus only be speculated to what extent people without second home access have chosen that position deliberately or whether they are constraint by e.g. economic situation or physical disabilities. Certainly, this would be a question for further research, although it is difficult to provide an answer based solely on the current survey.

Wolfe (1952) considered second home tourism to be a seasonal relocation of urban life into the countryside. The result of this study indicates that this might be the case. However, at least for Sweden, this means a seasonal relocation of a rather active group already interested in outdoor recreation even during everyday.

Figure 6.5   Cottage construction in Skeppsvik, Northern Sweden.

Acknowledgement

The data for this chapter was collected within the framework of the research program Outdoor Recreation in Change (www.friluftsforskning.se), a coopertion of six Swedish universities financed by the Swedish Environmental Protection Agency. Moreover, work on this chapter was also funded by the Swedish Research Council for Environment, Agricultural Sciences and Spatial Planning (Formas). The support of both funding organizations is hereby acknowledged.

References

Buller, H. and Hoggart, K. 1994. *International Counterurbanisation: British Migrants in Rural France*. Aldershot: Ashgate.
Chaplin, D. 1999. Consuming work/productive leisure: the consumption patterns of second home environments. *Leisure Studies*, 18, 41–55.
Fredman, P., Karlsson, S.E., Romild, U. and Sandell, K. (Eds.) 2008. *Besöka naturen hemma eller borta? Delresultat från en nationell enkät om friluftsliv och naturturism i Sverige*. Östersund: Friluftsliv i förändring.

Gallent, N., Mace, A. and Tewdwr-Jones, M. 2003. Dispellying a myth? Second homes in rural Wales. *Area*, 35, 271–84.

Gallent, N., Mace, A. and Tewdwr-Jones, M. 2005. *Second Homes: European Perspectives and UK Policies*. Aldershot: Ashgate.

Gallent, N., Shucksmith, M. and Tewdwr-Jones, M. 2003. *Housing in the European Countryside: Rural Pressure and Policy in Western Europe*. London: Routledge.

Haldrup, M. 2004. Laid-back mobilities: second-home holidays in time and space. *Tourism Geographies*, 6, 434–454 .

Hall, C.M. and Müller, D.K. (Eds.) 2004. *Tourism, Mobility and Second Homes: Between Elite Landscape and Common Ground*. Clevedon: Channel View.

Halseth, G. 1998. *Cottage County in Transition: A Social Geography of Change and Contention in the Rural-Recreational Countryside*. Montreal and Kingston: McGill-Queen's University Press.

Hartig, T. and Fransson, U. 2009. Leisure home ownership, access to nature, and health: a longitudinal study of urban residents in Sweden. *Environment and Planning* A, 41(1), 82–96.

Hörnsten, L. and Fredman, P. 2000. On the distance to recreational forests in Sweden. *Landscape and Urban Planning*, 51(1), 1–10.

Jaakson, R. 1986. Second-home domestic tourism. *Annals of Tourism Research*, 13, 367–91.

Kaltenborn, B.P. 1997a. Nature of place attachment: study among recreation home owners in southern Norway. *Leisure Sciences*, 19, 175–89.

Kaltenborn, B.P. 1997b. Recreation homes in natural settings: factors affecting place attachment. *Norsk Geografisk Tidsskrift*, 51, 187–98.

Kaltenborn, B.P. 1998. The alternate home: motives of recreation home use. *Norsk Geografisk Tidsskrift*, 51, 187–98.

Kaltenborn, B.P., Haaland, H. and Sandell, K. 2001. The public right of access: some challenges to sustainable tourism development in Scandinavia. *Journal of Sustainable Tourism*, 9(5), 417–33 .

King, R., Warnes, T. and Williams, A. 2000. *Sunset Lives: British Retirement Migration to the Mediterranean*. Oxford: Berg.

Lundmark, L. and Marjavaara, R. 2005. Second homel localizations in the Swedish mountain range. *Tourism*, 53(1), 3–16.

Marjavaara, R. 2007. Route to destruction: second home tourism in small island communities. *Island Studies Journal*, 2, 27–46.

Marjavaara, R. 2007. The displacement myth: second home tourism in the Stockholm archipelago. *Tourism Geographies*, 9(3), 296–317.

McHugh, K.E. 2006. Citadels in the sun, in *Multiple Dwelling and Tourism: Negotiating Place, Home and Identity*, edited by N. McIntyre, D. Williams and K. McHugh. Wallingford: Cabi, 262–77.

McIntyre, N., Roggenbuck, J.W. and Williams, D.R. (2006). Home and away: revisiting 'escape' in the context of second homes, in *Multiple Dwelling and*

Tourism: Negotiating Place, Home and Identity, edited by N. McIntyre, D. Williams and K. McHugh. Wallingford: Cabi, 114–28.

McIntyre, N., Williams, D. and McHugh, K. (Eds.) 2006. *Multiple Dwelling and Tourism: Negotiating Place, Home and Identity*. Wallingford: CABI.

Müller, D.K. 1999. *German Second Home Owners in the Swedish Countryside: On the Internationalization of the Leisure Space*. Östersund: Etour.

Müller, D.K. 2002a. German second home development in Sweden, in *Tourism and Migration: New Relationships between Production and Consumption*, edited by C.M. Hall, and A.M. Williams. Dordrecht: Kluwer, 169–86.

Müller, D.K. 2002b. Second home ownership and sustainable development in northern Sweden. *Tourism and Hospitality Research*, 3(4), 345–55.

Müller, D.K. 2004. Second homes in Sweden: patterns and issues, in *Tourism, Mobility and Second Homes: Between Elite Landscape and Common Ground*, edited by C.M. Hall and D.K. Müller. Clevedon: Channel View, 244–58.

Müller, D.K. 2006. The attractiveness of second home areas in Sweden: a quantitative analysis. *Current Issues in Tourism*, 9(4&5), 335–50.

Müller, D.K. 2007. Introduction: second homes in the Nordic countries: between common heritage and exclusive commodity. *Scandinavian Journal of Hospitality and Tourism*, 7(3), 193–201.

Paris, C. 2010. *Affluence, Mobility and Second Home Ownership*. London: Routledge.

Perkins, H.C., and Thorns, D.C. (2006). Home away from home: the primary/ second-home relationship, in McIntyre, D. Williams, and K. McHugh, *Multiple Dwelling and Tourism: Negotiating Place, Home and Identity* (ss. 67–81). Wallingford: Cabi.

Pitkänen, K. 2008. Second home landscape: the meaning(s) of landscape for second home tourism in Finland. *Tourism Geographies*, 10, 169–92.

Sievänen, T., Pouta, E. and Neuvonen, M. 2007. Recreational home users: potential clients for countryside tourism? *Scandinavian Journal of Hospitality and Tourism*, 7(3), 223–42.

Stedman, R.C. 2006. Places of escape: second-home meanings in northern Wisconsin, USA. in *Multiple Dwelling and Tourism: Negotiating Place, Home and Identity*, edited by N. McIntyre, D. Williams and K. McHugh. Wallingford: Cabi, 129–44.

Tuulentie, S. 2007. Settled tourists: second homes as a part of tourist life stories. *Scandinavian Journal of Hospitality and Tourism*, 7(3), 281–300.

Van Patten, S.R. and Williams, D.R. (2008). Problem sin palce: using discursive social psychology to investigate the meanings of seasonal homes. *Leisure Sciences*, 30, 448–64.

Williams, A.M., and Hall, C.M. 2002. Tourism, migration, circulation and mobility: the contingencies of time and place, in *Tourism and migration: New Relationships between Production and Consumption*, edited by C.M. Hall and A.M. Williams. Dordrecht: Kluwer, 1–52.

Williams, D.R. and Kaltenborn, B.P. 1999. Leisure places and modernity: the use and meaning of recreational cottages in Norway and the USA, in *Leisure/ Tourism Geographies: Practices and Geographical Knowledge*, edited by D. Crouch. London: Routledge, 214–30.

Wolfe, R.I. 1952. Wasaga Beach: the divorce from the geographic environment. *Canadian Geographer*, 2, 57–65.

Chapter 7

Vacation Homes in France since 1962

Jean-Marc Zaninetti

In France, a vacation home is a type of second home that is a property for seasonal, mostly recreational use. It is occupied for a maximum duration of five months a year;[1] otherwise, it would be the household's primary residence. There are two main different ways to spend vacations – travelling, or staying. Vacation properties are part of the second option. Summer homes have a long history in France (Cribier 1973), but we observe that vacation home development is a major feature of the second half of the twentieth century. At least four types of vacation homes can be identified according to their location: urban, countryside, coastal and mountain vacation homes. As the number of urban second homes is decreasing, countryside summer homes' total number has stabilized. However, this is the balance between two conflicting trends, decline in the vicinity of urban centres, and increase in peripheral isolated rural regions. Vacation properties develop mostly in mountain areas and along the coastline. This results in increasing regional disparities, as parts of the French territory display an increasing specialisation in a tourism-oriented economy.

This chapter mostly relies on quantitative analysis. It is based on first-hand analysis of the most recent census data that have been released by the French Statistics Institute (INSEE). GIS and spatial analysis are used to bring forward new insight on the changing geography of vacation homes in France, with a focus also on the economic significance of residential development in different regions. The environmental issues and planning challenges, raised particularly in coastal and mountain areas are discussed. This is finally illustrated by a detailed case-study of the coastal area of the department of Vendée, on the Atlantic Ocean, south of Nantes.

The Historical Development of Vacation Homes in France

According to the latest available Census of population and housing figures, there were 3.1 million vacation homes and occasional housing units in France[2] at the

---

1   INSEE definition, http://www.insee.fr/fr/methodes/default.asp?page=definitions/residence-secondaire.htm

2   This chapter considers only that part of the French territory that is located in Europe. We exclude the Overseas Regions from our study, despite the fact that second homes are also common in these regions.

beginning of the year 2008.[3] Second homes and occasional housing units make 9.7 per cent of the total housing stock. This is an important share in comparison with other developed countries. There are only 4.65 million vacant housing units for seasonal use in the USA according to the 2010 Census, 3.5 per cent of the housing stock. There were less than 1 million vacation homes in Metropolitan France in 1962.[4] Vacation home development took a very rapid pace from 1962 to 1990. Nowadays, vacation home growth persists, but it has cooled down. Second homes have developed along a long time span in France. This history can be retraced at least to the seventeenth century.

The Downward Imitation of an Aristocratic Custom

Many social practices spread by downward imitation along the social ladder. The development of tourism is a representative example of these imitation processes. There is a longstanding tradition of second homes in France. Before the French Revolution, the nobility used to have a family castle in the countryside and a *hôtel particulier*, a private house in Paris or Versailles for occasional use. After Louis XIV strengthened the royal power, the aristocracy shifted its principal residence to the Parisian *hôtel particulier* in order to attend the court for benefits from the King. The use of its ancestral castle became limited for relatively short summer stay only. As its wealth and power grew in the eighteenth century, the gentry imitated the nobility by buying mansions and castles in the countryside of the larger Paris region for recreational use during the summer season. These eighteenth century 'summer residences' are the forebears of modern second homes. This custom survived the Revolution and became a distinctive trait of the upper class. The mid-nineteenth century introduced the British-born innovation of sea bathing in France during the Second Empire (1852–70). The upper class built vacation properties along the coastlines in the vicinity of the newly built railway stations. Railways took an active role to attract customers to these newly developed coastal resorts. They advertised these destinations in order to attract more customers. The closer from Paris the best, the Channel attracted the larger number of visitors in Deauville, Trouville, Cabourg, etc. Short relaxation stays in hotels was eventually followed by the development of a vacation property. The role of literature (i.e. Marcel Proust) is also important to understand the expanding desire of the urban middle class to imitate the upper class. This opportunity was created during the years of high economic growth of the 1960s.

---

3 Since 2004, the renovated French census is based on yearly surveys. Census figures are moving estimates tallying 5 years of survey. The '2008' census is actually an estimation based on the 2006–2010 surveys.

4 The Census of 1962 is the first census of housing that counts second homes in France.

Expansion of Vacation Homes in the Second Half of the Twentieth Century

From its early beginning, the mass development of vacation homes took different guises. The old tradition of the countryside summer home has not disappeared, but it has been completed by the nineteenth century-born tradition of the seaside vacation property, and the new twentieth century-born trend of mountain winter vacation. Finally, there are still some urban *pied-à-terres* for occasional use in major urban cores, similar in their purpose to the early *hôtels particuliers* of the sixteenth and seventeenth centuries. Approximately six per cent of second homes are urban *pied-à-terre*.

Table 7.1   Vacation homes in Metropolitan France 1962–2008.

Years	Vacation homes and occasional residences* stock	Share of the total housing stock	Yearly average increase
1962	970 103	5.9%	
1968	1 258 940	6.9%	48 140
1975	1 676 655	8.0%	59 674
1982	2 265 672	9.6%	84 145
1990	2 814 291	10.7%	68 577
1999	2 902 093	10.1%	9 756
2008	3 098 999	9.7%	21 878

*   The occasional residences are counted separately since 1990 only. There were 0.4 million occasional residences in 1990 and there is only 0.2 million occasional residences left in 2008.

Source: INSEE censuses.

The development of vacation homes in France culminated in the late 1970s before to slow down since the 1990s. The relative share of second homes in the entire housing stock is down one per cent from its maximum of 1990. How can this change be explained? This points out first and foremost that the French middle class's real disposable income growth slowed down after 1990. This change is also probably related to changing lifestyles privileging travelling over staying vacation. An ageing population might also be an explanatory factor. Whatever the causes, this overall stabilization covers major regional disparities.

Increasing Regional Specialization

The comparison between 1968 and 2008 at the community level helps us to understand the major spatial and temporal change that occurred in France. Coastal and mountain areas have strongly specialized in the development of vacation properties, as second homes tend to decline elsewhere since 1990.

The housing density (Figure 7.1) is the best way to appreciate the absolute human impact on the land. Vacation home average density increased from 2.3/sq km in 1968 to 5.7/sq km in Metropolitan France in 2008. Higher than average densities were located in three types of regions in 1968:

1. Many second homes were concentrated in the vicinity of major cities, with two major clusters: the greater Parisian region and the Lyons – St Etienne – Grenoble triangle.
2. Coastal resorts were hotspots of vacation home development, with prominent concentrations in traditional resorts: The Azure Coast, La Baule, St Malo, Deauville, etc.
3. Some mountainous regions were already attractive for vacation property development: the Vosges Mountains, Savoy, and small enclaves in the Central Plateau and the Pyrenees.

Forty years later, the suburban clusters have decreased and moved away from city centres. Minor clusters are still to be found in western Burgundy and in eastern Normandy. Conversely, vacation home development multiplied along the coastlines, particularly the Atlantic and Mediterranean Coasts, and in almost every mountain area.

Vacation Home Development in Mountain Areas

The relative specialisation of mountain areas in tourism was already significant in the 1960s. Twenty per cent of the French territory is located above the elevation of 500 meters, which is the planning threshold of 'mountain' areas for French government agencies. Less than 225,000 vacation homes were located above 500 m in 1968. Mountain vacation properties formed 18 per cent of the existing stock of vacation home in France, a figure to be compared to the location of 7.5 per cent of the total housing stock in mountain areas. Vacation homes constituted already more than 16 per cent of the housing stock located above the threshold elevation of 500 meters in 1968.

Forty years later, the stock of mountain vacation properties has tripled, exceeding 740,000 in 2008. Mountain vacation properties form 24 per cent of the present stock of vacation home in France, a figure to be compared to the location of 7.8 per cent of the total housing stock in mountain areas. The relative specialisation of mountainous regions in tourism has increased significantly. According to the 2008 Census, vacation home constitute almost 30 per cent of

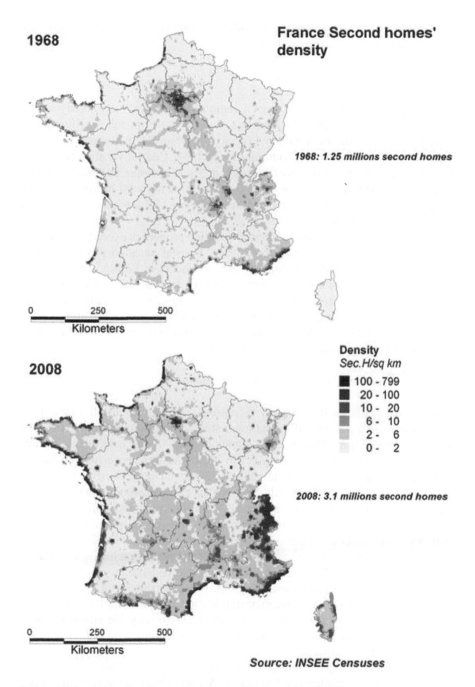

Figure 7.1   Vacation home density in France 1968–2008.

Source: INSEE Censuses.

the total housing stock above the elevation of 500 meters. The Alps display the largest concentration, particularly in Savoy.

Vacation Properties along the Coastline

The relative specialisation of coastal regions in tourism was already strong in the 1960s. Communities located less than 10 km from the coastline cover only 6.5 per cent of Metropolitan France. Less than 327,000 vacation homes were located in these coastal communities in 1968. Coastal vacation properties formed 26.5 per cent of the existing stock of vacation home in France, a figure to be compared to the location of 13.3 per cent of the total housing stock on the coastal strip. Vacation home constituted already 13.5 per cent of the coastal housing stock in 1968.

Forty years later, the stock of coastal vacation properties exceeded 1,287,000 in 2008, a fourfold increase in forty years. Coastal vacation properties form 41.5 per cent of the existing stock of vacation home in France, a figure to be compared to the location of 16.3 per cent of the total housing stock in coastal communities. The relative specialisation of coastal regions in seasonal residence is increasing. According to the 2008 Census, vacation home constitute almost 25 per cent of the coastal housing stock. The Mediterranean displays the largest concentrations, except in the Rhone River delta.

The Stable Number of Countryside Summer Homes

Non-coastal plain vacation homes increased +57 per cent in 40 years. They formed only 34.5 per cent of the national stock of vacation properties in 2008, versus 55 per cent in 1968. Second homes formed 4.7 per cent of the housing stock in non-coastal plain communities in 1968. Vacation property development in France tipped in relative term in 1990, when second home constituted 10.7 per cent of the national housing stock. Non-coastal plain vacation homes increased 82 per cent from 1968 to 1990 to reach 1.24 million units, including 0.4 million urban *pied-à-terre*. In 1990, 6.2 per cent of non-coastal plain housing was actually vacation homes or occasional residences. From 1990 to 2008, this figure dropped to 1.07 million non-coastal plains second homes including 0.2 million urban *pied-à-terre*, a 13 per cent decline. This now constitutes only 4.4 per cent of the housing stock in 2008 in these areas. Basically the decline of occasional residences explains this decline, their number has been halved. Recent changes display diverging regional patterns (Figure 7.2). Countryside summer homes have slightly increased in some peripheral rural regions, and declined in the vicinity of urban centres.

Except for a few peripheral enclaves (i.e. Vosges mountains), growth has concentrated in the southern half of France. Vacation home development clusters are visible in the Northern Alps, central Pyrenees, and along the Atlantic and Mediterranean coasts. Smaller enclaves are visible elsewhere, Périgord Noir,

Southern Ardèche, Corsica, etc. Decreasing density is visible in the major part of the country. The Parisian urban area displays the sharpest decline, but decreasing density can be observed around almost every major urban centre.

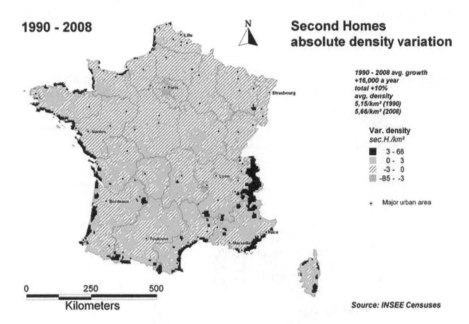

Figure 7.2   Change in vacation home density in France 1990–2008.
Source: INSEE Censuses.

The Economic Significance of Vacation Homes in Touristy Resort's Economic Specialisation

On average, 74 per cent of the French resident population spent 985 million nights per year outside of the principal residence in 2004, of which only 11 per cent abroad (INSEE 2008). Vacation homes accounted for 16 per cent of this total. In comparison, rented self-catering cottages totalled only 8 per cent and camping, 10 per cent. Vacation properties are an important part of the hospitality equipment in holiday resorts. A map of the relative share of vacation homes in the total housing stock (Figure 7.3 below) is very similar to a map of commercial hospitality facilities, hotels, camping, cottages to rent, and others.

This map displays the relative specialisation in tourism of coastal and mountainous areas. The coastline concentrates 41.5 per cent of French vacation homes, and the mountains 25 per cent. More than half the housing stock is made of vacation properties in numerous enclaves along the coastline, the Alps the Central Pyrenees, many rural sectors of Corsica and the south-eastern fringe of the Central

Plateau (Cevennes). Between one in four and one in two housing units are vacation homes in other mountainous areas. On average, 30 per cent of the housing units located above the elevation of 500 m are vacation homes. Vacation properties are between one in four and one in two housing units along most coastlines, except in highly urbanized areas. On average, 25 per cent of housing units located within 10 km from the coastline are vacation homes.

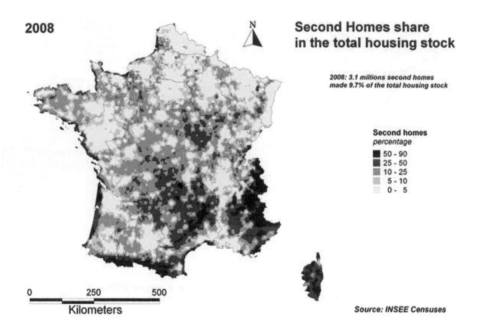

Figure 7.3   Vacation home share in France 2008.

Source: INSEE Censuses.

Foreign-owned Second-Homes

Foreigners own approximately 8.5 per cent of the total French vacation properties. Most of these foreign homeowners come from neighbouring European countries. British citizens form 24 per cent of these foreign second-home owners in France. British buyers invest along the Atlantic coastline and in South-western regions (Périgord, Quercy), as Dutch buyers do (8 per cent of foreign second-home owners). Swiss (14 per cent), Italians (14 per cent), Germans (13 per cent) and Belgians (11 per cent) invest preferably in border regions. Italians favour the Azure Coast, Germans prefer Alsace, Belgians the North coast, and the Swiss are present in the Jura Mountains and in Savoy.

Factors of Change

Identifying relevant explanatory factors of these major recent changes is a complicated task for at least two reasons. First, there are at least four different niche markets for vacation homes: coastal vacation properties, mountain vacation properties, countryside summer homes and urban *pied-à-terre*. Second, regional situations are strongly contrasted. Nevertheless, we may point out a few hypotheses.

1. **Macroeconomic factors** are prominent. During the 1968–90 period, strong inflation was associated with a sustained economic growth. Despite high interest rates, the actual cost of housing credit was limited by rising wages. Housing costs were limited, and many urban middle-class households were able to buy vacation properties, particularly along the coastlines or in mountain resorts, according to their preferences for winter or summer vacations. These conditions have changed since 1990. Inflation has been tamed, wages are stagnant and housing prices have skyrocketed. The actual cost of credit is soaring and economic conditions are no longer very supportive for a continued expansion of vacation home development.

2. **Retirement migration** is another likely factor in the current observed slowdown. The geography of retirement migrations is very similar to the geography of vacation homes (Zaninetti 2011). Retirees who move tend to quit major urban centres, and prefer to settle either in the countryside, or preferably along the coastline. Parisian retirees who move to neighbouring region prefer Burgundy, the Loire Valley and Normandy, three former clusters of vacation cottages. Farther from Paris, Brittany, the Atlantic and the Mediterranean coasts, and south-western regions to a lesser extent are favourite destinations for retirement migrations. In contrast, mountains are not attractive to retirees. At retirement age, some people have the option of switching the respective role of their Parisian residence and their former vacation home. They can either sell their former urban residence to benefit from the premium metropolitan market prices, or keep this apartment as a family heritage for their children and grandchildren. Such families gain access to better higher education facilities and better job opportunities. Recent retirement migrations contribute to mitigate an overspecialisation in second housing and tourism. Bringing permanent residents to coastal resorts, they stimulate the permanent development of services. As the proportion of retirees in the entire population is increasing, population ageing might further reduce the development of vacation homes in the incoming decades.

3. Developing **mountain and costal resorts** deliver plenty of cheap apartments on the market since 50 years, creating a booming market for vacation properties. Let us consider two characteristic examples; the department of Savoie located in the Northern Alps and the department of Hérault along the Mediterranean coast. Approximately 75 per cent of the local stock of

vacation homes located in the department of Savoy was apartments in condominiums in 1990, the national average being only 32 per cent at this period. A similar situation could be observed in the department of Hérault, where 64 per cent of vacation homes were apartments in 1990. This trend continues as the Vacation home stock expands in coastal and mountain resorts. In Savoy, vacation homes increased by 26 per cent from 1990 to 2008. Single detached houses used as vacation homes increased by 1,134 units only. Meanwhile, apartments used as vacation properties increased by 23,440 units. Nowadays, apartments make 78 per cent of vacation homes in the department of Savoy, as the national average has risen to 37 per cent (INSEE Census 2008).[5] The situation is more varied in the department of Hérault, where vacation home have increased by 32 per cent from 1990 to 2008. Nowadays, apartments make 57 per cent of vacation homes in the department of Hérault. The number of apartments used as vacation properties increased by 12,640 units, mostly along the coastline. Meanwhile, the number of detached houses used as vacation properties has risen by 10,050 units, mostly in the vast backcountry. These examples illustrate the importance of local development strategies in the increasing territorial specialisation in part of the country.

4. **Suburban development** is the main driver of the diminishing stock of second homes around urban centres. People commute over increasingly long distances to work. Most of the territory is now included in extensive commuter catchment's areas placed under urban influence. If you consider non-coastal areas, the role of urban influence on the second home market is clearly perceptible.

According to the latest census results, commuter pattern create extensive commuter catchment's areas, and the INSEE define as 'urban areas' at the minimum threshold of 40 per cent of commuters in the resident active population. These urban areas regroup 90 per cent of the resident population. Half the French territory is included into non-coastal urban areas. These regions regroup 76 per cent of the resident population, but only 27 per cent of the vacation home stock in 2008. The number of vacation properties and urban *pied-à-terre* is 16 per cent down from its 1990 figure. Since 1980, housing development is inferior to housing demand, particularly in large urbanized areas. In consequence, vacant homes, occasional residences and second homes are slowly converted into principal residences not only in the urban agglomeration, but also in its entire commuter catchment's area. The appeal of coastal areas towards vacation homeowners is so powerful that the urban factor is completely cancelled along the coastal strip.

---

5   Key figures about the French territory, its population and its economy are available at the INSEE website, http://www.insee.fr/fr/themes/.

Table 7.2   Metropolitan urban areas in France 2008.

Typology of urban areas	Share of the metropolitan territory (area)	Share of the metropolitan population (2008)	Share of the vacation home stock (2008)	Vacation home stock change 1990–2008 (%)
Coastal fringe* (any type)	7%	14%	42%	+35%
Non-coastal urban cores	11%	53%	13%	-15%
Non-coastal Inner Suburban commuter catchment's area (40% threshold)	39%	23%	14%	-17%
Non-coastal outer suburban commuter catchment's area (20% threshold)	19%	6%	8%	+2%
Non-coastal isolated rural communities (including most mountainous areas)	24%	4%	23%	+18%

*   The coastal fringe is defined at a distance of 10 km from the coastline.

Source: INSEE newly defined urban areas (2011), annual census survey 2006–2010.

Significance and Challenges of Vacation Homes for Regional Economy and Planning

Vacation home development is a major ingredient of the tourism economy. Its economic significance can be illustrated by the concepts of 'residential economy' or 'presential economy'. This industry generates a great part of the national income, but it raises severe environmental concerns and several planning issues.

The Economic Significance of Seasonal Daytime Population

'Presential economy' is a new concept that has been forged by Christophe Terrier and Laurent Davezies in recent years (Terrier 2006, Davezies 2008). Differently from the classical regional economic analysis, which focuses on industrial location (Benko 1998), Terrier and Davezies are interested in the actual location of income consumption. According to their theory, the importance of the service sector is not proportional to the 'permanent' resident population's income, but to the changing existing population, which fluctuates from day-to-day according to mobility, particularly during holidays. A report by the Direction of Tourism (Le Garrec 2005) mentions that an amount of 134 billion Euros was spent on the domestic market outside of the city of residence by French people during domestic leisure travels during the year 2004. In addition, spending by foreign visitors in France exceeded the expenditures of French residents abroad during the same year by 10 billion Euros. The total travel spending accounted for 8 per cent of the French GDP in 2004.

The French census survey takes place in March. The 'census population' is counted in its 'principal residence'. This is a good approximation of the year-round 'resident population'. The 'present population' at any time is the result of a simple calculation. The population present in any locality is the resident population, minus the absent persons who are away at this time of the year, plus the outside visitors who are present locally at this time of the year. Part of these seasonal residents is actually vacation homeowners. Theoretically, one could draw a yearly curve of the 'present population'. The French Ministry of Tourism conducted a large survey in 2005 to get better knowledge of this phenomenon (Terrier 2006). Let us illustrate these results by an example.

The department of Savoy contains many mountain resorts that fill in during the winter ski season. With an estimated 'resident population' of 392,300 in 2004, the survey estimated a peak of 'present population' of 747,000 in February 2005 and a minimum of 376,100 in November 2005. The 'present population' is a time-dependent concept. The number of visitors is calculated by the number of nights spend in Savoy. This can be converted into a 'permanent resident equivalent' by dividing the number of nights spend locally by 365. In Savoy, this 'permanent population equivalent' supplement has been estimated to 133,200 persons in 2005, including 66 per cent French and 34 per cent foreign visitors. In consequence, the potential consumption for many service businesses should be calculated on the basis of a 'present population' of 525,500 in 2005, a 34 per cent increase from the census population estimate. Figure 7.4 illustrates this calculation for the 96 departments of Metropolitan France.

There is a linear correlation ($r=+0.93$) between the relative specialisation in vacation homes and the relative gain or loss of population due to leisure travel. The economic potential for service businesses is larger than expected in most mountain and coastal departments, except in the departments with large metropolitan areas, i.e. Marseille in the Bouches-du-Rhône. The so-called 'presential economy' is closely related with the recent trend of development of the Southern and Western parts of the country, and the population redistribution leaving Northern and Eastern regions to resettle in Western and Southern France. On average, the balance between the 'present population' and the 'resident population' reduces the territorial imbalances between urban and rural areas, transferring income

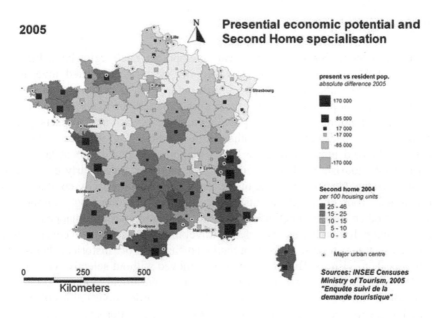

Figure 7.4   Correlation between vacation home specialisation and visitor-based economic potential.

Source: INSEE Censuses.

from heavily urbanized to mostly rural departments. However, this balance does not work for North Eastern France, which isn't attractive neither for tourism nor for vacation homes. This is a possible long-term consequence of the twentieth century wars that transformed the North-eastern region into a military buffer zone. Industrial legacy is another more important explanation factor. North-eastern France specialized in manufacturing from during the nineteenth century. This period has left a dull industrial landscape and the wars ruined a large part of their pre-industrial heritage. In contrast, Southern and Western France was mostly left rural and agricultural during the nineteenth century. They suffered rural to urban migrations during the nineteenth and early twentieth century, but this spared them the 'modernisation' of their built heritage. Mostly spared by the destructions of war, these regions turned to become attractive because of their pre-industrial landscape and heritage during the latter part of the twentieth century. They have become attractive both for tourism and vacation homeowners. Most mountains are located in the South – eastern part of France. As sea-bathing and water sports become major leisure activities, coastal resorts located along warmer seas are preferred over colder ones. Departments located along the Channel were forerunners in vacation home development in the nineteenth century because of the vicinity of Paris (Clary 1984), but recent development switched to the warmer Atlantic and Mediterranean coastlines after the Second World War (Barbier 1977).

Planning Issues

Presential economy is a dead end for local development. Each region has a different potential for 'commodification'. However, this requires dedicated investment in infrastructures and services. The infrastructure has to be proportionate not to the resident population, but to the peak population. In many coastal areas, that means doubling the census population. Soaring demand impacts public investment in roads, sewer and other infrastructures. Unfortunately, the cost of this oversized infrastructure has to be supported in part by local taxpayers. Furthermore, tourism tends to be an exclusive land-use, particularly in crowded coastal regions. Consumptive use of the landscape does not easily combine with production. Tourism tends to oust agriculture, mining, logging, commercial ports and other industrial activities, as industrial landscapes are unappealing for tourists and pollution harms severely the travel industry. The typical consumptive patterns of travel tend to recreate a sanitized 'natural' or 'historical' landscape that appeals to visitors. But this cannot be achieved without selective choices by the local stakeholders and a significant amount of investment (Judd 2003). Local stakeholders have been swift to understand the economic opportunities of the so-called 'presential economy'. High demand for land is a major source of unearned benefits, and that implies opportunistic and even corrupt policies (i.e. the Azure Coast). However, Davezies warns against this trend (2008). There is no possible gain in productivity in residential development. The most attractive places become more and more crowded, any further economic development will eventually be impossible because of exclusive land-use patterns. At some point, the residential development potential is saturated and excess urbanisation harms the natural and cultural heritage that created the region's attractiveness in the first place. This problem is paramount in coastal and mountain areas. Because it basically displaces income from one region to another, tourism is not a reliable economic development powerhouse. It is extremely sensitive to the overall economic volatility, as we can see it since 2008. Last but not least, tourism-related businesses tend to create low-skilled and low-wage jobs. This can be very damaging to the overall human capital of the regions that choose to overspecialize in the presential economy (Baumol 1967, Copeland 1991, Chao et al. 2006).

Environmental Issues for Coastal and Mountain Areas

Tourism infrastructure and vacation properties contribute strongly to increase the urbanisation pressure on the coastal environment. Including Lake Geneva, 898 communities which actually have a coastal stretch make up 4 per cent of France's area. More than 4.24 million housing units, 13.3 per cent of the national total, huddle in these coastal communities. The coastal average density, 193 housing units per square kilometer, is nearing urban level. Urban density in France is 230 housing units per square kilometer and the national mean density is 58 housing units per square kilometer. The coastal housing stock increased by nearly 440,000 from 1999 to 2008, an 11.5 per cent increase against a national average of 10.8 per cent. Vacation properties make more than 28 per cent of this coastal housing stock. The number of vacation homes (such as those illustrated in Figure 7.5) increased of 160,851 in coastal communities from 1999 to 2008, a 15.5 per cent increase, despite soaring costs (the national average is 6.7 per cent).

Figure 7.5   Isle of Ré, a high-end coastal resort in Charente Maritime. The village is 'Les Portes en Ré' (2012).

On the long run, France responded to demand for recreational access to the coastline by a continuous urbanisation process. From 1968 to 2008, the coastal housing stock multiplied fourfold. Vacation homes arc an important ingredient of this urbanisation recipe. There were less than 300,000 vacation properties in coastal communities in 1968, 15 per cent of the local housing stock. There were 1,200,000 vacation homes in coastal communities in 2008, 28 per cent of the local housing stock. Urbanisation includes hospitality infrastructures. There are 270,000 hotel rooms in coastal communities (22 per cent of the national stock). Fifty one per cent of the total French camping facilities (420,000 places) are concentrated in coastal communities. Nowadays, coastal campgrounds are considered urban enclaves of mobile homes (Loi Littoral 1986, article L 146–5). Operating year-round, some of them accommodate permanent residents. An estimated share of

60 per cent of the coastal vacation homes is actually rented during the summer season. In addition to their 6.188 million inhabitants, coastal communities can accommodate over 7 million visitors during the peak summer season, more than doubling their winter night-time population. These crowds defile the fragile coastal environment. The Coastal Conservancy Trust was established in 1975. Its mission is to restore remarkable natural habitats and protect the coastal strip from urbanisation. It acquired 73,500 ha of coastal land during its 30 first years of operation, only 3 per cent of the total area of coastal communities. Its rule extends to 660 km of coastline, 12 per cent of the coastal strip in France. It failed to meet its assigned goal of 25 per cent (Gérard 2009). With an annual budget of 25 million euros, it is hard for this government agency to resist the pressure of the tourism sector that is generating 20 billion euros of income a year in coastal areas.[6] Local governments are very supportive of tourism-oriented development and the coastline is becoming more and more urbanized. All this development increases the vulnerability to the sea-level rise. This point was sadly illustrated in February 2010 along the Atlantic coastline by the storm Xynthia, which caused 53 fatalities. Xynthia is only a harbinger of disasters to come. This stern perspective raises several issues. So far, France lacks preventive coastal evacuation plans. But Xynthia raised a more immediate insurance controversy as the government ordered 'black zone' maps of high hazard areas to be drawn, despite the infuriated resistance of home owners located into these areas. Local policymakers support a publicly funded structural protection policy, despite its obvious fairness issue and its possible adverse consequences in case of levee failure, as Xynthia precisely illustrated. This question will be discussed later.

As for mountain regions, problems are concentrated in ski resorts. Since 1968, France has developed 308 ski resorts that can accommodate 2 million visitors. Vacation homes make 54 per cent of these accommodations. Ski resorts raise severe environmental and planning concerns. First, mountain landscape is defiled by ski infrastructure. Second, climate change reduces the snow pack, raising concern about water supply. This threatens the future of lower elevation resorts, particularly in the Central Plateau and the Pyrenees. Third, skiing practice have declined since 1996. At this time 41 per cent of the French population attended ski resorts in the winter season, but the participation rate was down to 38 per cent in 2007, before the current economic crisis. Higher elevation ski resorts in the Northern Alps remain prosperous because of an increasing foreign attendance, but lower elevation stations have to rely more on the summer season to balance their budget. Moreover, vacation properties create a planning issue for ski resorts. Most first generation resorts were made of poorly built high-rise condominiums. The property is shared among many vacation property owners. As older buildings become dilapidated, private owners are increasingly reluctant to invest the adequate funds to maintain their property. The most fragile lower elevation resorts are threatened to crumble to ruin and create a new kind of 'tourism brownfield' (Amoudry 2002).

---

6   http://www.tourisme.gouv.fr/stat_etudes/etudes/territoires/rencontres/littoral.pdf.

Local policymakers turn to the government, claiming for subsidies. Our opinion is that both economic and environmental long-term prospect is bleak for many overdeveloped touristy resorts along the coastline and in mountain regions. But greedy local stakeholders voluntarily ignore any long-term consideration to reap the immediate benefits brought by residential development.

The Vendean Coast Case-study

The Vendean coast exemplifies the different points that have been discussed in this chapter. Located south of Nantes along the Atlantic Ocean, the department of Vendée boasts over 160 km of sandy beaches, including the islands of Yeu and Noirmoutier. As of 2007, its resident (winter) population numbered 607,430. A survey conducted by the Ministry of Tourism estimated the peak 'present population' at 1,120,000 summer residents in August 2003 (Terrier 2006), mostly concentrated along the 200 km long coastline (Figure 7.6 below). Two in three visitors stay in a vacation home. There were 108,633 vacation homes in 2007 in the department of Vendée, almost 29 per cent of the total housing stock; 90,786 vacation homes, 83.5 per cent of this stock, were concentrated in the 27 coastal communities alone, where they form 60 per cent of the local housing stock. The coastal communities are highly urbanized as a result of tourism. The average housing density of the 27 coastal communities reaches 182.5 units per square kilometer despite the absence of any major coastal cities. The departmental average housing density of 56 is very similar to the national average. Les Sables d'Olonne, the largest city along the Vendean coast, ranks only 152 on the list of French metropolitan areas with 45,455 winter residents in 2008. The highest concentrations of vacation homes extend north of Les Sables to Noirmoutier and south to La Faute-Sur-Mer. Vacation homes form the majority of the housing stock except in inland suburban communities.

Vacation home development has contributed greatly to urban development along the Vendean coastal region since 1968.

During this 39-year period, the total population increased by 51 per cent. Meanwhile, the total housing stock rose by 171 per cent and the number of vacation homes by 380 per cent. Development has concentrated along the coastline. Approximately 31 per cent of the population increase took place in the 27 coastal communities that make only 23 per cent of the total study land area. But nearly 60 per cent of the total housing development and 87 per cent of the vacation home development concentrated in the same 27 coastal communities to accommodate and expanding number of summer visitors. The relative specialisation in vacation homes of these communities rose from 39 per cent of the housing stock in 1968 to 60 per cent in 2007. Land scarcity pushed vacation home development away from the coastline to expand into the backcountry. However, vacation home development tries to keep within easy reach of the beaches, and the expansion zone is mostly limited to a 10 km distance from the sea. The relative specialisation in vacation homes of these immediate backcountry communities rose from 8 per cent of the housing stock in 1968 to 29 per cent in 2007. Small numbers of vacation homeowners have chosen to settle farther away from the sea. Communities located 10 to 30 km from the coastline have only 6 per cent of the local vacation home stock, and their relative specialisation in vacation home has risen from 4 per cent in 1968 to just 6 per cent in 2007. Urban pressure on the coast raises serious environmental concerns. The Coastal Conservancy Trust owns only 323 ha along the Vendean coast, mostly wetlands, less than 0.4 per cent of the land area of the 27 coastal communities.

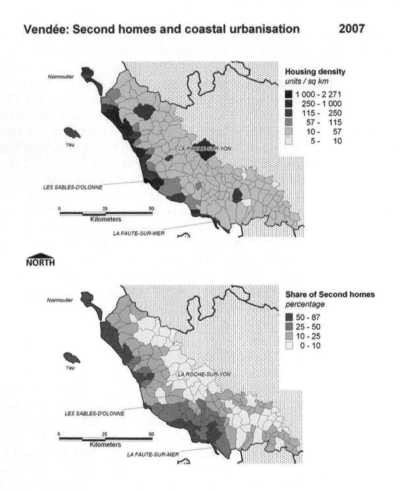

Figure 7.6   Vacation homes along the Vendean coast (2007).

Source: INSEE Census survey 2005–09.

Table 7.3   Vacation homes and urbanization of the Vendean coast 1968–2007.

Distance from the sea	Coastal communities	Immediate backcountry (<10 km)	Backcountry (10–30 km)	Total coastal Vendée (30 km limit)
Communities	27	26	92	145
Area (sq km)	825.8	568.5	2 157.4	3 551.7
Census population (winter) 1968	82,077	19,991	131,911	233,979
Census population (winter) 2007	118,506	39,594	195,004	353,104
Housing stock 1968	49,688	6,975	42,792	99,455
Housing stock 2007	150,697	24,561	93,861	269,118
Vacation homes 1968	19,418	583	1,655	21,656
Vacation homes 2007	90,786	7,194	6,028	104,008
Housing density 2007	182	43	44	76
Vacation home relative specialization 1968 (%)	39%	8%	4%	22%
Vacation home relative specialization 2007 (%)	60%	29%	6%	39%
Population average growth rate (yearly %)	0.95%	1.77%	1.01%	1.06%
Housing average growth rate (yearly %)	2.89%	3.28%	2.03%	2.59%
Vacation home average growth rate (yearly %)	4.03%	6.66%	3.37%	4.11%

Source: INSEE censuses.

The Vendean coast is a poster child of residential economic development. Nowadays, Census data shows that this coastal urbanisation attracts a growing number of permanent residents, including retirees, and creates jobs in the service sector that in turn attract an expanding workforce. The population has increased by 1.6 per cent per year from 1999 to 2007, against a national average of 0.68 per cent only. This rapid expansion is mostly due to domestic migrations; 1.55 per cent a year (national average 0.57 per cent). The population is ageing, 28 per cent was over 60 years old in 2007 (national average 22 per cent). This proportion reaches 38 per cent in the 27 coastal communities. The Vendean coast is attractive to retirees who used to spend their holidays in the region during their working life and eventually bought a vacation home there, particularly during the booming 1970s. The average yearly balance of domestic retirement migrations was 1,140 during the 2002–07 period, 1.1 per cent a year. This influx of seniors helps boost the service economy during the winter season. Local employment grew by 19 per cent from 1999 to 2007, a significant gain to be compared to the national average of 12 per cent during the same period. In consequence, the area is equally appealing to people of working age. The average yearly balance of domestic workers migrations was 1,066 during the 2002–07 period, 0.7 per cent a year. However, there is a clear spatial divide in this population of newcomers. Retired newcomers settle preferably directly in coastal communities. Active newcomers settle in backcountry communities, where they find affordable housing options.

This overspecialisation in tourism and vacation homes creates a rising vulnerability to coastal hazards. Let us illustrate this point with the sad story of the small city of La Faute-sur-Mer, where Xynthia's flood caused 27 fatalities in 2010.

There are an increasing number of aged urbanites living in the immediate vicinity of the coastline during the winter season. In past times, the coastline was mostly left to pasture and cropland. Shallow inland bays have been drained and converted to agricultural polders from the Middle Ages to the early twentieth century. To the south of the Vendean coast, the Poitevin Marsh is a large polder that has succeeded the former Gulf of Poitou. La Faute-sur-Mer is located on a sandspit of dunes that close this former gulf to the west. The spit is separated from the Poitou Marsh by the Lay river estuary. It used to be a tiny village of 328 inhabitants in 1936. This figure rose to 587 in 1968. La Faute had already turned into a family-oriented vacation resort (Figure 7.7).

Figure 7.7   Vacation homes in La-Faute-sur-Mer, Vendée (2012).

There were 1,030 housing units, 78 per cent of which being vacation homes in 1968. From 1968 to 2007, the peninsula was filled by new housing units with the successful development of new subdivisions. The total housing stock rose to 3,743, 86 per cent of which being vacation homes. A large majority of these newly built houses were detached one-storied cottages imitating the architectural style of the traditional Vendean farm. Meanwhile, the winter population was also on the rise, with 1,001 permanent residents estimated in January 2007, 48 per cent of whom retirees. The city gained 74 new retired residents from domestic migrations with the rest of France from 2002 to 2007. Up to 28 per cent of the retired population of La Faute were relative newcomers in 2007. Most victims of storm Xynthia were retirees (Vinet et al. 2011). As the weather alert recommended, people barricaded themselves at home. The storm came during the night. A storm surge of at least 5 meters rushed into the estuary of the Lay River, breaching the nineteenth century levees that were originally only protecting pastures. The most recent housing subdivisions, authorized in 2006 only, were flooded under more than 2 meters of water. Many elderly people were surprised during their sleep; some did not escape the rising water. This sad story illustrates the consequences of inadequate urban development in the former agricultural polders. Xynthia was not an unprecedented storm, nor an exceptional one. Historians mention the storm of 1935 in Noirmoutier, which caused a similar storm surge. There were no victims at this time. Local people, mostly peasants and fishermen, raised the alarm with bells, and the population retreated inland by their own means before the coming of the invading water. A strong sense of community, combined with an intimate knowledge of the local environment, spared many lives in 1935. Three quarter of a century later, urbanites have settled in dense and inadequately built subdivisions. The local authorities were taken by surprise. There was no ready evacuation plan, and the electricity shutdown added to the confusion. After the disaster, state officials have ordered 674 vulnerable houses, including 582 vacation homes, 18 per cent of the city's housing stock, to be bought back by the state and torn down in La Faute-sur-Mer. However, consecutive negotiation has finally reduced this total to 472 (Mercier and Chadenas 2012). Here, the storm has put the urbanisation pressure to a sudden halt. It is an indicator of the urgent need for integrated coastal management (ICM) to achieve sustainable urban development along the Vendean coast.

Conclusion and Perspectives

France appears as a large and diverse country in the European context. The diverging trends of vacation home development which have formed since the 1990s can be explained by vastly different regional conditions. We suppose, however, that these differences can be reconciled into a common economic analysis framework. A minority of regions are specializing in vacation home development. Regional contrasts are likely to become even more accentuated in the coming decades. As was demonstrated above, this is mostly because vacation home are actually tourist accommodation along the coastlines and in mountainous areas. A vacation home located in a tourist coastal resort or in a mountain ski resort is not simply a luxury; it is in the first place an investment that brings income to its owner. In contrast, a countryside summer home or an urban pied-à-terre is unlikely to generate rents, except perhaps in Paris-city proper. For this reason, the second home stock is likely to decrease in many regions. As the French society is becoming more polarized, and the disposable income gap *vis-à-vis* the affluent upper class is widening, statistical evidences suggest that since 1990 middle class households have been less likely to retain a second home ownership title that does not generate income. In the vicinity of urban centres, many second home are now being sold in the market and returning to the condition of main residences. The existing relative housing shortage and the correlated pressure of soaring housing costs help accelerate this transition in many metropolitan areas. This process combines with population ageing to further limit the national stock of second homes. As the bulging age classes of baby boomers retire in the next 25 years, the conversion of many second homes into main residences is likely, not only in the countryside, but also probably in coastal regions. On the opposite side of the same argument, there is still an obvious long-term development potential left for vacation homes in tourism-oriented areas. This raises planning and environment concerns in coastal and mountain settings. As the Vendean coast exemplifies, critical thinking about what is at stake should help policymakers to think forward to the needs of sustainable development of coastal and mountain areas.

References

Amoudry, J.P. 2002. L'avenir de la montagne : un développement équilibré dans un environnement préservé. Rapport d'information du Sénat n°15. [Online] Available at: http://www.senat.fr/rap/r02-015-1/r02-015-1.html [accessed April 2012].

Barbier, B. 1977. Les résidences secondaires et l'espace rural français. *Norois*, 1977, vol. 95 ter. 11–20.

Baumol, W. 1967. Macroeconomics of Unbalanced Growth: the Anatomy of Urban Crisis. *American Economic Review*, 57, 415–26.

Benko, G. 1998. *La science régionale*. Paris, Presses Universitaires de France.

Chao, C.C., Hazari, B.H., Laffargue, J.P., Sgro, P.M., and Yu, E.S.H. 2006. Tourism, Dutch Disease, and Welfare in an Open Dynamic Economy. *Japanese Economic Review*, 57(4), 501–15.

Clary, D. 1984. Le tourisme littoral: bilan des recherches. *Revue de géographie de Lyon*, 59(1), 63–72.

Copeland, B.R., 1991. Tourism, Welfare, and De-Industrialization in a Small Open Economy. *Economica*, 58, 515–29.

Cribier, F. 1973. Les résidences secondaires des citadins dans les campagnes françaises, *Études rurales*, *49–50*, L'urbanisation des campagnes, 181–204.

Davezies, L. 2008. *La République et ses territoires. La circulation invisible des richesses*. Paris: Seuil, La République des idées.

Gérard, Y. 2009. Une gouvernance environnementale selon l'état? Le conservatoire du littoral entre intérêt général et principe de proximité. *VertigO – la revue électronique en sciences de l'environnement*, 9(1), [Online] online 29 May 2009. Available at: http://vertigo.revues.org/8551; DOI: 10.4000/vertigo.8551 [accessed 24 April 2012].

INSEE. 2008. Le tourisme en France. Paris: INSEE. [Online] Available at: http://www.insee.fr/fr/publications-et-services/sommaire.asp?codesage= FRATOUR08 [accessed 20 April 2012].

Judd, D.R. (ed.). 2003. *The Infrastructure of play: Building the Tourist City*. London: M.E. Sharpe.

Le Garrec, M.A. (dir.). 2005. *Le tourisme en France*. Paris : INSEE – Références

Mercier, D. and Chadenas, C. 2012. La tempête Xynthia et la cartographie des «zones noires» sur le littoral français: analyse critique à partir de l'exemple de La Faute-sur-Mer (Vendée). *Norois*, 222-2012/1, 45–60.

Terrier, C. (dir.). 2006. *Mobilité touristique et population présente. Les bases de l'économie présentielle des départements*. Paris: Direction du Tourisme.

Vinet, F., Defossez, S. and Leclere, J.R. 2011. Comment se construit une catastrophe? *Place publique*, Hors série, 9–19.

Zaninetti, J.M. 2006. L'urbanisation du littoral en France. *Population et Avenir*, 677 mars-avril, 4–7 and 20.

Zaninetti, J.M. 2011. Les retraités en France : des migrations pas comme les autres. *Population et Avenir*, 703 mai-juin, 4–7 and 20.

Chapter 8

Second Home Tourism in Finland: Current Trends and Eco-social Impacts

Mervi J. Hiltunen, Kati Pitkänen, Mia Vepsäläinen and C. Michael Hall

Introduction

A growing number of people in the Western world have not only one fixed home, but share their lives between various places of dwelling. The decision to own leisure-oriented multiple dwellings is often related to a desire to spend time at a second home in a high amenity environment and pursue recreational interests and quality of life goals (Müller 2007, Marcoulier et al. 2011). In many countries second home ownership has become an elitist phenomenon (Hall and Müller 2004, Halseth 2004); however, in Finland (Pitkänen 2011a, Hiltunen 2009, Periäinen 2006), as in the other Nordic countries (Müller 2007, Hall et al. 2009), rural second home tourism has a long tradition and is today a fundamental part of cultural identity and modern life. The term 'second home tourism' refers here to second home ownership and the related mobile way of life between primary and second home. In Finland the term 'cottage' (*mökki*) is used as the equivalent to the term 'second home'. In Finnish this connotes the collective cultural conception of rural second home, regardless of the age, size or standard of the second home.

Finland, with its 5.4 million inhabitants, has one of the highest levels of second home access in the world. At the end of 2011 there were nearly half a million (493,000) registered second homes and approximately 3,600 new ones are built annually (Statistics Finland 2012a). In addition, there are around 170,000 rural vacant detached houses, of which over 70 per cent are used as second homes (Pitkänen et al. 2012). It has been estimated that every second family, i.e. around three million Finns, have access to a second home since cottages are also used by friends and relatives (Figure 8.1 below). Along with private ownership, rental and timeshare second homes are a growing business in the accommodation sector and are providing a new way of thinking about second home use. There are approximately 10,000 rental cottages in Finland (Nieminen 2010).

The geographical distribution of second homes is concentrated in the amenity rich rural regions of Lakeland, the west coast and the tourism centres of Lapland (Figure 8.2 below). Finland is a sparsely populated country which urbanized rapidly after the rural restructuring of the 1960s and 1970s. Second homes were traditionally built close to urban centres or close to family roots in sparsely populated areas and rural villages. Finnish second homes are typically built of wood on secluded private

Figure 8.1   **Outdoor activities close to nature fill the life at traditional summer cottages (2005).**

Source: K. Pitkänen.

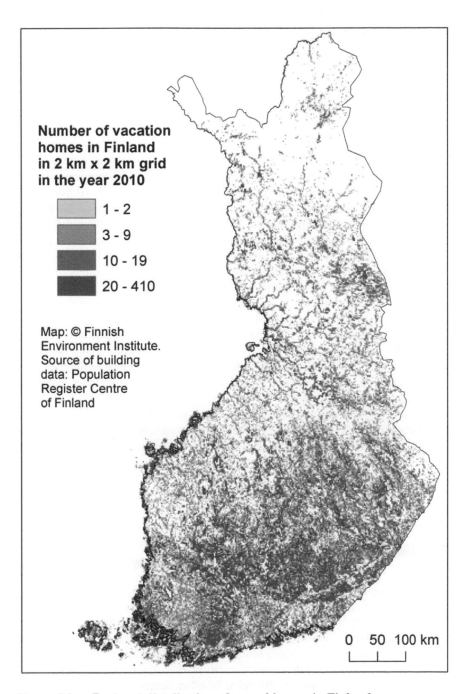

Figure 8.2   Regional distribution of second homes in Finland.

Source: Finnish Environment Institute.

properties, close to natural waters, and almost without exception have a sauna (Periäinen 2006). Second home tourism is often believed to be an obvious and unchanged part of Finnish culture and society. Traditionally rural second homes have provided urban dwellers with an opportunity to spend meaningful leisure time in peaceful environments as an alternative to urban daily life. Second home tourism in Finland is a form of escapism to the countryside, to nature, childhood nostalgia, family life and traditional gender roles (Pitkänen and Kokki 2005).

However, in reality both second home tourism and related images and lifestyles are continually changing. Even though much of the traditional summer cottage life still prevails, second home culture and lifestyles are currently diversifying (Pitkänen 2011a, Alasuutari and Alasuutari 2010, Hirvonen and Puustinen 2008, Pitkänen and Vepsäläinen 2008, Aho and Ilola 2006). The changes are induced by growing social welfare and wealth, increasing human mobility and leisure time, continuing urbanisation and population aging as well as by advances in technological development. Current trends in second home tourism indicate changes in the use, user groups, forms and locations of second homes. These have an impact on the regions and local communities of origin and destination as well as on the locations en-route, and, to a certain extent, culture and society at large. The overall changes in second home tourism affect land-use planning and infrastructure, political decisions and governance as well as public and private economies and service provision. These changes also have socio-cultural effects on second home owners' and users' lives and families, on local communities, on the patterns and practices of second home tourism as well as on related values and meanings.

Besides social impacts, second home tourism affects ecosystems and the environment too. Second home tourism is a phenomenon whose environmental impacts are often difficult to specify, pinpoint and calculate because of their dispersed nature. Nevertheless, the environmental impacts of second homes have been recognized in international research literature (Gallent et al. 2005, Müller et al. 2004, Gartner 1987, Mathieson and Wall 1982). In Finland researchers have recently highlighted environmental impacts (Hiltunen 2007), as well as changing human-nature relations (Vepsäläinen et al. 2011, Massa 2011), eco-efficiency (Rytkönen and Kirkkari 2010, Ahlqvist et al. 2008), and the use of natural resources (Salo et al. 2008) related to second homes.

Analysing Eco-social Impacts of Second Home Tourism

In previous research the impacts of second home tourism have usually been divided into three categories: environmental, economic and social/cultural (e.g. Marjavaara 2008, Hiltunen 2007, Müller et al. 2004, Gallent et al. 2005, Müller 1999, Visser 2006). Often the impacts are more descriptive than analytical, and are regarded as comparable or the equal to other forms of tourism. Yet, much analysis of the impacts of second homes fails to recognize that the phenomenon and its impacts are specific and quite often also profound because of property ownership and a longer lasting commitment to second home communities (Müller et al. 2004, Hall 2011).

The purpose of this chapter is to identify and discuss the environmental and social impacts of second home tourism in Finland. The aim is to highlight the relation and interaction between different impacts and reveal the underlying factors and processes inducing the impacts. The paper is primarily based on the results of the authors' recent works concerning Finnish second home development (Pitkänen and Kokki 2005); environmental impacts (Hiltunen 2007); future trends (Pitkänen and Vepsäläinen 2008, Hiltunen et al. 2013); human-nature relationship (Vepsäläinen et al. 2011); regional, cultural and social changes (Pitkänen 2011a); foreign second home ownership (Pitkänen 2011b); and regional patterns of second home tourism in Finland (Rehunen et al. 2013). Based on these studies, we have chosen three key trends in the current second home development that will have considerable impacts on the Finnish rural areas in the future.

The chapter provides a framework to assess the environmental soundness and social significance of second home tourism. We use the term eco-social to highlight the close relations and interactions between ecological and social systems (Eisto and Kotilainen 2010, Berkes and Folke 2000, Berkes et al. 2003). The ecological system is here approached from the research perspective of environmentally oriented social science and understood as part of a broader notion of environmental, with the main focus on the natural environment and human-nature relations. The social system, for its part, refers to human social and cultural environments, emphasising the second home owners' collective actions and influence on community and social development. The social impacts include community, cultural and economic impacts, as they are all parts of the social system and are oftentimes interconnected tightly for instance in regional development. The environmental impacts affect biotic and abiotic nature, which intertwine with the social system into an eco-social system.

In practice the eco-social system is more clearly understood as the physical and social living environment of second home owners. Broadly considered, this comprises the two dwelling environments, the urban home and rural second home, as well as the mobility between them. We use the concept of multiple-dwelling to emphasize the multiple place attachment and the mobile way of life of second home owners (e.g. Quinn 2004). Nevertheless, considering the eco-social impacts, a contextual and place specific focus on the living environment is of interest here. In this chapter, the individual perceptions and impacts on the personal social lives of second home owners are excluded. Instead, in analysing the eco-social impacts of second home tourism we focus especially on rural environments and local rural second home communities. Both the environmental and social impacts are considered at local level, yet they also resonate on the regional, national, and even global scales, especially in terms of climate change and mobile consumption patterns.

Current Trends in Finnish Second Home Development

Pitkänen and Vepsäläinen (2008) reviewed representations of the future of second home tourism in the Finnish media and policy discourse, and distinguished four themes that characterize the current development of second home tourism in Finland. These were: increasing year-round use of second homes, regional differentiation, internationalisation and the growing popularity of alternative forms of second homes. Similarly, in their study on the regional patterns of second home tourism Hiltunen et al. (2013) concluded that in recent years second home development has been concentrated in amenity rich hotspots with access to recreational activities. Based on these findings we propose, and analyse the impacts of three contemporary trends in second home development that also have international resonance (see, e.g. Visser 2006, Gallent et al. 2005, McIntyre et al. 2006). The first is the improvements in the building standard, fittings and décor of cottages, which is also associated with an increase in time spent at second homes (Pitkänen and Kokki 2005). Cottages are increasingly built or renovated to meet the needs and amenities of 'home' life which is here described as dual-dwelling. The second trend is the growing regional concentration of second homes in tourism centres. The third noticeable trend is the development of foreign second home ownership as a result of increased cross-border recreational mobility. Besides being topical, these trends describe well not only the ongoing changes in second home tourism in relation to both second home living and regional differentiation but also the emergence of new cottager groups.

Dual Dwelling

The term dual-dwelling implies the living at the primary home and second home alternately. Corresponding to dual-dwelling are such terms as semi-dwelling, part time living and dwelling, and semi-migration. These terms reflect the role of the second home as something more than merely a place to spend one's holiday (Pitkänen and Vepsäläinen 2008). Also in the Nordic region rural second homes are being increasingly equipped and built for year-round use (e.g. Kaltenborn 1998, Vittersø 2007). Retirement is a stage in life in which the use of second home increases. Furthermore, the proportion of households that wish to combine work and leisure in their second home is increasing and as a consequence the difference between first and second homes is becoming blurred (Müller and Hall 2004, McIntyre 2006, Sandell, 2006). The central features which characterize dual-dwelling in Finland are the high standard of second homes, year-round use, shoreline building and the frequent mobility between primary home and second home. These features and their eco-social impacts are analysed here.

High Standard of Second Homes

In Finland, the standard of second homes is changing from simple summer cottages into well-equipped and winterized villas (such as the ones in Figure 8.3 below). Old second homes are increasingly renovated and new ones built for year round use. For example, the average floor space of second homes built in the 1990s was 49 m², whereas second homes built in the 2000s were 64 m² on average. Nowadays over 90 per cent of second homes have electricity and a refrigerator, over 70 per cent a TV, over 20 per cent a shower and over 10 per cent a washing machine and dish washer. A fourth of all second homes is kept warm with electric heating also during winter months (Nieminen 2010).

The increasing popularity and amenity demands for second home living have initiated concerns on the environmental consequences of such development (e.g. Tress 2000, Dubois 2005, Hiltunen 2007, Massa 2011). In general, the increase in living space, equipment rate and year-round use of second homes also increases energy consumption and natural resource use (Rytkönen and Kirkkari 2010, Salo et al. 2008, Melasniemi-Uutela 2004). On the other hand, interest in investing in second home living can also mean that new second home constructions and renovations can be more energy efficient. Ahlqvist et al. (2008) believe that the use of ecologically designed technology and innovations at second homes is becoming more popular and will eventually help to reduce overall energy consumption. The positive environmental impacts also include the sustainable reuse and renovation of old cottages. Using old housing structures saves natural resources, employs embodied energy and maintains the site's footprint.

The positive social and community impacts of the higher standard of second homes include the increase in property values, which benefits the owner households in terms of equity, but also second home municipalities, since real estate taxes are paid according to the property value. Hence, the better and newer the second home stock, the more tax income the municipalities receive. Furthermore, the renovation and building of new cottages brings and maintains employment opportunities in the building sector. Nowadays second home owners are also ever more interested in using local maintenance services in cottage upkeep. However, the negative social impacts may include cultural changes in traditional rural landscape and built heritage, as modern and large second homes are built in rural environments (Periäinen 2006).

Year-Round Use of Second Homes

The contemporary Finnish second home is used relatively intensively. The average period of use is 75 days per year for all second homes, 103 days for winterized ones and 40 days for summer cottages. One-third of second homes are nowadays fit for winter use (Nieminen 2010). According to Pitkänen and Kokki (2005) the high usage rate is related to retirement, winterization, the minimum 60 m² floor space, and low travel time between primary and second home (less than three hours on average). In addition purchased cottages are used more often than inherited ones. In Finland,

Figure 8.3   Second homes today are built and equipped to meet the standard
of modern homes and year-round use (2005).

Source: K. Pitkänen.

retired second home owners are regarded as a potential source of new permanent residents in rural areas suffering from population decline (Aho and Ilola 2004). However, so far this potential has not realized. Instead, studies have shown that even though the time spent at second homes increases considerably after retirement, people are not willing to give up their permanent home but want to combine the best sides of both urban and rural living environments (Pitkänen and Kokki 2005).

The most positive environmental impact of the year round use of rural second homes is related to the regeneration of the human-nature relationship. The lived connection to rural nature and the environment is increasingly important with growing urbanisation. Second homes are in general used not only by the owner household but also by relatives and friends. The utilisation rate increases alongside the winterisation and higher standard of the building (Pitkänen and Kokki 2005). For second homes fit for winter the average number of users is 13 people annually, whereas in cottages used only in the summer the amount is half that (Nieminen 2010).

The year-round use of second homes may compel the owners to look for eco-efficient and energy saving technologies to reduce the costs of maintaining two homes. However, at the same time year-round use boosts the negative environmental impacts of second home tourism as a result of the more intensive use of the second home environment. The extension of human activities and tilling on second home plots may disturb vulnerable flora and fauna. Biodiversity may decline or change in second home plots where gardening and lawn keeping are quite common (Jokinen 2002), and may allow for the introduction of invasive species. Earth removal works increase as electricity, sewage system, water supply and broadband Internet connection become needed at the second home.

In dense second home areas along sensitive shorelines the harmful environmental impacts may cumulate over time if, for instance, wastewaters are not adequately managed. Negative impacts may include the air and noise pollution of motorized vehicles and equipments. The cottages and saunas are generally wood heated, which increases smoke emissions (Hiltunen 2007). Year round use also sets new demands on the infrastructure. For example, to construct roads to second homes and maintain them frost free and usable all year long consumes substantial resources. Private cottage roads leading from the main roads to shoreline plots fragment the natural environment and cut off species' ecological gateways. It is worth noting that as the whole road network in Finland covers the length of 454,000 km, the share of private and forest management roads is considerable at 350,000 km (Finnish Transport Agency 2013).

Interest in year-round living at second homes brings along many social benefits to rural areas. According to Sandell (2006), completely new types of rural residency have developed. Dual-dwellers are new 'locals' that are not necessarily involved in traditional rural industries like farming or forestry. Second home owners are potential rural migrants and part-time residents. The emergence of this new residential group reinforces social capital, and diversifies and revitalizes the social structure. It supports private and public services and provides employment opportunities in rural areas (Müller and Hall 2004, Sandell 2006). The year-round use of second homes also helps to maintain and develop infrastructure in remote areas. By renovating and maintaining rural housing stock, second home owners also contribute to preserving the rural cultural landscape and the market value of rural properties.

Long second home stays affect local service demand and supply. Dual-dwellers boost the local private economy not only seasonally but throughout the year. On the other hand, they use the public services of the second home municipality but are paying income taxes to the primary home municipality. The major costs of second home tourism for municipalities are road management, land use planning, rescue and emergency services, and library services (Nieminen 2010, Koski 2007). The aging of second home owners also increases the demand for public and private health care services.

Furthermore, in many municipalities second home related spatial and social planning measures lead to political conflicts. Second home owners' social rights and duties have been under public debate during the last decade. In many Finnish municipalities special committees have been established to deal with issues concerning second home owners as part-time dwellers. The second home owners invest not only economic but also strong emotional value in second homes and are therefore motivated to fight for their interests. Although this can be beneficial for the preservation of local nature and culture, it can also lead to conflicts between different interest groups and even to the displacement of local interests (Fountain and Hall 2002, Mottiar and Quinn 2003).

Shoreline Building

A significant factor causing eco-social impacts of dual-dwelling in Finland is building at sea, lake and river shorelines. Approximately 85 per cent of second homes are located less than 100 meters from the shoreline (Nieminen 2010). Around larger cities, especially in southern Finland, the best available building land and free shorelines are becoming rarer, which over the decades has led to the congestion of second homes in urban fringe areas. Furthermore, the shoreline ecosystems are very sensitive and vulnerable to environmental change. For instance, the problem of algal blooms, mainly caused by increased phosphorus and nitrogen from agriculture, is gradually worsening and affects the recreational use of shorelines.

The increased use of second homes and year-round dwelling at watersides contributes to landscape changes and shoreline disturbance. Shoreline development and private ownership also restrict the public right of access on shores (Jokinen 2002). On the other hand, locating second homes on a shoreline enables geothermal energy use and thereby reduces the need for non-renewable energy in second homes. A waterside location maintains nature-based outdoor activities, such as fishing, boating and swimming, which have traditionally been cottage leisure pursuits, as well as Finnish sauna culture and traditions. Therefore, there are many positive social impacts to second home tourism which support health and cultural traditions.

Second homes at shorelines are built detached from the traditional rural settlement structure and villages, leading to dispersal of the community spatial structure. In some areas, parallel to the intensified use of shorelines, rural villages and population centres are marked by unoccupied and rundown dwellings. This can be considered ecologically and socially unsustainable development. Koski (2007) suggests that the costs for municipal economies of the scattering and growing settlement structure induced by second homes are considerable. Second home owners also burden the municipal planning and building authorities. Many new second homes are built with exceptional permits despite detailed planning and zoning at shorelines.

Frequent Mobility

Travelling to second homes has increased considerably during the past decade. In 2000, Finns made around three million trips to own second home while in 2011 5.9 million (Statistics Finland 2012b). The majority of all second home trips were short weekend trips (4.4 million), which in 2009 were on average 21 per cottage owner household (Nieminen 2010). The distance between primary and second home was 118 km on average, although half of second home owners have less than 50 km and 60 minutes of driving to their second home (Nieminen 2010). The short distance between home and second home induces frequent routine travelling.

The vast majority (95 per cent) of second home trips are made by private cars and the car is also used for trips in the second home region (Nieminen 2010). It has been estimated that the energy consumption related to second home mobility is at least twice as much as electricity consumption at second homes (Lahti 2010). The negative environmental impacts of second home mobility are related to air pollution and greenhouse gas emissions, noise levels and the use of natural resources for road building and management (Hiltunen 2007). The negative impacts caused by second home mobility are also increasing as more than one car is often used by cottage users for access.

The more time is spent at the second home the more second home owners move around in the second home region. In rural areas and especially at remote second home locations services are seldom accessible by foot, bicycle or public transport. The closing of rural public and private services has raised questions about the regional and social equality of mobility (Vepsäläinen and Hiltunen 2001). The private car is often a necessity in rural areas as distances are long and services located in population centres (Rehunen et al. 2012). Also motorboats are commonly used during the summer time. The year-round use of second homes increases the mobility in the second home area, whereas longer stays at the second home might reduce the mobility between the permanent and second home, and decrease the negative environmental impacts caused by long haul second home mobility.

It is often believed that remote work done at home decreases daily mobility. The interest of doing remote work at second homes has increased in recent years, and in 10 per cent of second home households remote work is done on average 23 days a year. However, two-thirds of second home owners do remote work less than 10 days a year (Nieminen 2010). The increasing possibilities and interest in doing remote work may decrease the frequent mobility between primary and second home in the longer term. However, commuting from second home is also on the rise. In 20 per cent of second home households, trips were made to work on average on 14 days a year. The distance between primary and second home in those cases was an average of 77 kilometres (Nieminen 2010). The positive (+) and negative (-) environmental and social impacts of dual-dwelling are summarized in Table 8.1.

Second Homes in Tourism Centres

The concentration of second homes in certain amenity-rich hotspot areas is an increasingly typical regional pattern (Müller 2004, 2005; Kauppila 2009). Such areas with dense second home agglomeration have emerged as a result of growing interest in nature based, sport and wellness tourism and reflect the change towards a more touristic orientation in the motives for second home ownership in the Nordic countries (Müller 2004, Tuulentie 2007). Previous generations' main motives for having a second home had been the appeal of peaceful nature and return to childhood and family rural environments. For today's generation, family roots do not necessarily play as central a role anymore, and the second home motives become increasingly more diversified, often related to hobbies and outdoor activities (Jansson and Müller 2004, Pitkänen and Vepsäläinen 2008).

Currently, the construction of new second homes in Finland is fastest in the tourism centres of Lapland (Tuulentie 2007, Hiltunen et al. 2013). The building of second homes in tourist centres has grown exponentially from the 1980s on (Kauppila 2006). Besides privately owned detached cottages, time share cottages and apartment houses have also been built in response to the burgeoning demand for both domestic and inbound winter tourism, as illustrated in Figure 8.4 below. In other parts of Finland, second home hotspots are also emerging around wellness and spa tourism locations. On a more moderate scale, the growing concentration of second homes appears in many traditional second home areas at shore sites, especially in the Finnish Lakeland. In some rural municipalities, special second home areas are being built close to villages and population centres in accordance with community zoning and planning. Often such planned second home areas are built for aging people who wish to spend their second home life close to services in rural environments. Increasingly second home villages or agglomerations are also promoted for families with small children and for foreign second home owners.

The central features characterising second homes in tourism centres are the high turnover of second home tourists, long-haul travelling and spatial concentration of second homes within a small area. These features and their eco-social impacts will be analysed in the following sections.

Table 8.1   Eco-social impacts of dual-dwelling.

Key Features	Environmental impacts	Social impacts
High standard of second homes	+ Eco-efficient new second homes and technology − Growing consumption of energy and natural resources	+ Increase in property values and taxes + Employment opportunities in construction and building maintenance − Changes in cultural landscape
Year-round use of second homes	+ Regeneration of human nature relationship + More users per cottage and higher usage rate + Use of energy saving technologies − Intensive use and tilling of the environment − Scattered loading to natural waters − Air, waste and noise emissions − Road building and maintenance	+ Advance local economy + Reinforces social capital + Maintains infrastructure, services and cultural landscape + Potential rural migration − Changes in local power relations and structures − Public costs of infrastructure and social service provision
Shoreline building	+ Use of geothermal energy from lakes − Overdevelopment of shores − Landscape change and disturbance of nature − Restricts public access − Dispersal spatial structure	+ Retention of traditional nature based activities and sauna tradition − Costs of scattering spatial structure
Frequent mobility	+ Frequent travelling may decrease − Increase in daily mobility in second home region − Increase in frequent travelling	+ Urban-rural interaction + Liveliness in rural areas − Public costs of road management

High Tourist Turnover

Kauppila (2006) has pointed out that second homes in tourism centres are different from second homes in general. The ownership of second homes is quite often based on timeshare and many of the second homes are owned by companies and used by company employees. Rental cottages are a typical form of accommodation in tourism centres as well. Hence, second homes are in intensive use especially during the high season, when tourist turnover is high. These features have important environmental and social impacts. Environmentally, co-ownership reduces the

Figure 8.4   Second homes in ski resorts of Lapland are often in time share or rental use during the winter holidays (2012).

Source: M. J. Hiltunen.

need for new developments, but at the same time increases energy consumption and mobility falling upon each second home. Socially the high utilisation rate of second homes brings income and business opportunities to the area. However, the high tourist turnover means that they might not be very committed to the local community. This, added to the large amount of tourists, might entail the negative social side-effects of tourism. Moreover, many tourist centres are lively and full of activity only part of the year, and as the high season is over, many of them might turn into 'ghost towns'. Seasonality is one of the main problems for sustaining local living conditions and employment. Seasonality also brings many seasonal workers and temporary occupants to tourism centres who do not necessarily commit themselves to local community life or interact with permanent dwellers as much as with each other. Finally, many tourism resorts typically have a life-course of a growth period followed by stagnation and even decline. However, according to Kauppila (2006) in some cases second homes have remained in the area even though the tourism business has declined. This indicates the place attachment and commitment of second home owners to rural tourism centres.

Long-haul Travelling

In relation to environmental impacts, an important feature of tourism centres is that they are often located far from urban areas and reached by private cars which then are used in local mobility as well. On the other hand, many tourist centres are also reached by plane, train and long haul buses, in which case people may move locally in an environmentally responsible manner by walking, skiing, bicycling or by using public transportation. When assessing environmental impacts, it is essential to take into account the modes of action and mobility both at the destination and while travelling. In terms of the effects of emissions, the ecological burden of tourism consists largely of travelling between the areas of origin and destination (Salo et al. 2008, Gössling and Hall 2006, Scott et al. 2012). Socially, long-haul second home tourism may help even out some of the economic differences between urban and rural regions by attracting investments to the more peripheral rural areas. On the other hand, because of external business ownership and seasonal labour force, it is also fairly common that a part of the business profits and income taxes flow out of the region.

Spatial Concentration of Second Homes

Intense second home and tourism development differentiates tourism centres from the surrounding rural areas (Hall 2011). From environmental perspectives tourism centres change the local landscape significantly. Besides construction, tourist activities such as downhill skiing or golfing are energy and land use consuming. On the other hand, second home tourism in tourism centres can be considered ecologically efficient due to comprehensive planning and dense spatial structure. The environmental benefits of a dense spatial structure are today seen as unquestionable as it minimizes the need for mobility, favours public transport and enables eco-efficient building and use of infrastructure. Furthermore, a dense spatial structure saves resources while a housing structure can be directed so that natural areas remain unbuilt on (Heinonen et al. 2002).

In rural communities with tourism centres the local attitudes of permanent dwellers and entrepreneurs towards the tourism industry and second home development are largely positive (Leinonen et al. 2008a). Tourist flows and related investments create jobs and income, and help to sustain local infrastructure and services. Second homes also bring important revenues for the community in the form of yearly property and real estate taxes. Nevertheless, while assessing economic benefits, attention should be paid not only to the tourism centre itself, but also to the interaction between the centre and the surrounding countryside, and the relations between the new and traditional sources of livelihood (Leinonen 2008a, 2008b). For example, the economic benefits of tourism centres in Finnish Lapland are often limited to the core areas (Kauppila 2004), and regional differences in municipal and communal development may vary widely. Furthermore, tourism centres might displace other local land use and business activities.

Tourism development brings employment and services to remote areas, yet the benefits for local people are not always only positive. Tourism activities, services and second homes often agglomerate to certain areas separate from the local villages. Often tourism structures also gradually displace the old village structure, and local inhabitants may even be forced to move away as older buildings are torn down or converted for other purposes than housing. The price level of goods in local shops and municipal payments such as property taxes may also become too high for low-income earners including retirees.

Finally, second homes in tourist centres do not necessarily replace traditional second home tourism elsewhere in the country. As a consequence of the general growth in welfare and wealth, people increasingly have the possibility to purchase more than one second home. Such third homes are seen as a weak signal of second home tourism (Tuulentie 2007, Pitkänen and Vepsäläinen 2008) and are becoming progressively more popular. It is already relatively common for people to have the possibility to use more than one second home. Some 38 per cent of the respondents in a questionnaire survey (Hirvonen and Puustinen 2008) stated that they used more than one second home during the previous year. In another survey (Pitkänen and Kokki 2005) 19.5 per cent of cottage owner households stated that they own more than one second home either in Finland or abroad. Having several second homes simultaneously, and travelling between these multiple places, brings about a completely new set of environmental and social impacts. Table 8.2 summarizes the positive and negative eco-social impacts of second home densification in tourism centres.

Foreign Second Home Ownership

Improved access to communication and transportation, general opening of borders, and growth in income and private financial resources have enabled certain people to seek desirable environments or cheaper and available properties abroad (Williams and Van Patten 2006, McCarthy 2007, Woods 2009). A well-known example of the internationalisation of second home ownership are 'snowbirds', people who migrate seasonally to sunnier and warmer locations within or across national boundaries (Karisto 2000, Williams et al. 2000, 2004, Timothy 2002). Examples of nationalities crossing borders in search of second homes include the British in rural France (Buller and Hoggart 1994, Hoggart and Buller 1995, Chaplin 1999), Americans in Mexico and Canada (Timothy 2004) and Germans in Denmark (Tress 2002) and Sweden (Müller 1999, 2002).

In Finland transnational second home tourism has become an issue in recent years alongside the growing number of Russians and, to a lesser extent, Norwegians interested in buying leisure properties on the Finnish side of the border (Tuulentie et al. 2012). However, the history of Finnish cottage culture has long been marked by its international character. The origin of second home ownership dates back to the eighteenth century and the time under Swedish rule. Later at the beginning

Table 8.2   Eco-social impacts of second homes in tourism centres.

Key features	Environmental impacts	Social impacts
High turnover of second home tourists	+ Eco-efficient co-ownership − Growth of energy consumption and mobility per cottage	+ High utilisation rate, more income to local economy − Low commitment to local community − Social side-effects of tourism − Seasonality, unemployment
Spatial concentration of cottages	+ Systematic land use planning + Dense and eco-efficient spatial structure − Landscape change − Energy intensive tourism (down hill skiing, golf)	+ Income for tourism industry + New employment and business opportunities + Sustenance of infrastructure and services + Taxes and payments − Uneven regional development − Displacement of other forms of land-use and business activities − Rising price level
Long-haul second home tourism	+ Use of public transport (buses and trains) − Use of air traffic and private cars	+ Investments to the area − Business profits and income taxes flow out of the region

of the nineteenth century the Russian occupation made Finland a destination of Russians (Jaatinen 1997, Lovell 2003). However, since independence in 1917, second home ownership has remained almost entirely a domestic phenomenon. This has been partly due to national legislation that restricted property ownership by foreigners. Foreigners have been allowed to buy properties since 2000, after the accession of Finland to the European Union (EU) in 1995. The 2000s have gradually witnessed an increase in the number of foreign second home purchases. This is mainly due to increases in the number of Russian second home owners. Since 2007 Russians have bought over 400 properties annually (National Land Survey 2011). The foreign demand has mostly focused on properties in the ski centres in Lapland and Finnish Lakeland in south-eastern Finland, within a few hours reach from St. Petersburg (Tuulentie 2006, Pitkänen and Vepsäläinen 2008, Pitkänen 2011b) (Figure 8.5 below).

Lipkina (2013) has studied Russians' second home purchases in eastern Finland and found that the Russians' demand focuses mainly on large and high-priced properties or unbuilt lakeside plots. However, old farmhouses, vacant dwellings and old public buildings are increasingly sold to Russians. In eastern Finland, second homes owned by Russians are mostly geographically scattered. However, due to real estate agents' offers and available properties there are a few

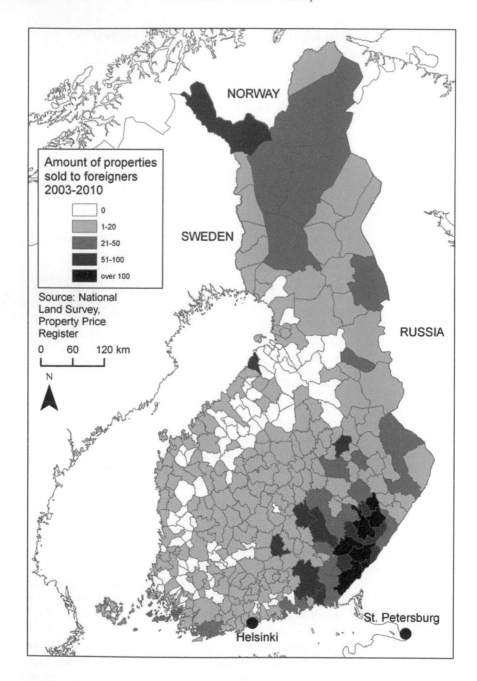

Figure 8.5   Foreign second home ownership in Finland.

Source: National Land Survey.

concentrations where the Russian outnumber the Finnish second home owners and local residents. Lipkina (2011, 2013) conducted interviews with Russian newcomers and found that those buying second homes in Finland are typically families in managerial positions from the metropolitan areas of St. Petersburg and Moscow. They are attracted by the positive image of Finland which emphasizes untouched and clean nature as well as friendly people. Finland is considered to be close even to Moscow, and the price level of suitable properties is considered to be lower than in the urban fringe areas of Moscow and St. Petersburg. However, one of the most important pull factors is that the Russians consider their investment to be physically, economically and politically safe in Finland. Buying safe and attractive lakeside properties in Russia is considered nearly impossible.

The central features characterising foreign second home ownership are the emergence of a completely new second home market in certain regions, and also socio-economic, cultural and language differences between the newcomers and the local population as well as cross-border mobility. These features and their eco-social impacts will be analysed in the following sections.

Emergence of New Second Home Market in Eastern Finland

There is no research on the environmental consequences of transnational second home tourism in Finland. In the case of Russians, the environmental impacts can be considered similar to modern domestic second home development. However, some features of Russian second home tourism give rise to a specific set of impacts. First of all, the emergence of completely new second home market has, in some areas, led to the intense development and landscape change of shorelines that would otherwise not have occurred. Not only large and well-equipped second homes, but also gardens, jetties and roads have been built on areas with no previous settlement structure. Besides environmental impacts, the emergence of this new market has had a number of social impacts. Pitkänen (2011b) has studied how foreign second home tourists have been discussed in the media. In media reports, the Russian buyers have been associated with dubious real estate business, land-jobbing and money laundering. As the Russians are speculated to be capable and willing to pay more than Finns, this has been seen to lead to the creation of selective market, i.e. to the marketing available properties only for Russians. A frequent fear in the Finnish media has been that the external demand will raise property prices and thereby affect Finns' possibilities to acquire second homes. Media has also reported on the displacement of locals from the housing market and the change of residential areas to vacation use as the Russians have been interested in buying houses and plots in amenity-rich locations meant for permanent residence and provided with municipal engineering (Pitkänen 2011b).

On the other hand, the positive impacts of the new second home market have also been highlighted. Such newcomers bring considerable revenues for the local economy and have revitalized the rural real estate market in areas suffering from rural decline and out-migration. Maintenance services, especially the building and retail sectors, as well as real estate and tourism businesses have been seen to benefit from Russian investments in second homes. In response to the claims that Russians displace Finns, it has been reported that the Russians have bought especially large and high-priced properties and old public properties that are hard to sell in the domestic market (Pitkänen 2011b).

Figure 8.6   Newly built Russian owned second homes in Eastern Finland (2010).

Source: K. Pitkänen

Social Differences

The cultural, language and socio-economic differences between the Finnish host society and the Russian newcomers brings about a set of eco-social impacts. The cultural and language differences have caused concern in the media (Pitkänen 2011b). In relation to environmental impacts, it has been feared that the Russians will not follow Finnish regulations for the building and use of natural areas. Similarly, concerns have been voiced that differences in environmental consciousness and nature use traditions will lead to negative environmental impacts and restrict access to natural areas. On the other hand, although this has not been mentioned in the media, the language and economic differences can also have positive environmental consequences. The Russian newcomers have the assets to invest in ecologically sound technology and instead of doing everything by themselves they resort to professionals in building and landscape design.

Cultural and language differences can lead into the creation of ethnic enclaves. Internationally there are examples of instances when the growth of newcomers raised concerns over the preservation of local culture, language and forms of land use (Buller and Hoggart 1994). However, concentrations of foreign second home tourists can also attract foreign service providers and businesses (Timothy 2002). In this respect, a significant role is played by the real estate agents operating across borders, who have the power to direct demand to certain areas (Williams et al. 2004, Müller 1999, 2002).

Furthermore, in remote and declining communities the increase in the number of socioeconomically advantaged second home owners can cause tensions between the locals and the newcomers. The differences between the two groups are intensified if the newcomers are perceived as complete outsiders with no previous place affiliation (Pitkänen 2011b, Buller and Hoggart 1994). Research on Russian second home tourism in Finland emphasizes that, from the perspective of the Finnish society, the phenomenon is a complicated and emotional matter with the potential to raise opposition and even conflicts (Pitkänen 2011b). This is partly a result of the troubled political history between the countries, which reflects as a certain kind of negativeness towards Russia among the civil society (Paasi 1999). In the media it has been emphasized that the local people are against selling properties to Russians. On the other hand, international leisure migration has also been perceived as positive and helping to revitalize the Finnish countryside, including new business opportunities. Due to language problems, the Russians need special services such as legal help, translations and mediators helping with Finnish institutions, banks and insurance companies. For this purpose a number of new businesses, often run by Russian migrants, have sprung up.

Cross-border Mobility

A feature that makes Russian second home tourism different from the domestic is the related cross-border mobility. In the global context, the environmental impacts of transnational second home tourism are especially induced by long distance travelling. Often the decision of acquiring a second home or a time-share apartment abroad is dependent on low-cost air connections. Transnational second home tourism thereby contributes to the emissions caused by air traffic. In the case of Russians, the second home destinations in Finland are accessible mainly by private car. Some of the second home owners from Moscow drive 16–18 hours to their property in Finland (Lipkina 2011). Due to the long distances and poor road quality in Russian rural areas, the Russians prefer large four-wheel drives that might not be as fuel efficient as smaller cars. In general, the mobility related to Russian second home ownership in Finland is similar to other tourism mobility from Russia. Besides the trip to the second home, the Russian second home owners often make shorter trips from the second home to do shopping, visit tourist attractions and buy services.

Hence, the ecological impacts of an average Russian second home owner's mobility can be estimated to be above an average Finnish second home owner.

Compared with social and cultural impacts, the ecological impacts caused by transnational second home traffic are international. So are also the political impacts of such mobility. The Finnish-Russian border is the longest border the European Union has with Russia, and cross-border mobility has increased during the past decade. The Finnish Border Guard counted 12 million border crossings in passenger traffic on the Finnish-Russian border in 2012, an increase of 13 per cent from the year 2011 (Finnish Border Guard 2013). Increasing cross-border mobility has meant greater foreign investments and income from tourism, but the capacity of border crossings as well as embassies and consulates granting visas has also been exceeded. Furthermore, the difficulties of border crossing and visa regulations restrict Russian second home owners' ability to visit their second homes. Table 8.3 summarizes the positive and negative environmental and social impacts of Russian second home tourism in Finland.

Factors and Processes Causing Eco-social Impacts of Second Home Tourism

In this chapter the eco-social impacts of second home tourism in Finland have been discussed by highlighting the interrelatedness of environmental and social impacts. Hiltunen (2007) has previously concluded that environmental impacts of second home tourism in Finland derive from (1) firstly, dwelling, living and consuming at second home, (2) secondly, motorized mobility between home and second home, and (3) thirdly, building cottages dispersed in the countryside and on shoreline locations. Here the impact assessment was widened to include social, community and cultural impacts by scrutinising impacts through current trends in second home tourism. The eco-social impacts of second home development were analysed according to the central features of each trend (see Tables 8.1–8.3).

The aim has been to identify the underlying factors and processes leading to these impacts. After identifying current trends and their impacts, we looked for their similarities and noticed that the eco-social impacts were induced by certain factors which could be further categorized into two main dimensions: location and agency. These categories summarize the key processes and basic factors in second home tourism that have the potential to induce the most significant eco-social impacts of the phenomenon (Figure 8.7 below).

Location Factors

The category of location refers to the spatial dimension and processes of second home tourism. For example, shoreline building and frequent mobility between the primary and the second home were among the key features related to/affecting eco-social impacts of dual-dwelling. Similarly, in relation to second homes in tourism centres the spatial concentration of second homes and long-haul mobility

Table 8.3   Eco-social impacts of Russian second home tourism in Finland.

Key features	Environmental impacts	Social impacts
Emergence of completely new second home market	– Intensive shoreline development (buildings, jetties, gardens, roads) – Landscape change	+ Revenues for local economy + Revitalisation of rural real estate market + Demand of properties that would be hard to sell on domestic market – Real estate speculation – Displacement of local population and Finnish second home buyers – Change of residential areas to vacation use
Socio-economic, cultural and language differences	+ Assets to invests in ecologically sound technology + Use of professionals in building and landscape design – Difficulties with environmental regulations – Differences in environmental awareness and nature use traditions – Restricts public access of nature	+ New business opportunities and jobs + Multiculturalism and revitalisation of local communities – Suspicions, prejudices, hostility – Conflicts – Ethnic enclaves
Cross-border mobility	– Long haul travelling – Intensive use of private cars	+ Foreign investments and tourism income – Congestion of border crosses and consulates – Low utilization rate of second homes

were important. As regards foreign second homes, the emergence of a regionally specific new market of second homes and cross-border mobility were emphasized. Hence, the eco-social impacts of second homes can be seen to be caused by spatial factors and processes which include (1) regional patterns and distribution of second homes, (2) relative distance between primary and second home, and (3) second home location in the rural spatial structure.

The spatial distribution of second homes is not geographically even, but the strongest demand focuses on amenity rich rural areas fulfilling certain criteria (e.g. proximity to urban centres, shorelines, tourism centres, border areas). These areas are marked by concentrations of second homes and second home tourists, which also

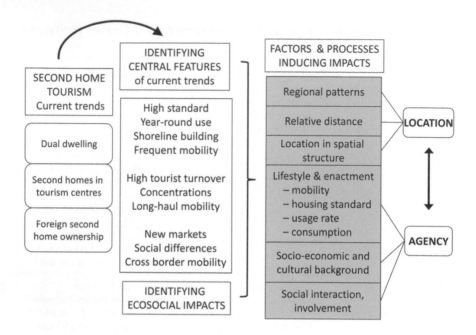

Figure 8.7   Assessing eco-social impacts of second home tourism.

intensify the eco-social impacts of the phenomenon. In less marketable areas there are fewer second homes whereby the eco-social impacts, too, can be very different.

A crucial factor in relation to eco-social impacts is the relative distance between the primary and the second home. The distance has an effect on the amount and frequency of second home trips as well as on the modes of travelling and length of stays. In general, the shorter the travel distance the more frequently trips are made. On the other hand, with longer distances people spend longer periods at the second home. The absolute distance, however, is not always decisive since also travel time, economic costs and experienced distance matter in travelling (Hall 2005). For example, in terms of cross-border mobility the distance might be relevant only when measured after the border, and in terms of long-haul mobility the travel modes are crucial, as the distant second home may be reached time-efficiently by plane or night train.

The location of second homes within the spatial structure is important. In Finland, second homes are usually built in secluded surroundings along lake and sea sides, detached from rural population centres and villages. This leads to a dispersed spatial structure of second homes with environmental and social impacts e.g. in the form of increased public and private economic costs. However, second homes are increasingly built in dense spatial form in areas with municipal engineering (foreign second home owners) and even contribute to the creation of new population centres (tourism centres).

Agency Factors

The category of agency refers to the human dimension and enactment of second home tourism. In our review, regarding dual-dwelling, high standard and year-round use of second homes had important environmental and social consequences. With respect to second homes in tourism centres, the high turnover of tourists leads to impacts that are different from traditional types of second home tourism. In foreign second home tourism, in turn, the socio-economic, cultural and language differences between the Russians and the Finns were emphasized. In other words, the eco-social impacts can be seen to be related to (1) second home lifestyle and enactments (including mobility, standard of second home, usage rate and consumption), (2) the socio-economic and cultural background of second home owners, and (3) the social interaction and networks of second home owners, their local commitment, as well as their involvement in rural communities.

The length of stays and standard of second homes are central lifestyle related aspects which lead to eco-social impacts of second home tourism. Activities, consumption decisions and usage rate all contribute to the way the second home owner household's lifestyle is composed in the interaction between the two homes. This affects the impacts second home tourism has on the environment and local communities.

The socio-economic and cultural background of second home owners reflects eco-social impacts as well. Urban second home owners' nostalgic images of the countryside can differ greatly from the meanings the countryside has for permanent rural dwellers. Rural images affect changes second home owners are ready to accept in the cottage environment and how they want to participate in rural change. Such background factors may also play significant roles in local conflicts caused by cultural differences. In remote and declining communities the increase in the number of socioeconomically advantaged second home owners can cause tensions between the locals and the newcomers.

Finally, an important role is also played by social interaction. In this respect it is important how the interaction between second home owners and local dwellers evolves. Local interaction may develop quite differently depending on the form of second home tourism. For example, for many dual-dwellers living and residing in the second home community comes close to permanent residency characterized by active local social and often also local family relations, whereas for timeshare cottagers short visits to tourism centres and a lifestyle emphasising leisure and outdoor activities involve less social contact with the local people. The involvement in the local community has not only important social, but also environmental consequences in the sense that it affects how second home owners use the environment and commit themselves to its protection.

It is often thought that the positive economic and social advantages even out the negative environmental impacts of second home tourism (Nieminen 2010). By juxtaposing the positive and negative impacts we have aimed to reveal that the distinction is not that simple and straightforward. The impacts of second home tourism may thus be environmentally sound and then again socially harmful. Consequently, instead of making an unproblematic comparison, it is important to identify and understand what factors and processes cause and explain the different impacts of second home tourism in specific locations in time and space. The eco-social impacts of dual-dwelling are similar in many respects to those of permanent dwelling, while the impacts of second home growth in tourism centres and transnational second home tourism are often more comparable to tourism impacts generally.

Concluding Remarks

Current developments in Finnish second home tourism are strongly influenced by new forms and patterns of production and consumption, advances in mobility and technology, and increases, for some, in personal wealth and leisure time. The growing number of retirees in the Western world with disposable spare time and purchasing power is especially important in the part-time dweller segment in rural areas with potential for yet further growth. Second home ownership often follows a path of cyclical migration beginning with holiday visits, progressing towards the purchase of a second home and later to an extended residence upon retirement (Tuulentie 2007, Müller and Hall 2004, Quinn 2004, Timothy 2004). However, in countries such as Finland, with a long second home tradition, cottages are often inherited and second home ownership goes on within the family. It is likely that both of these ownership mechanisms will be important in the foreseeable future.

The interplay between amenity factors and affordability is increasingly affecting access to second homes in both locational and socio-economic terms. The high demand for coast and lake shoreline second homes has a significant effect on the price and supply of plots and real estate values. Second home property prices have continued to increase in recent decades. To purchase a second home for a reasonable price has become difficult, especially near larger urban centres in southern Finland, where the value of shoreline sites may easily exceed the value of suburban plots. This situation appears to have caused a displacement effect on demand, and new second home development seems to have moved further north to the more peripheral and less populated areas or even to cheaper destinations abroad (Pitkänen and Vepsäläinen 2008). Furthermore, prices for second homes and shoreline plots have risen in the amenity rich regions of rural Lakeland and Lapland as well. This situation means that, despite the place of second homes in Finnish culture and identity, second home tourism as a socially accessible and equitable form of recreational living seems to be getting increasingly out of the financial reach of younger generations.

From a planning and policy perspective, the reason for recognising eco-social impacts of second home tourism and identifying underlying factors and processes is the need to enhance the positive impacts and mitigate the negative impacts caused by second homes. Positive impacts are generally related to local social benefits and community development. Enhancement measures of second home tourism typically aim at maximising economic benefits. Alongside their contribution to the regional economy, second home owners are also expected by policy makers to add to local social capital and the overall liveliness of rural areas. Such concerns are extremely important for those more peripheral areas of Finland that have been affected by out-migration and economic restructuring in recent years. The negative impacts caused by second home tourism are most often related to issues concerning the natural and built environments. Land use planning and environmental governance measures aim to prevent or reduce the harmful impacts on landscape and nature. Indeed, second home tourism reflects the same difficulties long recognized in tourism in general regarding the need to manage or even limit development in order to maintain the very qualities that attracted people in the first place.

In this chapter the key focus was on the eco-social impact assessment of second home tourism. It is apparent that environmental impacts cannot be understood if they are disconnected from social and community development. Both environmental and social impacts interweave in eco-social impacts that affect the relational system of second home environment, and broadly concern policy and decision making in different levels in society. As suggested at the outset of this chapter this means that, despite being a 'traditional' form of leisure mobility in the Finnish context, second home tourism has changed and will continue to change. In particular this means that researchers, along with planners and policy makers, will need to develop new ways of understanding the extent to which past and current second home decisions may lock in regions to certain futures or at least limit options. In this the major societal trends identified in this chapter, such as retirement populations, impacts of technology, and, for some, growing wealth and leisure time and the blurring of work and leisure, will all affect the future of second homes in Finland and, in light of increased mobility, in other countries in which Finns may purchase second homes. The future of second homes in Finland is likely to be marked therefore as much by increasing Finnish international second home mobility as it is by the purchase and development of second homes in Finland by people from elsewhere in Europe. Such potential change only reinforces that second homes need to be understood not just in space but also over time and within new forms of relationality that frame multiple-dwelling and the eco-social system.

Acknowledgment

The chapter was written during a research project funded by Academy of Finland, project number SA 255424.

References

Ahlqvist, K., Santavuori, M., Mustonen, P., Massa, I. and Rytkönen, A. 2008. *Mökkeily elämäntapana ja ekotehokkaiden käytäntöjen hyväksyttävyys.*

Vapaa-ajan asumisen ekotehokkuus. TTS tutkimuksen raportteja ja oppaita 36. Nurmijärvi: Työtehoseura.

Aho, S. and Ilola, H. 2004. *Maaseutu suomalaisten asenteissa, toiveissa ja kokemuksissa.* Lapin yliopiston kauppatieteiden ja matkailun tiedekunnan julkaisuja. B. Tutkimusraportteja ja selvityksiä 2. Rovaniemi: Lapin yliopisto.

Aho, S. and Ilola, H. 2006. *Toinen koti maalla? Kakkosasuminen ja maaseudun elinvoimaisuus.* Lapin yliopiston kauppatieteiden ja matkailun tiedekunnan julkaisuja B. Tutkimusraportteja ja selvityksiä 6. Rovaniemi: Lapin ylipisto.

Alasuutari, P. and Alasuutari, M. 2010. *Mökkihulluus. Vapaa-ajan asumisen taika ja taito.* Rovaniemi: Lapin yliopistokustannus.

Berkes, F. and Folke, C. 2000. *Linking social and ecological systems. Managementvpractices and social mechanisms for building resilience.* Cambridge: Cambridge University Press.

Berkes, F., Colding, J., and Folke, C. 2003. *Navigating social-ecological systems.* Cambridge: Cambridge University Press.

Buller, H. and Hoggart, K. 1994. The social integration of British home owners into French rural communities. *Journal of Rural Studies*, 10(2), 197–210.

Chaplin, D. 1999. Consuming work/productive leisure: The consumption patterns of second home environments. *Leisure Studies*, 18(1), 41–55.

Dubois, G. 2005. Indicators for an Environmental Assessment of Tourism at National Level. *Current Issues in Tourism*, 8(2&3), 140–54.

Eisto, I. and Kotilainen, J. 2010. Resilienssitutkimus ja sosioekologiset järjestelmät, in *Luonnonvarayhdyskunnat ja muuttuva ympäristö – resilienssitutkimuksen näkökulmia Itä-Suomeen*, edited by J. Kotilainen and I. Eisto. Publications of the University of Eastern Finland. Reports and Studies in Social Sciences and Business Studies No 2. Joensuu: University of Eastern Finland, 14–30.

Finnish Border Guard 2013. Suomen itärajan rajaliikenne kasvoi 13% vuonna 2012. [Online]. Available at: http://www.raja.fi/tietoa/tiedotteet/1/0/suomen_itarajan_ rajaliikenne_kasvoi_13_vuonna_2012 [accessed: 15 January 2013]

Finnish Transport Agency 2013. Liikenneverkko, Tiet. http://portal.liikennevirasto. fi/sivu/www/f/liikenneverkko/tiet, [accessed: 16 January, 2013].

Fountain, J. and Hall, C.M. 2002. The impacts of lifestyle migration on rural communities: A case study of Akaroa, New Zealand, in *Tourism and Migration: New relationships between production and consumption*, edited by C.M. Hall and A.M. Williams. Dordrecht: Kluwer Academic Publishers, 153–68.

Gallent, N., Mace, A. and Tewdwr-Jones, M. 2005. *Second Homes: European perspectives and UK policies.* Aldershot: Ashgate.

Gartner, W.C. 1987. Environmental Impacts of Recreational Home Developments. *Annals of Tourism Research*, 14(1), 38–57.

Gössling, S. and Hall, C.M. 2006. An introduction to tourism and global environmental change, in *Tourism and Global Environmental Change*, edited by S. Gössling and C.M. Hall. Bodmin: Routledge. 1–33.

Hall, C.M. 2005. *Tourism: Rethinking the Social Science of Mobility*, Harlow: Prentice-Hall.

Hall, C.M. 2011. Housing tourists: Accommodating short-term visitors, in *Rural housing, exurbanization and amenity driven development: Contrasting the haves and the have nots*, edited by D. Marcoullier, M. Lapping and O. Furuseth. Farnham: Ashgate, 113–28.

Hall, C.M. and Müller, D.K. 2004. *Tourism, Mobility and Second Homes: Between elite landscape and common ground*. Clevedon: Channel View Publications.

Hall C.M., Müller D.K. and Saarinen J. 2009. *Nordic Tourism: Issues and Cases*. Bristol: Channel View Publications.

Halseth, G. 2004. The 'cottage' privilege: Increasingly elite landscapes of second homes in Canada, in *Tourism, Mobility and Second Homes: Between elite landscapes and common ground* edited by C.M. Hall and D.K. Müller. Clevedon: Channel View Publications, 35–54.

Heinonen, S., Kasanen, P. and Walls, M. 2002. *Ekotehokas yhteiskunta – Haasteita luonnon ja ihmisen systeemien yhteensovittamiselle*. Publications of the Ministry of the Environment, Environmental Policy, The Finnish Environment 598. Helsinki: Edita Prima Oy.

Hiltunen, M.J. 2007. Environmental impacts of rural second home tourism: Case Lake District in Finland. *Scandinavian Journal of Hospitality and Tourism*, 7(3), 243–65.

Hiltunen, M.J. 2009. Second Homes in Finland, in *Nordic Tourism: Issues and Cases*, edited by C.M. Hall, D.K. Müller and J. Saarinen. Bristol: Channel View Publications, 185–88.

Hiltunen, M.J., Pitkänen, K., Vepsäläinen, M., Shemeikka, P. and Rehunen A. 2013. Vetovoimaiset kylämökkeily-ympäristöt. In *Vapaa-ajan asuminen maaseudun kylissä ja taajamissa*, edited by Rehunen, A. et al. Ministry of the Environment, The Finnish Environment. Forthcoming.

Hirvonen, P. and Puustinen, S. 2008. *Vapaa-ajan asumisen uudet tuulet. Suomalaisten näkemyksiä vapaa-ajan asumisesta*. Yhdyskuntasuunnittelun tutkimus- ja koulutuskeskuksen julkaisuja B 94. Espoo: Teknillinen korkeakoulu.

Hoggart, K. and Buller, H. 1995. Geographical differences in British property acquisitions in rural France. *The Geographical Journal* 161(1), 69–78.

Jaatinen, S. 1997. *Elysium Wiburgense: Villabebyggelsen och villakulturen kring Viborg*. Bidrag till kännedom av Finlands natur of folk 151. Helsinki: Finnish Society of Sciences and Letters

Jansson, B. and Müller, D.K. 2004. Second home plans among second home owners in Northern Europe's periphery, in *Tourism, mobility and second homes: Between elite landscape and common ground*, edited by C.M. Hall and D.K. Müller. Clevedon: Channel View Publications, 261–72.

Jokinen, A. 2002. Free-time habitation and layers of ecological history at a southern Finnish lake. *Landscape and Urban Planning*, 61(2–4), 99–112.

Kaltenborn, B.P. 1998. The alternate home: Motives of recreation home use. *Norsk Geografisk Tidsskrift*, 52(3), 121–34.

Karisto, A. 2000. *Suomalaiselämää Espanjassa*. Kansanelämän kuvauksia; 49. Jyväskylä: Suomalaisen Kirjallisuuden Seura.

Kauppila, P. 2004. Matkailukeskusten kehitysprosessi ja rooli aluehkehityksessä paikallistasolla: esimerkkeinä Levi, Ruka, Saariselkä ja Ylläs. *Nordia Geographical Publications 33:1.*

Kauppila, P. 2006. Matkailukeskukset, vapaa-ajanrakennukset ja kehitysprosessi: tarkastelussa Levi, Ruka, Saariselkä ja Ylläs. in *Matkailukehityksestä aluekehitykseen*, edited by R. Leinonen et al. *Naturpolis Kuusamo, koulutus-ja kehittämispalvelut, tutkimuksia* 1/2006, 69–109.

Kauppila, P. 2009. Resorts' second home tourism and regional development: a viewpoint of a Northern periphery. *Nordia Geographical Publications* 38: 5, 3–12.

Koski, K. 2007. *Vakituisen ranta-asutuksen kuntataloudelliset vaikutukset.* Publications of the Ministry of the Environment. Land Use Department. The Finnish Environment 38.

Lahti, P. (2010). Mökkiliikenteen energia- ja ekotehokkuus. In Vapaa-ajan asumisen ekotehokkuus, edited by Rytkönen, A. & Kirkkari A-M. Publications of the Ministry of the Environment, Department of the Built Environment. The Finnish Environment 6, 64-88. Helsinki: Edita.

Leinonen, R., Kauppila, P. and Saarinen, J. 2008a. Maa- ja metsätalous, matkailu vai teollisuus? Koillis-Suomen julkisen ja yksityisen sektorin toimijoiden näkökulmia matkailuun ja aluekehitykseen. *Alue ja Ympäristö* 37(1), 29–40.

Leinonen, R., Kauppila, P. and Saarinen, J. 2008b. Kestävä matkailusuunnittelu ja syrjäseutujen aluekehitys. *Terra 120(2)*, 107–12.

Lipkina, O. 2011. *Dacha perspectives: reasons to have a second home in Finland.* Paper to the Nordic Geographers Meeting: Geographical Knowledge, Nature and Practice, Roskilde, Denmark, 24–27 May 2011.

Lipkina, O. 2013. Motives for Russian second home ownership in Finland. *Scandinavian Journal of Hospitality and Tourism.* Submitted.

Lovell, S. 2003. *Summerfolk: A history of the dacha 1710–2000.* Ithaca: Cornell University Press.

Marcoulier, D., Lapping, M. and Furuseth, O. 2011. *Rural Housing, Exurbanization, and Amenity-Driven Development: Contrasting the Haves and the Have Nots.* Aldershot: Ashgate.

Marjavaara, R. 2008. *Second home tourism: The root to displacement in Sweden.* Gerum Kulturgeografi 2008:1. Kulturgeografiska institutionen. Umeå: Umeå universitet.

Massa, I. 2011. Mökkeily muuttuu, ympäristökuorma kasvaa, in *Ihminen ja ympäristö*, edited by J. Niemelä et al. Helsinki: Gaudeamus, 299–304.

Mathieson, A. and Wall, G. 1982. *Tourism: Economic, physical and social impacts.* Harlow: Longman Scientific & Technical.

McCarthy, J. 2007. Rural geography: Globalizing the countryside. *Progress in Human Geography*, 32(1), 129–37.

McIntyre, N. 2006. Introduction, in *Multiple Dwelling and Tourism: Negotiating place, home and identity*, edited by N. McIntyre, D.R. Williams and K.E. McHugh. Wallingford: Cabi, 3–14.

McIntyre, N., Williams, D.R. and McHugh, K.E. 2006. *Multiple Dwelling and Tourism: Negotiating place, home and identity*. Wallingford, Cabi.

Melasniemi-Uutela, H. 2004. Suomalaisen mökkikulttuurin suunta? in *Ihanne ja todellisuus. Näkökulmia kulutuksen muutokseen*, edited by K. Ahlqvist, and A. Raijas. Tilastokeskus. Helsinki: Edita Prima, 145–63.

Mottiar, Z. and Quinn, B. 2003. Shaping leisure/tourism places – the role of holiday home owners: A case study of Courtown, Co. Wexford, Ireland. *Leisure Studies* 22(2), 109–27.

Müller, D.K. 1999. *German second home owners in the Swedish countryside: On the internationalization of the leisure space*. Kulturgeografiska institutionen. Umeå Universitet, Umeå.

Müller, D.K. 2002. Reinventing the countryside: German second-home owners in Southern Sweden. *Current Issues in Tourism* 5(5), 426–46.

Müller, D.K. 2004. Second homes in Sweden: Patterns and issues, in *Tourism, Mobility and Second Homes: Between Elite Landscape and Common Ground*, edited by C.M. Hall and D.K. Müller. Clevedon: Channel View Publications, 244–60.

Müller, D.K. 2005. Second home tourism in the Swedish mountain range. In *Nature-based Tourism in Peripheral Areas: Development or Disaster?*, edited by C.M. Hall and S. Boyd Clevedon: Channel View Publications, 133–48.

Müller, D.K. 2007. Second Homes in the Nordic Countries. Between common Heritage and Exclusive Commodity. *Scandinavian Journal of Hospitality and Tourism*, 7(3), 193–201.

Müller, D.K. and C.M. Hall 2004. The Future of Second Home Tourism. *In Tourism, Mobility and Second Homes. Between Elite Landscape and Common Ground*, edited by C.M. Hall and D.K. Müller. Clevedon: Channel View Publications.

Müller, D.K., Hall, C.M. and Keen, D. 2004: Second home tourism impact, planning and management, in *Tourism, Mobility and Second Homes: Between Elite Landscape and Common Ground*, edited by C.M. Hall and D.K. Müller. Clevedon: Channel View Publications, 15–34.

National Land Survey 2011. *"Ne vie meitin maat" – Kiinteistökauppa vapautettiin vuonna 2000*. [Online: National Land Survey]. Available at: http://www.maanmittauslaitos.fi/tiedotteet/2011/02/ne-vie-meitin-maat-kiinteistokauppa-vapautettiin-vuonna-2000 [accessed 21 February 2012].

Nieminen, M. 2010. *Kesämökkibarometri 2009*. Publications of the Ministry of Employment and the Economy, Regional Development, 12. Helsinki: Edita. [Online]. Available at: http://www.tem.fi/files/27185/tem_12_2010_web.pdf [accessed 16 January 2013].

Paasi, A. 1999. Boundaries as social practice and sicourse. The Finnish-Russian border: *Regional Studies*, 33(7), 669–80.

Periäinen, K. 2006. The summer cottage: A dream in the Finnish forest, in *Multiple Dwelling and Tourism: Negotiating place, home and identity*, edited by N. McIntyre, D.R. Williams and K.E. McHugh. Wallingford: Cabi, 103–13.

Pitkänen, K. 2011a. *Mökkimaisema muutoksessa: Kulttuurimaantieteellinen näkökulma mökkeilyyn*. Publications of the University of Eastern Finland, Dissertations in Social Sciences and Business Studies. Joensuu: University of Eastern Finland.

Pitkänen, K. 2011b. Contested cottage landscapes: Host perspective to the increase of foreign second home ownership in Finland 1990–2008. *Fennia* 189(1), 43–59.

Pitkänen, K. and Kokki, R. 2005. *Mennäänkö mökille? Näkökulmia pääkaupunkiseutulaisten vapaa-ajan asumiseen Järvi-Suomessa*. Savonlinnan koulutus- ja kehittämiskeskuksen julkaisuja n:o 11. Savonlinna: Joensuun yliopisto.

Pitkänen, K. and Vepsäläinen, M. 2008. Foreseeing the future of second home tourism. The case of Finnish media and policy discourse. *Scandinavian Journal of Hospitality and Tourism*, 8(1), 1–24.

Pitkänen, K., Sikiö, M. and Rehunen, A. 2012. *Hidden life of rural north. A study on empty dwellings in rural Finland*. Paper to the 2nd Nordic Conference for Rural Research: Rural at the Edge, Joensuu, Finland, 21–23 May 2012.

Quinn, B. 2004. Dwelling through multiple places: A case study of second home ownership in Ireland, in *Tourism, Mobility and Second Homes: Between Elite Landscape and Common Ground*, edited by C.M. Hall and D.K. Müller. Clevedon: Channel View Publications, 113–32.

Rehunen, A., Rantanen, M., Lehtola, I. and Hiltunen, M.J. (eds). 2012. Palvelujen saavutettavuus muutoksessa. Maaseudun vakituisten ja vapaa-ajan asukkaiden palveluympäristön kehityssuunnat ja uudet mahdollisuudet. Helsingin yliopisto, Ruralia-instituutti, Raportteja 88. Available at: http://www.helsinki.fi/ruralia/julkaisut/pdf/Raportteja88.pdf

Rehunen, A., Vepsäläinen, M., Hiltunen, M. and Pitkänen, K. 2013. *Vapaa-ajan asuminen kylissä ja taajamissa.*. Ministry of the Environment, The Finnish Environment. Forthcoming.

Rytkönen, A. and Kirkkari, A-M. 2010. *Vapaa-ajan asumisen ekotehokkuus*. Ministry of the Environment, Department of Built Environment. The Finnish Environment 6. Helsinki: Edita.

Salo, M., Lähteenoja, S. and Lettenmeier, M. 2008. *MatkailuMIPS – Matkailun luonnonvarojen kulutus*. Publications of the Ministry of Employment and the Economy, Employment and entrepreneurship, 8. Helsinki: Edita.

Sandell, K. 2006. Access under stress: The right of public access tradition in Sweden, in *Multiple Dwelling and Tourism: Negotiating place, home and identity*, edited by N. McIntyre, D.R. Williams and K.E. McHugh. Wallingford: Cabi, 278–349.

Scott, D., Hall, C.M. and Gössling, S. 2012. *Tourism and Climate Change: Impacts, Adaptation and Mitigation*. London: Routledge.

Statistics Finland 2012a. *Rakennukset ja kesämökit 2011*. Official statistics of Finland. [Online]. Available at: http://www.stat.fi/til/rakke/2011/ rakke_2011_2012-05-25_fi.pdf [accessed: 14 January 2013].

Statistics Finland 2012b. *Suomalaisten matkailu 2011*. Official statistics of Finland. [Online]. Available at: http://www.stat.fi/til/smat/2011/smat_2011_2012-05-30_fi.pdf [accessed: 16 January 2013].

Timothy, D. 2002. Tourism and the Growth of Urban Ethnic Islands, in *Tourism and Migration: New relationships between production and consumption*, edited by C.M. Hall and A.M. Williams. Dordrecht, Kluwer Academic Publishers.

Timothy, D.J. 2004. Recreational second homes in the United States: Development issues and contemporary patters. In *Tourism, mobility and second homes. Between elite landscape and common ground*, edited by C.M. Hall and D.K. Müller. Clevedon: Channel View Publications, 133–48.

Tress, G. 2000. *Die Ferienhauslandschaft: Motivationenen, Umweltsauswirkungen und Leitbilder im Ferienhaustourismus in Dänemark*. Forskningsrapport nr. 120, Publikationer fra Geografi, Institut for Geografi og Internationale Udviklingsstudier, Roskilde Universitetscenter.

Tress, G. 2002. Development of second-home tourism in Denmark. Scandinavian Journal of Hospitality and Tourism 2(2), 109-122.

Tuulentie, S. 2006. Tourists making themselves at home: Second homes as a part of tourist careers. In *Multiple Dwelling and Tourism: Negotiating place, home and identity*, edited by N. McIntyre, D.R. Williams and K.E. McHugh. Wallingford, Cabi. 145–58.

Tuulentie, S. 2007. Settled tourists: Second homes as a part of tourist life stories. *Scandinavian Journal of Hospitality and Tourism*, 7(3), 281–300.

Tuulentie, S., Lipkina, O. and Pitkänen, K. 2012. Do borders matter? Norwegian and Russian second home owners' relation to their leisure places in Finland. In Developing Tourism – Sustaining Regions. Book of Abstracts, edited by M. Ednarsson, F. Hoppstadius, L. Lundmark, R. Marjavaara, D. Müller, K. Pitkänen and U. Åkerlund, The 21st Nordic Symposium in Tourism and Hospitality Research. Umeå University, Umeå.

Vepsäläinen, M. and Hiltunen M.J. 2001. *Liikkumisen arkea Muu-Suomessa. Liikenteen ja tienpidon sosiaalinen ja alueellinen tasa-arvo*. Department of Geography. Publications No. 9. Joensuu: University of Joensuu,

Vepsäläinen, M., Pitkänen K. and Hiltunen, M. 2011. Mökkeilijöiden muuttuvat luontokäsitykset, in *Ihminen ja ympäristö*, edited by J. Niemelä et al. Helsinki: Gaudeamus Helsinki, 305–9

Visser, G. 2006. South Africa has second homes too! An exploration of the unexplored. *Current Issues in Tourism* 9(4&5), 351–83.

Vittersø, G. 2007. Norwegian cabin life in transition. *Scandinavian Journal of Hospitality and Tourism* 7(3), 266–80.

Williams, A.M., King, R., Warnes, A. and Patterson, G. 2000. Tourism and international retirement migration: New forms of an old relationship in Southern Europe. Tourism Geographies 2(1), 28–49.

Williams, A.M., King, R. and Warnes, T. 2004. British second homes in Southern Europe: Shifting nodes in the scapes and flows of migration and tourism. In *Tourism, Mobility and Second Homes: Between elite landscape and common ground,* edited by C.M Hall and D.K. Müller. Clevedon: Channel View Publications. 97–112.

Williams, D.R. and Van Patten, S.R. 2006. Home and away? Creating identities and sustaining places in a multi-centred world. In *Multiple Dwelling and Tourism: Negotiating Place, Home and Identity,* edited by N. McIntyre, D.R. Williams and K.E. McHugh. Wallingford: Cabi, 32-50.

Woods, M. 2009. The local politics of the global countryside: boosterism, aspirational ruralism and the contested reconstitution of Queenstown, New Zealand. *GeoJournal,* 76(4), 365–81.

PART III
Leisure Housing Expansion:
Driving Forces and Policy Choices

Chapter 9

Historic, Symbolic Aspects and Policy Issues of the Second Home Phenomenon in the Greek Tourism Context: The Cyclades Case Study

Olga Karayiannis, Olga Iakovidou and Paris Tsartas

Introduction

The aim of this chapter is to present basic aspects of the second home phenomenon within the framework of tourism studies, focusing on the Greek reality and particularities. The chapter is organized in four sections and their sub-sections.

In the first section an attempt is made to approach the broader historical framework of the second home phenomenon, in conjunction with its evolving symbolic, functional and spatial dimensions. The second section discusses the theoretical foundations of the phenomenon within the tourism context. The third section, which constitutes the main body of the chapter, aims to present an outline of the historical evolution of the second home phenomenon in Greece. Taking into account the particularities and parallel historical and developmental framework of the country, a literature review is presented and research shortcomings are identified, thus suggesting contemporary approaches needed. The present extent of the phenomenon in Greece is examined in light of basic data, and an effort is made to present the spatial concentration of second homes in one of the most touristic prefectures of Greece, the insular complex of the Cyclades. This is done by presenting and commenting on basic data available that reflect parameters of interest. Given that secondary data were inadequate to respond to crucial questions, basic results of primary research conducted on one island of the Cyclades are also presented, potentially allowing extrapolations.

The chapter is completed with a synopsis of basic conclusions and concerns prominent for the management of the second home phenomenon and its current prospects in the Greek reality.

Historical, Symbolic, Functional and Spatial Aspects

By definition, maintaining and using second homes presupposes a movement from the *primary* residence and doubtlessly takes on different meanings in various times and throughout the history of human social organization. With the exception of the nomadic hunter-gatherer period of social organization, where constant movement was the norm and residence was circumstantial, the second home phenomenon comes into existence during the period of permanent residency (Agricultural Revolution), while the use of the second home is more probably related to employment.[1]

The maintenance, use of and travel to second homes due to motives similar to those generating tourism (i.e. rest, recreation, change of environment), run a parallel course to the tourist phenomenon, with respect to basic parameters. Thus, such second homes, similarly to tourism journeys, were a privilege of the elite, their main characteristic being that they belonged to the nobility of the eighteenth century, and were maintained in spa towns or coastal areas, in order to enable seasonal escape from city life (Hall and Müller 2004, reference to Coppock 1977).

Increasingly intensified urbanization, a consequence of the successive stages of the Industrial Revolution, attributed a partially different content to the second home phenomenon. On the one hand, the countryside was gradually abandoned by its residents, thus transforming, in many cases, *first* homes into *second* homes. On the other hand, due to the gradual improvement of living standards in developed countries, second homes turned from a luxury into an accessible option for a wider range of social strata (NCSR 1998, Hall and Müller 2004). This trend reached its peak in the final decades of the last century and mainly from 1960 onwards.

In relation to function-utility, symbolism and location, second homes can be roughly categorised, albeit with frequent overlaps. Thus, second homes have become symbols of present-day desertion of rural areas or of their *post-production* phase and transformation into *consumption landscapes* (Gallent et al. 2005, Muller et al. 2004), in other words simultaneously expressing change and operating as agents of change in rural areas (Muller 2004). In some cases, second home maintenance concerns owners or users with origins from the corresponding locality, especially in rural areas. In other cases, choices concerning second homes are determined by investment criteria, such as in Northern European countries (mainly France, Germany, the Netherlands, and Scandinavian countries), where the aim is to earn income from letting one's main home for limited periods of time (Ball 2005). Often second homes can be related to changes in one's life cycle (retirement migration or having a family), or can even be retained for educational or professional training purposes. Sometimes second homes are gradually transformed into primary homes, rendering the initially seasonal or circumstantial migration to these homes permanent (Hall & Müller 2004). In other cases, the two homes might be used concurrently especially when their use is related to the distance of one's employment location (Ball 2005), or with specific working circumstances, due to which one may need to share time between two residences, for more or less definite periods of time. As far as the contemporary spatial aspect

---

1   The Agricultural Revolution came about with the simultaneous development of farming and animal breeding. For animal breeding, people often moved seasonally, e.g., from colder to warmer and more fertile lands during the winter months, for grazing, making use of some kind of second home.

of the phenomenon is concerned, there are cases where second homes are located in urban centres, belong to permanent rural residents, are used for city recreation or in order to use services not available in the area of one's permanent residence. Furthermore, second home location may be geographically related to one's main residence, in cases where travel time and distance make it easy to visit them often (e.g. as weekend homes), or, alternatively, can be independent from one's main abode, at more distant, attractive and often touristic destinations, i.e. holiday-seasonal migration homes.

Finally, if, apart from the particularities and different cases above, one takes into account the plethora of mobile forms of second homes (i.e., mobile homes, such as caravans, boats) (Hall and Müller 2004), the multifaceted and complicated character of this phenomenon becomes apparent.

Conceptual Limitations and Reflections

Tourism and Second Home Connections

As discussed in the previous section, apart from cases where maintaining a second home is related to one's employment, education and, in part, investment reasons, the second home phenomenon presents successive overlapping aspects with the tourism phenomenon in terms of space, morphology and time-frame. However, it has also become apparent that the second home phenomenon, even in cases where it is activated and shaped by tourism motives, still 'escapes' the narrow framework of understanding and interpreting tourism. This fact may justify the view of certain researchers who, as mentioned by Paris Tsartas, consider the phenomenon as 'not conceptually belonging to the category of tourism' (Tsartas et al. 2001: 45–6).

An attempt to present the similarities and differences between the two phenomena is put forward in Table 9.1 below.

Attempting to correlate the different forms of tourism and the expression of the second home phenomenon, one can see that the latter can be 'produced' locally, in the sense that the subjects seeking second homes are indigenous, with or without origins in the reception area, stimulating flows of visitors to the reception area, more closely related to domestic tourism. Alternatively, second home stimulation may occur by foreign subjects, initiating flows of foreign visitors to the second home destination which are more closely related to incoming tourism.

If examined in the light of spatial development, the second home phenomenon, when located in rural areas, may be referred to as a form of rural tourism, while when located in urban centres may be referred to as urban tourism. In any case, as far as location is concerned, second home complexes, in the sense of newly-built, organized 'villages', or even golf resorts, condo hotels and/or time-shares, seem to be sharing more qualities with the tourist phenomenon.

Table 9.1   Comparison between second home and tourism phenomenon.

Fields of comparison	Similarities	Differences
Infra & super structure usage/ *Services generation in the host area*	Regarding infra and superstructures of transportation, food, free time activities and others that support tourism	Concerning the accommodation The reinforcement of the local building sector and real estate market in the second homes case
Travel motives	Vacation, recreation-free time, getting closer to nature, better climate, special interests, etc.	Many times, demand for second homes is motivated by investment aspirations or retirement needs
Organisation and repeatability of travel	It is possible in both cases that the trip is privately organised and repetitive	The trip to second homes is almost exclusively privately organised and is by definition repetitive
Impacts in the host areas	Common impacts, socioeconomic and spatial-environmental	In the case of second homes it is possible that the impacts have a more permanent character
Expenditure	Regarding travel expenses, food and recreation	Regarding the expenses that concern the purchase and maintenance of second homes
Space and time coexistence (tourism and SH) *Tourism development paradigms*	-usual coexistence or connection *(in some areas second homes prevail and more conventional forms of tourism follow and in some areas second homes phenomenon develops in mature tourism areas)* -national and international factors influence the development paradigms	In the second home paradigm national, regional and local factors are more influential

Source: Adapted from Tsartas 1998.

Using travel motives as a criterion, if a second home is used for summer holidays, it is possible to refer to the phenomenon as 'holiday tourism' or 'summer tourism', while if it is related to a special interest, e.g. maintaining a second home at a ski resort for the purpose of practising mountain activities, it may be referred to as 'special interest tourism'. Additionally, if second homes are maintained in people's places of origin, the phenomenon may be partially referred to as 'visiting friends and family' and/or as belonging to the 'roots travel' category (Hall 2005).

Tourism and second homes coexist at particular places and times, and the relevant literature available mainly originates from users of the 'area tourism life cycle' theory. According to their findings, there are times when an area initially dominated by second home tourism develops more conventional forms of tourism (Wolfe 1952, Butler 2005), while, at other times, second home tourism may develop into what are already mature tourist resorts, usually at saturation stages, functioning as stabilizing factors (Strapp 1988, Lagiewski 2005). Time-place correlation and interaction between second home tourism and mass tourism as well as special interest tourism is significant, interesting and, definitely, inadequately researched.

For the purposes of this chapter, the focus is on tourism-related second homes.

Second Homes within the Tourism Context

Both tourism and second homes constitute multidisciplinary sectors of research, attracting the interest of scientists from related fields, such as regional planning – mainly rural, rural and human geography (Hall and Müller 2004), environment, urban planning, land-planning, land use, infrastructure project planning, mobility, migration, etc. However, the scientific incorporation of second homes into the tourism framework was only marginally pursued (if at all) for most of the period of second home study. The official consideration of the second home phenomenon, from the point of view of official tourism study organizations, was initially expressed in the 1995 manual of basic tourism concepts by the World Tourism Organization. The criteria by which a movement was characterized as 'touristic' structured a limiting framework that marginalized or even disputed the view that considers second homes as part of the tourism phenomenon.[2] Besides, the term 'tourist' for owners-users of second homes seems to be rendered inappropriate (Svenson 2004), taking into account that the concept of a tourist is connected to the presence of a flow of anonymous visitors or strangers arriving among the reception population, while the concept of a second home owner is very often identified with the presence of known visitors to the reception population, whether they originate from the area or have become known over time.

However, the globally increasing importance of the phenomenon seems to be gradually enhancing the recognition of the need to study second homes in the context of tourism. According to the joint publication by the WTO and the UN Statistics Division, in 2008, the corresponding definition of one's usual environment ceases to include trips to second homes (UNSC 2007). As characteristically mentioned in the publication: 'Trips to vacation homes are usually tourism trips. Recognizing the growing importance of these trips in an increasing number of countries, and

---

2   A place which one visits, e.g. on a weekly basis, is considered part of their usual environment according to the manual, and, therefore, any such trip is excluded from the characterisation of 'tourism trip'. Although second homes are seen as tourist lodgings in this manual, it is obvious that such a belief tends to exclude many second homes that are used by their owners on a weekend basis.

because of the specificities of the corresponding expenditures and activities, tourism statistics compilers are encouraged to measure them separately for analytical purposes and cross country comparisons.' (UNSC 2007: 13). Similarly, at European level, EU directive 95/57/EC of the Council on the statistics of tourism, considers second homes 'tourism accommodation' and visits to them by owners and guests as 'tourist activity' (EC 1999).

There is the exception of important publications moving towards the direction of correlating the two phenomena during the last decade (Muller et al. 2004, Gallent et al. 2005, McIntyre et al. 2006, McWatters 2009). Two broader and related scientific debates advance simultaneously, both raising the issue of sound research practices regarding tourism and, essentially, the limits of where the phenomenon starts and ends. Drawing on second home study, the first debate refers to the correlation between the fields of migration and tourism, arguing that tourism is unjustifiably studied separately from migration, which is both a broader and older phenomenon (Williams and Hall 2000, Bell and Ward 2000, O'Reilly 2003, Müller 2004, Hall 2005, Illes and Gabor 2008), proposing terms related to the second home phenomenon, such as 'amenity migration', 'seasonal migration' or 'retirement migration'. The second scientific debate raises the issue of including all the above phenomena, as well as others that presently comprise distinct fields, under the umbrella of the 'mobility' concept (Bell and Ward 2000, Uriely 2001, O'Reilly 2003, Müller 2004, Hall 2005, Urry 2000).[3] In the context of this debate, taking into account the unprecedented current mobility, Urry, highlighting its dominant role in modern life, expressed the need to redefine the research unit of sociology from society, in the sense of a static territorial settlement, to that of mobility (sociology of mobilities) (Urry 2000). As for the phenomena of abode, tourism and second homes, the concept of 'multiple dwellings' is introduced, while Lash and Urry support that our era is characterized in a sense by the end of tourism, considering that most people are tourists most of the time, either because they are physically moving or because they participate in a kind of a virtual mobility, through the consumption of visual points and images (Lash and Urry 1994, Nazon 2004).

Taking into account the connections and overlaps between tourism and second homes, in combination with the above mentioned current research trends of tourism, it is possible to characterise second homes as a tourism frontier issue (Karayiannis 2007b). Placing second homes in the conceptual framework of tourism consequently legitimizes the term 'second home tourism'. The adoption of the term, however, is not made light-heartedly, but consciously realizing the presented drawbacks that characterize the interconnection of these two phenomena.

In short, it seems necessary for researchers involved in this field to adopt theoretic flexibility, putting into practice the currently desirable aim of multidisciplinary or interdisciplinary approaches.

---

3   Mobility in the broader sense, of people, goods, commercial goods, capital, ideas, technology, etc. (Urry 2000)

Second Home Tourism in the Greek Reality

Housing Policy and the Second Home Phenomenon in Greece: Brief Historical Overview

The development of second homes and second home tourism in Greece (since 1970) is definitely linked to housing issues, as well as land and urban planning policies, which in turn doubtlessly reflect broader socioeconomic conditions and shifting priorities.

The Greek housing policy implemented in the post-war period and for decades to follow, consisted of encouragement and mobilization of small private capital in the direction of satisfying housing needs (Himonitis 2001, Getimis 2000, Sifounakis 2005), a fact that led to a boom of construction activity. It should be noted that the residential capital of Greece is mainly post-war, taking into account that approximately 85 per cent of residential buildings were constructed after 1946 (2001 census). It is widely accepted that the policy implemented in the post-war period, at least until the mid-1970s, was characterized by the lack of adequate state intervention (Himonitis 2001) and was based on the tradition of small land ownership and its exploitation for purposes of political party interests, while serving macroeconomic goals[4] (Himonitis 2001, Getimis 2000).

The Greek particularity of 'off-plan building' was, and still is, decisive for the land and urban planning of the country. The term was first introduced through the planning legislation of 1923, which legitimised construction activity either within or outside city plans (Bladou 2008). The main purpose of off-plan regulation was to exclusively serve agricultural uses, or others not appropriate within cities (at least by the 1928 revision). But through the following legislative revisions of 1978 and 1985, off-plan building was expanded to all other categories of uses, from homes up to commercial stores. As Blandou mentions, 'For every activity that was not possible to be placed within city plans, the State, instead of revising the land planning terms and uses, foreseeing and attending for the planning of socioeconomic activities, resorted to the easy solution of off plan building. The relevant legislation evolved to be a substitute of the rural planning of the country, each time accordingly adjusted' (Bladou 2008).

The phenomenon of second homes with tourism content picked up momentum in Greece between 1970 and 1990 (Getimis 2000). While already in the 60s–70s

---

4   It is widely accepted that in the post-war era construction activity served as a general growth lever (Getimis 2000), while the dynamism of the sector and its modern significance for the Greek economy is well known. As mentioned by Aggelidis (Aggelidis 2000), Greece's growth options after the war, and under the influence of the American Mission in Greece, gave priority to financial sectors that did not demand significant investment of fixed capital and technology to ensure growth, and which could more quickly contribute towards fighting unemployment, such as construction, farming and light industry for consumer goods.

there arose intense investment interest in real estate in rural areas of growing tourism significance (Getimis 2000), second – and mainly – holiday homes were mass albeit mostly individually produced in the last three decades (since 1980s). This fact is considered to be a consequence of a series of parameters, such as intense urbanization and the deterioration of quality of life in unplanned developed cities, improvement of transport (infrastructure and transport means), increased leisure time[5], increased income, influence by consumption standards of the lower and middle social strata of the developed world, as well as national policy trends (NCSR 1998).

Monitoring the second home phenomenon at the national, regional and local levels, through national statistics, remains partially problematic to this day. For several decades since the appearance of the phenomenon the main index has been the number of vacant homes through the population and housing census[6] (NCSR 1998, CPER et al. 1998, Himonitis 2001, 2005). The number of vacant homes during the 50s–60s in rural areas, where most of them were situated, reflected social decline due to abandonment (domestic and external immigration). As of the 1970s, justifiably, the increasing number of vacant homes in these areas was connected to the spread of holiday home ownership, mainly by natives (Himonitis 2001, 2005). It was only in the 2001 census that vacant homes were classified as second or holiday residences. The picture is, therefore, still unclear, considering that even this category occasionally includes cases where second homes are maintained for reasons other than recreation, such as, i.e. professional, educational, etc. In the most recent population and housing census (2011), this category was further divided, differentiating holiday from second homes, thus for the first time more effectively facilitating the monitoring of the second home phenomenon, more closely related to tourism [Hellenic Statistical Authority (HSA) 2011].

Research into Second Homes and Second Home Tourism in Greece

The two best known nationwide studies on second holiday homes in Greece were carried out by recognized national bodies and institutions: the National Centre for Social Research (NCSR), the Centre for Planning and Economic Research (CPER), the National Technical University (NTUA), the University of Thessaly (UTh) in cooperation with the Ministry of the Environment, Energy and Climate Change (MEECC). The studies were completed in 1998 and mainly focused on spatial aspects, addressing the growth of the phenomenon. The main participants in these studies were land and urban planners, topographers, as well

---

    5   Weekend days off were instituted in the 1970s, a fact that positively influenced the growth of the holiday home phenomenon (Himonitis 2005).
    6   The problem is that a percentage of vacant homes concerns both second homes serving non-tourism needs of the household (such as work or studies), and residences for lease or sale, as well as others in need of repair or deserted (Himonitis 2001).

as economists, sociologists, and, to a limited extent, specialists in the tourism field. The phenomenon was studied on the basis of secondary statistical data at national level, followed by case studies from suburban and coastal areas, as well as tourism destinations.

The basic subjects examined in these studies, were:

> – *Second holiday home 'production'.* The prevailing way to achieve this was by segmenting land, buying a land plot, undertaking the construction privately or assigning to a contractor, whether the lot was located within or outside city plans (NCSR 1998). At the same time, there were also other, albeit secondary cases, of renovation-maintenance-extension of existing homes, construction of blocks of flats in already established holiday resorts and the construction of organized holiday-tourist compounds (NCSR 1998).
>
> – *Land planning allocation and models for the spatial development of second holiday homes – Impacts on the urban and natural environment,* per second home model. Dominant types and corresponding impact presented as follows:

1. Non-organized continuous expansive second holiday home growth into zones of broader suburban areas, mainly coastal and independently of pre-existing residential districts. Impacts: arbitrary urbanization and expansive degradation of the urban and natural environment. Hemming in of residents within a gradually urbanized environment, with frequently intense problems due to lack of infrastructure.

2. Disorganized growth spatially focused on the periphery of existing settlements, extending their borders, into touristic areas. Impacts: High density construction and excessive concentration of activities. Seasonal failure of networks and infrastructure. Insufficient protection of cultural and socio-economic structures from tourism and second holiday homes use. Issues of conflicting use and negative impact on farming.

3. Growth within the limits of already existing settlements, whether traditional or not, in the broader insular or mainland area, with impacts similar to those of type 2.

4. Scattered growth over broader holiday or tourism interest zones. Impacts: Overconsumption of natural resources and problems in protecting certain sensitive scenic zones, shores, ecological habitats and cultural sites. Issues of conflicting use and negative impact on farming.

5. Lastly, growth outside existing settlements, in the form of organized complexes of an entrepreneurial or partnership nature, as part of intense construction off (outside city) plan, focused on touristic areas. Impacts similar to those of type 4.

According to the study by the CPER and the UTh, more recently a trend is being observed concerning a shift in preferred locations, from coastal regions to more mountainous ones, towards traditionally agricultural and forest districts, historically and culturally important sites, and areas of special interest (CPER et al. 1998 phase C). Simultaneously second holiday homes are no longer an exclusive need of major urban centre residents; residents of smaller semi-urban and rural areas own them as well (CPER et al. 1998 phase C).

> – *Second holiday home demand features.* Major determining factors for second holiday home demand are the phenomena of urbanization and environmental degradation in urban centres, the economic status of households, the level of education and ownership of a private car (CPER et al. 1998 phase C). As far as travelling to second holiday homes is concerned, it should be noted that trips are becoming more frequent but shorter in duration. There are also increasing trends in illegally renting second holiday homes for limited periods of time, as well as their conversion into permanent residences, mainly in suburban areas, upon retirement. The two studies seem to differ in their conclusions regarding social access to second holiday homes. The NCSR and NTUA consider the belief that second holiday homes are commonplace to be a myth, since they mainly belong to middle and upper social income classes, while the lower and lower-middle strata are excluded (NCSR 1998 phase B). On the contrary, according to the study by the CPER and UTh, second holiday homes have turned from once 'a privilege of high and middle income owners into a gained right by the rest of the socio-economic strata' (CPER et al. 1998 phase C: 11).

Both studies conclude with policy directions and measures proposed for managing the spatial expansion of the phenomenon and its consequent impacts.

The Need for Contemporary Research Approaches in Greece

The studies on second holiday homes in Greece presented above are recognised nationwide but are more than a decade old. Justifiably for the time they were undertaken, they gave priority to dealing with an uncontrollably growing phenomenon and its spatial-environmental impacts and dealt less with its operational aspects and the evaluation of its local socioeconomic implications. The NCSR study (2001), on *Qualitative Characteristics and Demand Trends in Domestic Tourism*, is indicative of the overlap between domestic tourism and second holiday home tourism. Suggestively, it points out:

- the lack of systematic assessment of the impacts of second holiday homes and their contribution to economic development at a national, regional and local level, and to a greater extent,
- the inadequate assessment of the social and cultural dimensions of the

presence of a significant number of native tourists at a given destination, as well as the correlation of domestic tourism and more so second holiday home tourism, with special and alternative forms of tourism (Tsartas et al. 2001, Iakovidou and Turner 1995).

Although spatial and growth issues of the phenomenon have not been solved to date, the need to explore its effects on host regions regarding its operational dimension has become obvious. Such an enquiry should take place in a holistic manner, co-assessing socioeconomic and environmental impacts, taking into account the given local historical-growth context and any existing particularities. Important evaluation parameters should be, amongst others: the level of the revival of local economies due to the phenomenon, its duration perspectives, as well as the extent of potential gradual dependence of the local economy on its activities (monoculture), the strengthening of local productive sectors or service provision industries, the replenishment of human capital through the creation of jobs, the long term local growth dynamics, the broader effects on local life quality (Partalidou and Iakovidou 2008).

As far as local environmental systems are concerned, an evaluation of second holiday homes in relation to sustainability criteria is essential, looking into the soundness of natural resource management (water, land, biodiversity, landscape, etc.), waste and sewage treatment, changes in land use, etc. Taking into account that, in Greece, second holiday homes often develop in rural regions, which in the post-war period were decimated in population and fell into decline, the phenomenon's contribution to local sustainability should be evaluated within this framework.

Finally, the study of second home tourism in Greece should take into account current developments and the internationalized framework within which the phenomenon is developing. Second home tourism is now possibly both domestic and international, stimulates visitor flows during both the summer season and other periods, while it gradually evolves, to some extent, to permanent migration. Contemporary research and evaluation of the phenomenon in Greece should evidently be carried out in these revised contexts.

Contemporary Reality in Greece

Comparative Position in the European Framework

According to the latest data of the housing census (2001) the total number of homes characterised as second or holiday homes in Greece, comes to 924,877, in a total of 1,441,690 vacant homes and 5,465,167 conventional homes, representing 17 per cent of the latter (as compared to 15 per cent in 1991, based on the estimates of the CPER and the UTh). In any case, Greece is included in the European countries with the highest volume of second and vacant homes (Ball 2005, 2007), both as a percentage of its habitable housing stock as well as in relation to its total population. According to the 2001 census, Greece is among the three top-ranking countries in the European Union (25), along with Portugal and Spain, on the basis of these indices (Karayiannis 2007b).

Second Home Depiction at a Prefecture Level

According to our previous publication, (Karayiannis et al. 2010), a crude spatial representation of the second home phenomenon at prefecture level in Greece can only shed some light on the multifaceted and complex character of the phenomenon and its spatial specificities and impacts. A typical example appears to be the concentration of second homes near coastal areas, both in major urban centres, such as Attica (around the capital of Greece), and Thessaloniki (second biggest city nationally), highlighting the demand for second homes near the coast as well as in suburban areas (Figure 9.1). High concentrations are found in prefectures that are only a short distance from peri-urban areas and always coastal, e.g. the prefectures of Euboia and Corinth (both near Athens), Chalkidiki (near Thessaloniki), as well as prefectures with tourist attractions and development, e.g. Chalkidiki, the Cyclades islands, Herakleion, Laconia, etc.

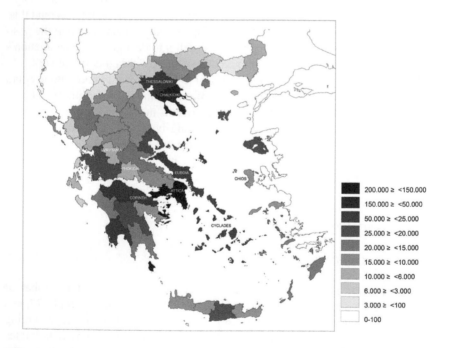

Figure 9.1   Second home distribution at prefecture level (2001 data).

Source: Housing census data processed by the authors, map data provided by geodata.gov.gr.

For newly built houses, the year of construction being the criterion, the picture does not change significantly, and coastal suburban and/or tourist prefectures top the rankings.

As to the relative scale of the phenomenon, i.e. in relation to the number of permanent population per region, the picture has somewhat changed, although some factors remain constant. The phenomenon still appears most intense in coastal and tourist prefectures, e.g. Chalkidiki and the Cyclades, while, quite possibly, the incorporation into that category of prefectures such as Evrytania, Phokida and Chios, is due to the post-war population decline of these areas, and the high stock of houses that were converted from primary to second homes.

Taking into account GDP per capita, both municipalities above the national average and those with lower scores seem to have higher SHs concentrations, indicating, once again, that the phenomenon may appear both in attractive areas of affluent urban prefectures, declining rural areas, as well as in financially powerful tourism destinations.

In any case, the absolute or relative significance of the phenomenon for each prefecture is affected by parameters of the broader socioeconomic development of the regions in question, which dictates that further analysis per prefecture should be undertaken.

Data on Domestic Second Home Tourism

The only available data on the number of trips to second homes, overnight stays and expenditure in Greece, concern domestic demand and are traced in the annual tourist demand (holiday) survey by HSA since 1999. According to the survey for the years 1999–2009 (HSA 1999–2009):

– 31–40 per cent of the trips taken by Greeks with more than 4 overnight stays, both within and outside Greece, were to a second home, while for 2009 this figure came to 2.8 million trips, reaching 37 per cent of the total number of trips taken;
– the corresponding percentage of trips with more than 4 overnight stays taken by Greeks inland only, came to 33–45 per cent and in 2009 reached 41 per cent of total trips taken;
– overnight stays at second homes on total trips with more than 4 nights within and outside the country came to 37–49 per cent during the 1999–2009 period, and in 2009 they came to 46.3 million overnight stays or 51 per cent of the corresponding total;
– the corresponding percentage of Greek overnight stays, on trips with more than 4 nights, undertaken in second homes within Greece, ranged from 41 to 57 per cent, while in 2009 this figure reached 57 per cent of the overall number of overnight stays;

– 29–31 per cent of expenses undertaken by Greeks on trips with more than one overnight stay, either within or outside Greece, concern trips involving second homes (the corresponding percentage for trips within Greece is 27–42 per cent). This figure came to €1.2 billion for 2009, reaching a percentage of 28 per cent of overall expenses of trips taken by Greeks for more than 1 overnight stay, within and outside Greece, and 34 per cent of trips only within Greece.

Consequently, the great importance of second homes for Greeks is reflected in these indices too. Unfortunately, there is no equivalent information available for international second home demand in Greece.

Second Homes in a Popular Greek Tourism Destination: The Cyclades Archipelagos

Geography

The Cyclades Archipelagos is one of Greece's 51 prefectures and along with the Prefecture of the Dodecanese islands, they compose the Southern Aegean Region, one of the 13 Regional Districts of Greece and one of the four that are exclusively insular. The region's long distance from European development centres, the lack of borders with other European states and its geographic fragmentation make the Southern Aegean 'an exceptionally isolated and unusual European Region' (MDCS 2007). With a total of 9,837 insular areas (islands, islets, rocks), the Prefecture ranks first on the number of islands it includes, with 2,242 (Mergos et al. 2004), of which 24 are inhabited islands.

Regarding the natural and man-made environment of the broader area of the Southern Aegean, its significance and uniqueness lie among other things in its natural landscape, from the point of view variety, scale and form, in its flora and fauna, as well as in the architectural wealth of the islands, from both an aesthetic and a historical perspective.

Human Presence

Human presence in the Cyclades has been continuous since antiquity, a fact confirmed by the wealth of archaeological findings as well as by the noteworthy rural landscape typical of the area (Mendoni and Margaris 1998). The basic productive activities since antiquity have been farming, animal raising and shipping. The Cyclades island complex, upon their accession to the Greek state in 1830, were densely populated and showed an important shipping and commercial activity (Mergos et al. 2004, Spilanis and Kizos 2004). After the Second World War, the prefecture suffered a significant population decrease, although over the last few decades there has been a notable recovery, attributed to the employment opportunities that the gradual expansion of the tourism phenomenon brought about (Spilanis 2000, MDCS 2007, Mergos et al. 2004).

Society, the Tourism Sector and Current Issues

The current population of the Cyclades, according to the 2011 census, comes to 117,840 people (little more than 1 per cent of Greece's population) recovering already from the 1980s and approaching the levels of the 1950s, when its greatest decline was noted (Mergos et al. 2004, Spilanis 2000). However, approximately 70 per cent of the prefecture's population resides in just 6 of the 24 islands, highlighting the significant inequalities within the prefecture.

According to the most recent census data available at the time of writing this chapter (2001), the majority of the Prefecture's active population is employed in the tertiary sector (approximately 60 per cent), which has been constantly growing over the last few decades. The secondary sector comes second (approximately 29 per cent) and next, at a steadily declining rate, exceeding the corresponding national average, comes the primary sector (approximately 11 per cent) (HSA 2001). According to its most recent GDP data available (2008), the Prefecture ranks second nationally, and it exceeds the national average by 29 per cent. Nonetheless, significant variations between islands comprising the Prefecture are still observed.

According to the Regional Operational Program of Crete and the Aegean Islands 2007–2013, the common problems of the Region of 'Crete and the Aegean Islands' include (MDCS 2007: 18–19):

- The existence of significant inter- and intra-regional inequalities, amongst which is the uneven distribution of human and natural resources and infrastructure;
- High seasonality of activities, which are, to a great extent, based on tourism;
- A fragile economy mainly dependent on the tertiary sector;
- Intense pressures due to construction, both residential and for tourism;
- Over-exploitation of the limited natural resources.

As far as the main risks of island landscape deterioration in the broader region of Southern Aegean are concerned, they include *de facto* urbanization caused by arbitrary tourism development (expansion of urban centres, construction of tourism superstructure and infrastructure, mainly along the coastline), extensive and/or localized quarrying activities, construction of infrastructure projects as well as uncontrollable disposal of solid waste (MDCS 2007).

At present, the Cyclades Archipelago is one the most popular tourist destinations in Greece, for domestic as well as foreign tourists. It is considered to be one of the prefectures with a specialisation in tourism (Spilanis 2000). Concerning its tourism *product*, beach tourism prevails (GNTO 2003: phase B). A series of parameters, such as the significant contribution of domestic tourism in the area, small accommodation size, a significant number of small and medium enterprises with rooms to let and low dependence on tour operators (Papanikos 2000), indicate a tourist development paradigm of a less organized character. A significant role seems to have been played in the area by the second homes phenomenon (Karayiannis 2005), inflating the hospitality potential as well as the flow of regular visitors-tourists to the area, often unnoticed by usual tourism statistics.

The basic problems that concern the tourist sector are related to the low quality of the product offered, as well as of public infrastructure and services, the inability to shift away from the typical 3S model (sea, sand, sex), which is, in any case, reflected in the low tourist expenditure and high seasonality (MDCS 2007).

Second Home Data

Based on the most recent available housing census data (2001), the Cyclades rank third among Greece's prefectures as to the total number of second homes. They exhibit the highest stock of second homes amongst the insular prefectures of Greece, as well as one of the highest percentages of second homes compared to total population and housing stock (Karayiannis, et al. 2010, 2007b).

More specifically, the prefecture ranks second in the country in the index number of second homes per resident with a value of 0.38 units, and third based on the percentage of second homes over the total of its conventional homes, with a percentage of 45 per cent. For newly built second homes, year of construction being the criterion, as compared to other Greek prefectures, the Prefecture of the Cyclades again ranks high. More specifically, it ranks second in number of second homes built in the 1971–2001 period per 100 residents, at a value of 22.42 units, while it holds first place as to the number of second homes built prior to the 1970s, at a value of 25.60 units (Karayiannis et al. 2010). This index possibly indicates the relative significance of primary homes that were converted to second, due to intensified urbanization in the post-war period throughout Greece.

The following tables, map and diagram illustrate an attempt to monitor the phenomenon on the islands, comprising the prefecture in absolute, historical and relative terms, as well as to estimate its development over the last ten years.

In table 9.2, basic data concerning the islands are presented (area, coastline, population in 2001 and 2011) as well as the numbers of main, second and vacant homes (2001 census). Additionally, the percentage of second homes per island built in recent decades and the percentage of second homes in relation to primary homes (with some omissions due to unavailability of data) have been calculated. The 24 islands are ranked according to the number of second homes, from highest to lowest, and are then grouped into 3 categories divided by the dotted lines, based on this division.

The following should be noted:

- The second homes stock comes to approximately 41,500 units (slightly less than half the conventional homes), of which more than one third (36 per cent) have been constructed relatively recently, more specifically

Table 9.2 Basic data and second home indices per island.

islands	surface in km2	coasline in km	population 2001	population 2011	conventional homes 2001	second homes (SH) 2001		% SH built between 1981-2001	% SH/conventional homes 2001	vacant homes 2001
NAXOS	390.00	133.00	17,357	19,440	14,197	6,584		34%	46%	7,627
PAROS	194.52	111.00	12,514	13,710	10,582	4,964		46%	47%	6,009
SANTORINI	75.79	67.00	13,725	15,250	10,933	4,086		36%	37%	5,794
TINOS	194.21	114.00	8,115	8,590	8,309	3,968		35%	48%	4,965
MYKONOS	85.48	89.00	9,274	10,190	7,275	3,624		54%	50%	4,418
ANDROS	379.67	176.00	9,285	9,170	7,423	2,992		26%	40%	3,621
SYROS	83.63	84.00	19,793	21,390	12,581	2,880	75%	42%	23%	4,382
MILOS	150.60	139.00	4,736	4,960	4,377	2,066		23%	47%	2,427
TZIA	132.00	88.00	2,162	2,420	2,869	1,787		42%	62%	2,021
SERIFOS	73.23	83.00	1,262	1,480	2,329	1,673		20%	72%	1,724
KYTHNOS	99.26	111.00	1,538	1,310	2,370	1,593		39%	67%	1,707
SIFNOS	73.18	75.00	2,574	2,570	2,370	1,392		20%	59%	1,450
KIMOLOS	35.71	45.00	838	920	1,223	792		18%	65%	808
AMORGOS	120.67	126.00	1,852	1,940	1,526	663		31%	43%	746
IOS	107.80	87.00	1,862	2,030	1,567	629	22%	32%	40%	867
ANTIPAROS	34.83	49.00	1,011	1,190	954	567		70%	59%	600
FOLEGANDROS	32.07	42.00	676	780	693	355		19%	51%	412
SIKINOS	41.03	40.00	238	260	429	311		31%	72%	316
ANAFI	38.35	38.00	272	240	382	244		19%	64%	256
DONOUSA	13.48	31.00	166	*	201	90		32%	45%	119
HERAKLEIA	17.60	29.00	133	*	142	83		48%	58%	90
SHINOUSA	7.78	8.51	197	*	173	78		60%	45%	95
KOYFONHSIA	5.70	*	376	*	217	70		26%	32%	98
THIRASIA	9.30	17.00	268	*			3%			
TOTALS			*110,224*	*117,840*	*93,122*	*41,491*		*36%*	*45%*	*50,552*

Sources: Authors' own data processing from population and housing census HSA 2001, 2011 and Salfo et al. 2003.

between 1981–2001. If the number of vacant homes is considered to be a more dependable index of the phenomenon, then its scale is readjusted to approximately 50,500 second homes, which is more than half the conventional homes in the prefecture;

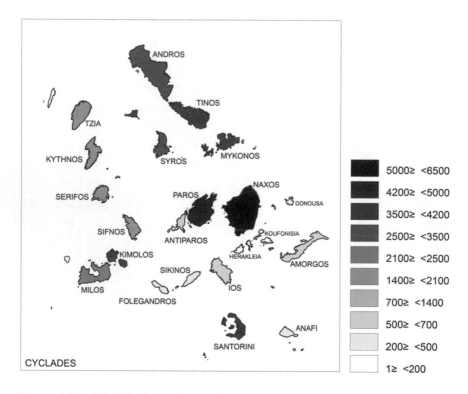

Figure 9.2   Distribution of second home stock in the Cyclades (2001 data).

Source: Housing census data processed by the authors, map data provided by geodata.gov.gr.

- In absolute numbers, the phenomenon is more intense on islands with dynamic tourism sectors, which have reached stages of tourism saturation (a great decrease in their physical carrying capacity and over-intensive use of natural and cultural resources) (Salfo et al. 2003). These islands include Paros, Santorini, and Myconos, but also second holiday home islands more closely related to domestic tourism, which serve demand from the Attica peninsula, such as Tinos, Andros, Tzia (Kea), and Kythnos. Although Syros falls into this last category of second holiday home islands, since it is the administrative centre of the prefecture, and more urbanised, it is likely to present other particularities, i.e. internal demand for second holiday homes or demand from nearby islands, besides demand from Athens. Seventy-five per cent of the second homes stock of the entire prefecture is found on the islands above, as well as on the island of Naxos;

- Other islands that fall into the same category of development as Naxos, and, more specifically, islands that are characterized by population growth, environmental pressures, shortages in basic resources (i.e. water) and differentiated touristic development, as far as accommodation is concerned (class, category, official – illegal beds), (Salfo et al. 2003), such as Melos, Serifos, Sifnos, Amorgos, Ios, receive lower ratings with regard to absolute numbers of second homes.
- Lower in ranking, as expected, are islands of smaller size and population, which until recently retained a relatively high level of self-sufficiency, based on a farming economy. This feature is gradually being reversed, with the increasing demand for small tourist units and rooms to let, on such islands as Folegandros, Sikinos, Anafi, Donousa, Heraklia, Shinousa, and Koufonisia. Important here is the exception of Kimolos and Antiparos, which, due to their small distance from Melos and Paros (bigger and more developed islands), respectively, are possibly under more significant pressures – in absolute terms – due to demand for second homes, and therefore rank higher;
- However, when observing the percentile indices of second home participation per island vis-à-vis the total of habitable stock, as well as the share of more recently constructed second homes, initial rankings seem to change slightly. The lead seems to be taken by several islands which are smaller in size and population, and less developed (see grey shading), such as Shinousa, Heraklia, Sikinos, Anafi, Serifos, indicating qualitative differentiation as far as pressures from the phenomenon are concerned. High rankings are also held with respect to these indices by Paros, Syros, Kythnos and Syros.

Taking into account that there is no reliable source of information monitoring second homes stock development for the period between the two censuses, i.e. before a decade has elapsed each time, discussion on the development of the phenomenon seems troublesome at present. Nonetheless, for cases of prefectures such as the Cyclades due to the relatively limited population fluctuation, it may be possible to utilise the only source of information regarding construction activity, which is the monthly record of legitimate construction activity.[7]

More specifically, although there is no distinction between the use of primary homes or second homes in these data, in the case of the prefecture of the Cyclades it is quite plausible to accept that due to the relatively small population increase between the last two censuses (2001 and 2011), a significant percentage of homes that received building permits very possibly fall into the category of second homes. Although there is no knowledge as to whether these homes have indeed been

---

7   It is interesting to point out that the term 'legal' is being used in the official data of HSA, insinuating that illegal construction activity is also possible in Greece.

constructed, this construction activity is deemed indicative for the period between the last two censuses and, therefore, the relevant data should be presented.

Making another hypothesis, i.e. that a residence requires on average 1-4 years for completion from the date the building permit has been issued, the building activity in number of homes in the Prefecture of the Cyclades, per island, from 1997 (3–4 years prior to the 2001 census) until 2010 (the last year for which there are data available on an annual basis) are presented in the following diagram (Figure 9.3).

Prior to commenting, it should be pointed out that the number of homes presented concern permits for new buildings as well as extensions to already existing homes, not repairs or restorations, as such data were not available (although they would also be of great interest). For these later categories only the number of permits is known and not the actual number of homes they account for. Based on these data, the percentage of new buildings and extensions represents the majority of permits issued, specifically 70–80 per cent of permits for the entire period in question, while repairs and restorations represent a percentage ranging from 10 to 14 per cent. Thus permitted construction activity seems to concern mainly new constructions and extensions and significantly fewer repairs and restorations of existing buildings.

As can be seen in the upper section of Figure 9.3 the development of construction activity as recorded in the index of the number of homes receiving permits reveals significant construction activity, exceeding 40,000 for the 1997–2010 period for the whole of the prefecture, which corresponds to almost half the residential stock of 2001 and almost double the second home stock.

The islands that seem to be under the greatest pressure in absolute figures are still, with small differences in ranking, the highly touristic islands, i.e. Paros, Myconos, Santorini, as well as the so called second holiday home islands, such as Syros, Tinos, Andros, and Tzia, with Naxos once again ranking high, as in the presentation of the census data. The ranking of the previous census table has more or less been maintained, with small differences, also for the rest of the prefecture's islands.

However, the lower section of Figure 9.3 presents the islands ranked with regard to the higher percentage of estimated change between their overall home stock since 2001 (assuming that constructions licensed in the 1997–2010 period have indeed been completed) and the corresponding population growth rates. Significant changes are observed in the ranking of islands regarding the relative pressures from the phenomenon, with smaller islands, such as Koufonisia, Shinousa, Antiparos, Folegandros, Sikinos and Heraklia, rising in the ranking, and in contrast islands with larger population, such as Naxos, Santorini, Syros, Tinos and others, coming lower.

Some islands, though, such as Paros and Tzia, as well as Myconos, retain their high position in this ranking, reflecting the intense pressures they, too, are under, in relative terms.

In any case, the great divergence between percentage shifts of homes on each island and their demographic changes, which does not seem to justify a demand in construction activity of such intensity for the permanent population, reinforces

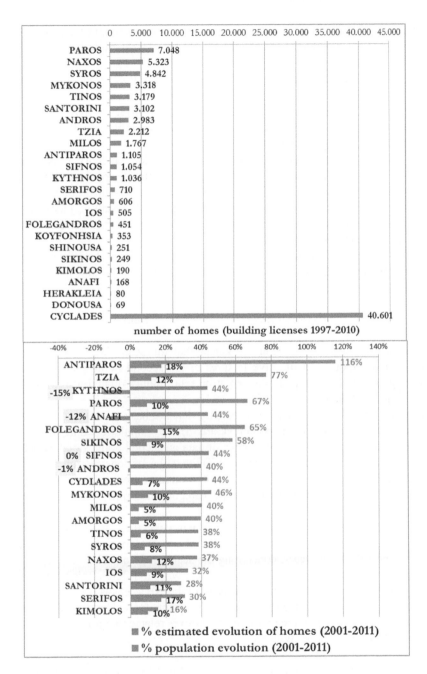

Figure 9.3   Building activity of homes in the Cyclades.

Source: Authors' own processing of legitimate construction activity tables HSA 1997–2010, population census 2001, 2011.

the initial assumption that a large percentage of this activity should be attributed to the second home phenomenon (also possibly to informal tourist lodgings which are not far from the phenomenon in question).

Adopting a series of indices, in order to assess environmental and social pressures caused by the phenomenon, as estimated to be reflected in home construction activity per island per km² of coastline and surface – environmental indices – per resident – social index – are examined in Table 9.3. Results are presented per island in descending order.

Based on the two environmental criteria above the known touristic islands, i.e. Santorini, Myconos, Paros, the capital of the prefecture, Syros, as well as the largest island of the Cyclades, Naxos, are under the greatest pressure. Tzia remains in the high pressure group, as does Tinos, with regard to its coastline index. The differnce here derives once again from the inclusion of smaller islands, such as Koufonisia, Antiparos and Shinousa.

Finally, as far as social pressures are concerned, they seem highest for smaller islands, i.e. Shinousa, Antiparos, Koufonisia, Folegandros, and Anafi, while Tzia and Kythnos maintain their high rankings.

Pending Questions – Indications Based on the Island of Andros

From the above analysis, it becomes evident that the scientific record of the magnitude and important qualitative features of the second home phenomenon presents significant difficulties in Greece at present, mainly related to retrieving adequate statistical data.

In an attempt of a synopsis, it can be said that the basic current sources of information concerning the phenomenon are: i) the population and housing census, essentially the only credible source of data on second home stock, which, however, is carried out every ten years; and ii) the tables of construction activity, which are updated monthly for intended activity (but not necessarily undertaken), compiling data on the number of permits (per category of permit), m² and homes, from Greece's regional authorities.

As far as the former source of information is concerned, in order for any management of the phenomenon to be possible, it is necessary to have data more often than once every ten years, concerning the scale of development and spatial concentration of second homes. Furthermore, there was no differentiation between holiday homes and secondary homes, at least until the 2011 census (second homes were classified in a unified category). This does not allow for a clear differentiation of second homes, the use of which is driven by motives related to tourism and not to professional or other reasons. With regard to the latter source of information, while data for permits for construction activity could bridge this gap, unfortunately they do not provide information specifically on second homes. Instead, they provide information on the total number of homes. Furthermore, no information is given on the number of homes created through repairing or restoring existing premises. What is not made known either is what percentage of construction activity issued with

Table 9.3   Environmental and social pressures of second homes per island.

islands	number of homes (permits 1997–2010)/surface (m2)	islands	number of homes (permits 1997–2010)/coastline (m)	islands	number of homes (permits 1997–2010)/redident (2001)	number of homes (permits 1997–2010)/redident (2011)
KOUFONHSIA	61.93	PAROS	63.50	SHINOUSA	1.27	*
SYROS	57.90	SYROS	57.64	ANTIPAROS	1.09	0.93
SANTORINI	40.93	SANTORINI	46.30	SIKINOS	1.05	0.96
MYKONOS	38.82	NAXOS	40.02	TZIA	1.02	0.91
PAROS	36.23	MYKONOS	37.28	KOYFONHSIA	0.94	*
SHINOUSA	32.26	SHINOUSA	29.49	KYTHNOS	0.67	0.79
ANTIPAROS	31.73	TINOS	27.89	FOLEGANDROS	0.67	0.58
TZIA	16.76	TZIA	25.14	ANAFI	0.62	0.70
TINOS	16.37	ANTIPAROS	22.55	HERAKLEIA	0.60	*
SIFNOS	14.40	ANDROS	16.95	PAROS	0.56	0.51
FOLEGANDROS	14.06	SIFNOS	14.05	SERIFOS	0.56	0.48
NAXOS	13.65	MILOS	12.71	DONOUSA	0.42	*
MILOS	11.73	FOLEGANDROS	10.74	SIFNOS	0.41	0.41
KYTHNOS	10.44	KYTHNOS	9.33	TINOS	0.39	0.37
SERIFOS	9.70	SERIFOS	8.55	MILOS	0.37	0.36
ANDROS	7.86	SIKINOS	6.23	MYKONOS	0.36	0.33
SIKINOS	6.07	IOS	5.80	AMORGOS	0.33	0.31
THIRASIA	5.48	AMORGOS	4.81	ANDROS	0.32	0.33
KIMOLOS	5.32	ANAFI	4.42	NAXOS	0.31	0.27
DONOUSA	5.12	KIMOLOS	4.22	IOS	0.27	0.25
AMORGOS	5.02	THIRASIA	3.00	SYROS	0.24	0.23
IOS	4.68	HERAKLEIA	2.76	KIMOLOS	0.23	0.21
HERAKLEIA	4.55	DONOUSA	2.23	SANTORINI	0.23	0.20
ANAFI	4.38	KOUFONHSIA		THIRASIA	0.19	*

Source: Authors' own processing of legal construction activity HSA 1997–2010, population census 2001, 2011, Salfo et al. 2003.

permits has materialized. Furthermore, taking into account the particularities of the Greek institutional framework (lack of land use legal framework, lack of protection of natural and cultural resources and off plan construction), there are no answers to significant questions, such as, what is the scale of off-plan construction, and construction within or very close to regions of high ecological value (coastlines, Important Bird Areas, NATURA sites, among others).

The situation is made worse for monitoring relevant figures on the islands, because data available are not provided on an island level, but on a Prefecture and community level, requiring further processing in order to record and compare construction per island.

Finally, a particularly interesting piece of information which is not available is the origin of second homes' owners and, basically, to what extent they constitute part of domestic or international tourism demand.

It becomes obvious that meeting the need for data, as described above, is only possible through extensive primary research. An attempt to answer these questions, in combination with significant others of a qualitative nature, has been made through a case study on one of the Cycladic islands with a significant number of second homes, namely Andros. The study was carried out during the 2010–2011 period, recording quantitative and qualitative data from home construction permits from the relevant Andros authorities and carrying out interviews with civil engineers for purposes of cross-referencing data of interest. Initial indications (Karayiannis forthcoming) reveal:

- Overwhelming participation of the second homes phenomenon in the region's construction activity. More specifically, this reflects either pure demand for second homes (privately constructed) or foreseen demand for second homes (homes for sale or to let, built in small groups of usually 2–4 homes by contractors). These come to approximately 90 per cent of licensed homes during the 1998–2009 period, and their number is estimated at 2,252 units;
- Significant percentage of actually completed second homes amounting to more than 60 per cent of second home licensed activity, with 1,370 homes for the same period. However, 15 per cent of permits issued remain unused and 7 per cent unfinished, quite often for several decades, leaving long-lasting scars on the island landscape (see Figure 9.4);
- High proportion of new second home construction activity: almost 81 per cent as compared to 8 per cent of repairing existing homes;
- High percentages of off-plan construction at approximately 60 per cent of permits issued for the period;
- Prevalence of domestic demand for second homes: 95 per cent of demand comes from Greeks, of whom more than half from Athens and less than half from people who originated from Andros. Only 5 per cent express international demand, mainly from European countries.

Figure 9.4   Incomplete construction of second homes on the island of Andros.

Source: O. Karayiannis

It should, however, be pointed out that there are also transactions concerning existing homes to be used as second homes, which would reflect cases of reusing the existing stock. These have not been covered in this study.

Second Home Policy, Management and Prospects in Greece

Policy Aspects and Management

A plethora of papers highlight the unfavourable impacts of housing policy and development in Greece. The policy implemented 'intentional[ly] rather than by delinquency', according to Himoniti (2001: 65), might have resulted in housing sufficiency,[8] but it never stopped functioning in a fragmented manner rather than as an integral part of a long-term sustainability policy. This gradually led to the major problems of degradation of urban and non-urban settlements currently encountered (NCSR 1998, CPER 1998, Sifounakis 2005). As already mentioned, for many decades the Greek particularity of building off-plan allowed profiteering

---

8   Albeit with some shortcomings, as noted by Himoniti (2001: 65).

from land exploitation. Requests are being made, with increasing frequency in public and scientific discourse, to limit, if not completely abolish, such a practice.

Demand for second homes, for recreation in better quality environments, due to the troublesome expansion of urban areas, in combination with the construction of second homes in the problematic context mentioned above, has often merely conveyed exactly the same land-planning and environmental problems from urban to the peri-urban, rural, coastal and insular areas (Aggellidis 2000, Getimis 2002 Himoniti 2001, 2005, Sifounakis 2005, CPER et al. 1998, NCSR 1998). As mentioned in the study of CPER for second homes:

> (...) from an exceptionally profitable subject, in the first post-war decades, mainly due to the lack of alternative options for investment, it [second home] has evolved into a durable consumption good of disputable return, in terms of benefit versus cost. It appears to be a much more harmful activity for the environment, as compared to permanent residence, due to its illegal construction, poor aesthetics, and absence of construction quality and land-planning, inadequate provision of infrastructure and services. (CPER 1998 3rd phase: 12)

Amidst the above problematic context, the demand for the construction of organized mass second home complexes, of the condo hotels, golf and spa resorts types, was added. In 2009, the Greek authorities, in their effort to satisfy the demand for these investment options in Greece – if not under lobbying pressure by interested investors – prepared a special land-plan for tourism, providing favourable regulations and naming these second homes 'touristic' (MEECC 2007). However, following intense organized protests by environmental organizations and other social stakeholders, including some from the tourism industry, the Greek authorities were forced to withdraw that law draft (HSPECH 2009). The recent acute fiscal crisis in Greece has brought back to the limelight the pressing need to attract foreign investments to the country, and reopened the debate for creating favourable legislative framework for this type of investment.

Regardless of any strategic directives concerning the development of individual or organised second homes expansion in Greece, in accordance with the study by NCSR and CPER, as well as numerous other national documents, a fundamental prerequisite condition is to prepare land use plans. This is an issue that has been pending in Greece for decades, and clearly illustrates the fragmented and short-term perspective towards the management of national space and commons.

Prospects of Second Home Tourism in Greece in the Light of Sustainability

As for the prospects of the phenomenon and the data concerning demand within the broader European framework, according to the European Housing Review (Ball 2005, 2007):

– second home markets in Europe are now much more internationally oriented than in the past and a significant market share seems to include retired Europeans;

– all influencing factors currently seem to be having a reviving effect on the development of the second home sector (Ball 2005);

– a significantly enhancing factor seems to be the role of low fare flights, which, however, so far does not apply to the Greek islands, Portugal and Turkey;

– a series of durable consumer's goods that concern home construction and operation have become significantly cheaper;

– the very enlargement of the European Union provides its citizens with the opportunity to buy a home within the European territory, while at the same time support to less developed regions through regional programs improves general living conditions in potential second home areas, rendering them more accessible and attractive.

Elements of the profile of potential buyers as well as their international criteria, according to construction companies' data, indicate (Droussioti 2006):

– demand comes from Northern Europeans, middle-to-high income investors, usually over 50 years old, with a family, several of whom are pensioners;

– decisive factors affecting their preferences include: the climate, security, the broader environment, accessibility, types of residences available and services provided (telecommunications networks, infrastructure, etc.);

– second home buyers attracted to Greece, in particular, are British, German, Italian and Swedish, while significant flows of Northern Europeans are expected by 2014 in search of permanent or seasonal homes in attractive rural areas.

However, all data presented above are in direct interaction with the unfavourable international and Greek economic circumstances and any long-term forecasts are highly uncertain.

In any case, any evaluation of all data presented above, regarding the prospects of the phenomenon in Greece, ought to be based on the principle of sustainability as the supreme criterion for assessing any human activity (WCED 1987, UNEP 2002).

The international study of the phenomenon reveals not only negative but also positive social and economic, local and national implications (Wallace et al. 2005, Gallent et al. 2005, Hall and Müller 2004, McIntyre et al. 2006). These are determined by the particular legislative frameworks controlling the growth and operation of the phenomenon.

Conclusively, the unfavourable impacts from the evolution of the phenomenon in Greece so far, in combination with the prospect for large-scale investment, both by investors from abroad and by major Greek construction companies, invite essential reflection and policy reforms, incorporating past national experiences and international knowledge.

Summary and Discussion

Greece constitutes one of the European countries with the highest percentage of second homes, both expressed as a total of its housing stock as well as per inhabitant.

The majority of second homes are located in coastal suburban areas, in attractive areas of the countryside, other coastal areas, islands, etc., often of touristic interest, whether socioeconomically robust or not.

Evidently second homes in Greece comprise a domestic phenomenon considering the steadily high numbers of trips, overnight stays and expenditure for the 1999–2009 period as well as trips within the country Although foreign ownership definitely exists in Greece, its extent is not formally known, due to the lack of relevant statistical data. In any case, although precise estimates are not possible, it can be reasonably assumed, due to the historical socioeconomic development of the country, that a significant number of second homes in the Greek countryside resulted from the conversion of inherited primary homes into second, as a consequence of extensive post-war urbanization.

However, the case of the Cyclades demonstrated significant levels of recent authorized construction activity. This may be attributed, to a great extent, to the demand and foreseen demand for second homes, considering that population in this area at that time was relatively stable. If such authorized construction activity did truly take place in recent years, it will result in a more-than-twofold increase of the second home stock in this prefecture, which will probably be confirmed by the data of the recent 2011 census.

In any case, this mass construction activity concerns in its vast majority new constructions and extensions and much less repairing or restoring existing buildings, thus demonstrating that the Greek second home sector is new-construction intensive.

This estimated construction activity for second homes seems to exert more pressure on the more touristic islands of Mykonos, Paros and Santorini, as well as the so-called second home islands, i.e. Syros, Tinos, Andros, Tzia and Naxos. In any case, what is of great interest is the significant deviation between the increase in homes on each island and their respective population growth. The latter cannot actually justify such a (high) demand for construction activity for the needs of the resident population, supporting the initial assumption that a significant percentage of this activity could well be attributed to the second home phenomenon.

The scientific depiction of the scale and important quality features of the second home phenomenon is quite difficult in our country nowadays, due to the lack of adequate statistical data, as already shown in the previous analysis. Crucial questions regarding the development of second homes between censuses, the

Figure 9.5   Scattered construction of second homes in small complexes in off-plan areas, in inclination, on ridges and within systematically burned graze lands.

Source: O. Karayiannis.

rates of construction outside the formal urban plan – off plan – and the sources of demand for second homes remain unanswered on the basis of the available statistical data recording systems.

Indications after extensive primary research on second homes on the island of Andros suggested a significant number of newly built second homes between the last two censuses, a high percentage of off-plan constructions and the prevalent role of domestic demand, with high percentage of owners originating from the area.

The tourism-related second home phenomenon in Greece is a contemporary, widely spread, still evolving phenomenon. Sustainability-led research approaches for second homes in Greece are considered to be necessary, assessing the phenomenon holistically, considering not only its growth, but also its function, co-estimating economic, social and environmental aspects, as well as its international dimension.

The second home development is not a 'positive' or 'negative' phenomenon *per se*, since its impact may undoubtedly vary depending on national and regional frameworks, local socioeconomic and environmental particularities, as well as the existence and implementation of appropriate policies (Muller et al. 2004, Gallent et al. 2005). Regarding the particularities of the spatial expansion of the phenomenon in Greece, encouraging the reuse – through restorations – of existing houses, instead of continuously building new, could be a good practice.

In any case, the following issues are foreseen to become most significant concerning the contemporary development of second homes in Greece:

- the reform of the Greek institutional framework (land-use and planning);
- the political handling of the Greek particularity of off-plan building, which has led to significant degradation of natural and cultural resources of great national and local importance; as well as
- the pressure exerted by international construction companies and real estate agents concerning the construction of organized complexes of the condo hotel and golf resort types.

The preparation and implementation of a complete land-use plan in Greece has been pending for decades, clearly reflecting the fragmented and short-term approach to national space and commons. According to experts and scientists, the environmentally necessary measures for the restriction of non-organized construction in Greece, which should have been put into practice long ago, are going to be accompanied with a significant political cost for any institution or person attempting to implement them. The traditionally high dependency of the Greek economy on the local construction sector, combined with the current international financial crisis and the crucial fiscal deficit of the country are expected to further hold back the adoption of necessary legislative measures.

To conclude, in order to achieve effective and sustainable management of second homes in Greece, either in their traditional individual development or in their more recent organized form, it is necessary to integrate the knowledge acquired so far from national experience with the international scientific study of the phenomenon.

References

Aggelides, M., 2000. *Land planning and sustainable development*, Athens, Greecce, Symmetria Publications

Ball, M. 2005. *European Housing Review 2005*, London: Royal Institution of Chartered Surveyors-RICS.

Ball, M. 2007. *European Housing Review 2007*, London: Royal Institution of Chartered Surveyors-RICS.

Bell, M. and Ward, G. 2000. Comparing temporary mobility with permanent migration. *Tourism Geographies*, 2(1), 87–107.

Bladou, Al. 2008. *Off-Plan Building: The Greek Particularity of Urban Planning-Causes and Proposals*. Nomos and Physis. [Online] Available at: http://www.nomosphysis.org.gr/articles.php?artid=3342&lang=1&catpid=2, (Accessed: February 2012).

Butler, R.W. 2005. *The Tourism Area Life Cycle Vol. 1, Applications and Modifications*, Aspects of Tourism, UK, Channel View Publications

CPER-Centre for Planning and Economic Research, University of Thessaly. 1998. *Summer Homes and Housing Development in Greece, Phases B and C*, Greece, CPER.

Droussioti Th. 2006. The second homes, Proceedings of the Rhodes Tourism Forum, Greece, (Cybarco-Civil Engineering Building Construction Property Development) Available at Available at: http://www.rhodesforum. gr/PRESENTATIONS%202/DAY%202/%CE%94%CE%A1%CE%9F %CE%A5%CE%A3%CE%99%CE%A9%CE%A4%CE%9F%CE%A5- %CE%97%20%CE%91%CE%93%CE%9F%CE%A1%CE%91%20 %CE%A0%CE%91%CE%A1%CE%91%CE%98%CE%95%CE%A1%CE %99%CE%A3%CE%A4%CE%99%CE%9A%CE%97%CE%A3%20%CE %9A%CE%91%CE%A4%CE%9F%CE%99%CE%9A%CE%99%CE%91 %CE%A3.pdf (Accessed : 24 February 2012)

EC-European Commission. 1999. (E(1998)3950): *Decision of the 9th December 1998 regarding the implementation procedures of the directive 95/57/EK of the Council in relation to the statistical data collection in tourism*, Official Newspaper no. L 009 15/01/1999 0023 – 0047, [Online] Available at: http:// eur-lex.europa.eu/LexUriserv/LexUriserv.do?uri=CELEX:31999D0035:EL:H TML (Accessed : May 2008).

Getimis, P. 2000. *Housing policy in Greece, the limits of reform*, Athens: Ulysses

HSA-Hellenic Statistical Authority 2011. Population cencus. Available at: http:// www.statistics.gr/portal/page/portal/ESYE/PAGE-themes?p_param=A1602&r_ param=SAM01&y_param=2011_00&mytabs=0 and http://web.statistics.gr/ tenders/2010_OCR/p1_1_set.pdf (Assessed: February 2012).

HSA-Hellenic Statistical Authority. 2001. Population and housing cencus. Available at : http://www.statistics.gr/portal/page/portal/ESYE/PAGE-themes?p_param=A1604 (Assessed: March 2009).

HSA-Hellenic Statistical Authority. 1999–2008, Tourism Demand Side Statistics (Holiday Survey). Available at: http://www.statistics.gr/portal/page/portal/ESYE/ PAGE-themes?p_param=A2001 (Assessed: February 2012).

HSA-Hellenic Statistical Authority. 1997–2010. Building activity data. Available at : http://www.statistics.gr/portal/page/portal/ESYE/PAGE-themes?p_param=A1302 (Assessed: February 2012).

HSPECH-Hellenic Society for the Protection of the Environment and the Cultural Heritage. 2009. Actions for Sustainable Tourism. [Online] Available at: http:// www.diktioaigaiou.gr/contents/draseis.php?kkid=78&kid=63&action=show &m1=4&lang=1 (Accessed: 5 March 2009).

Gallent, N., Mace, A. and Tewdwr-Jones M. 2005, *Second Homes, European Perspectives and UK Policies*, UK, Ashgate.

Hall, M.C. and Müller, D.K. 2004. *Tourism, Mobility and Second Homes, Between Elite Landscape and Common Ground*, Aspects of Tourism, Clevedon: Channel View Publications.

Hall, M.C. 2005, Reconsidering the Geography of Tourism and Contemporary Mobility, *Geographical Research*, 43(2), 125–39

Himonitis-Terovitis, S. 2001. *Research on the evolution of housing the last decades: comments and proposals for the urban areas*, Athens: Centre for Planning and Economic Research (CPER).

Himonitis-Terovitis, S. 2005. *Developments in the housing market*, Athens: Centre for Planning and Economic Research.

Iakovidou, O. and Turner, C. 1995, The female gender in Greek agrotourism. *Annals of Tourism Research.* 22(2), 481–4

Illes, S. and Gabor, M. 2008, Relationships between International Tourism and Migration In Hungary: Tourism Flows and Foreign Property Ownership, *Tourism Geographies*, 10(1), 98–118.

Karayiannis, O. 2005, 'New' tourism 'products': Sociological research of views and beliefs on Andros, Unpublished Thesis for Postgraduate Degree in Tourism Management, Patras: Greek Open University.

Karayiannis, O. 2007a. Construction activity in the Cylcades the last decades: attempting a survey and assessment, *EYPLOIA-electronic newspaper of the Network of Ecological Organizations of the Aegean*, 13th issue. [Online] Available at: http://old.eyploia.gr/modules.php?name=News&file=article&s id=816. (Accessed: 10 February 2010).

Karayiannis, O. 2007b, Second homes tourism: Assessing the phenomenon impacts in the Greek insular areas, Unpublished PhD proposal, Aegean University, Interdepartmental Post Graduate Department of 'Planning, Management and Tourism Policy', Chios Island.

Karayiannis, O., Iakovidou O. and Tsartas P. 2010, Il fenomeno dell'abitazione secondaria in Grecia e suoi rapporti con il turismo, in *Il turismo residenziale-Il fenomeno della mobilità turistica residenziale in Europa: nuovi stili di vita e sviluppo sostenibile del turismo*, edited by Tullio Romita, Milano: Franco Angeli, pp. 91–109.

Lagiewski, R.M. 2005, The Application of the TACL Model: A Literature Survey, in *The Tourism Area Life Cycle Vol.1, Applications and Modifications*, Aspects of Tourism: 28, edited by R.W.Bulter, Clevedon: Channel View Publications, pp. 27–50.

Lash, S. and Urry, J. 1994. *Economies of signs and space*. London: Sage Publications.

McIntyre, N., Williams, D. and McHugh, K. 2006. *Multiple Dwelling and Tourism, Negotiating Place, Home and Identity*, Wallingford: CAB International.

McWatters, R.M. 2009. *Residential Tourism-(De)Constructing Paradise, Tourism and Cultural Change Series*, Clevedon: Channel View Publications.

Mendoni, G.L. and Margaris, N. 1998. *The history of landscape and local stories, Centre of Greek and Roman Antiquities-National Research Foundation, Aegean University, Department of Environment*. Athens: Hellenic Ministry for the Environment, Physical Planning and Public Works.

Mergos, G., Papadaskalopoulos, Th., Christofakis, M., Arseniadou, Ir. and Kalliri, Ag. 2004. *Economic characteristics and development strategy for insular*

Greece- Department of Economic Surveys, Survey no 1. Athens: Athens Academy.

MDCS-Ministry for Development, Competitiveness and Shipping, Prefecture of Aegean Islands and Crete. 2007. *Regional Operational Program of Crete and the Aegean Islands 2007–2013*, [Online] available at http://www.espa.gr/elibrary/Episimo_Keimeno_Kritis-Nison_Aigaiou.pdf (Accessed: February 2012)

MEECC-Ministry of Environment, Energy and Climate Change. 2007. Special Land-planning Framework for Tourism. [Online] Available at: http://www.minenv.gr/download/2008/kya.tourismos.esxaa.pdf and http://www.minenv.gr/4/42/00/kya.xorotaksiko.tourismou%20final.pdf (Accessed: February 2012).

Nazou, D. 2003. *Multiple Identities and its Representatives in a Tourist Island of the Cyclades: 'Entrepreneurship' and 'Locality' in Mykonos Island.* Unpublished PhD. Thesis. Department of Social Anthropology and History, University of the Aegean, Lesvos, Greece (in Greek).

NCSR-National Centre for Social Research, NTUA-National Technical University. 1998. *Summer Homes and Housing Development in Greece*, edited by Panagiotatou E., Greece: EKKE and NTUA.

Papanikos, G. 2000. *Greek Small and Medium Size Hotel Companies*, Athens: Institute of Tourism Research and Forecasts.

Partalidou, M. and Iakovidou, O. 2008, Crafting a policy framework of indicators and quality standards for rural tourism management, *International Journal of Tourism Policy*, 1(4), 353 – 367.

Salfo and Colleagues Surveys SA., Enviplan-G.T.Tsekouras and Associates, Fotopoulos, S. and K., Zaharatos, G., and Tsartas, P. 2003. *Tourism Development Survey of the Southern Aegean Region, Phases A & B*. Athens: Greek National Tourism Organization.

Sifounakis, N. 2005. *Policy: Planning and Action, the experience of the period 2000–2004*, Athens: Kastaniotis Publications.

Spilanis G. 2000. Tourism and Regional Development : the Aegean islands case study, in *Tourism Development, Multidisciplinary approaches*, edited by Tsartas P. Athens: Exantas, pp. 149-180.

Spilanis, G. and Kizos, Th. 2004. *Notes for the Subject: Islands Geography-Geography Department* Mytilene: Aegean University.

Strapp, J.D. 1988. The Resort Cycle And Second Homes, *Annals of Tourism Research*, 15, 504–16.

Tsartas, P. 1998. Tourism and second homes tourism: Conceptual limitations, in EKKE and NTUA, *Summer Homes and Housing Development in Greece*, edited by E. Panagiotatou, Greece: EKKE and NTUA, pp. 195–239.

Tsartas, P., Manologlou, E. and Markou, A. 2001. *Internal Tourism Qualitative Characteristics and Demand Trends*, Athens: National Centre for Social Research (EKKE).

UNEP-United Nations Environmental Programme. 2002. *Integrating the Environment and Development: 1972–2002.* [Online] Available at: http://www.unep.org/geo/geo3/pdfs/Chapter1.pdf (Accessed: February 2012).

Uriely, N. 2001. Travelling Workers' and 'Working Tourists': Variations across the Interaction between Work and Tourism. *International Journal of Tourism Research*, 3, 1–8

Urry, J. 2000. *Sociology Beyond Societies, mobilities for the twenty-first century,* London: Routledge.

UNSC-United Nations Statistical Commission, UNWTO-World Tourism Organization. 2007. *International Recommendations for Tourism Statistics 2008.* UNSC, UNWTO. [Online] Available at: http://unstats.un.org/unsd/statcom/doc08/BG-TourismStats.pdf (Accessed: February 2012).

Wallace, A., Bevan, M., Croucher, K., Jackson, K., O'Malley, L. and Orton, V. 2005. *The Impact of Empty, Second and Holiday Homes on the Sustainability of Rural Communities : A Systematic Literature Review,* York: The Centre for Housing Policy, The University of York. [Online] Available at: http://scholar.googleusercontent.com/scholar?q=cache:cv0s2aqlZHAJ:scholar.google.com/+The+Impact+of+Empty,+Second+and+Holiday+Homes+on+the+Sustainability+of+Rural+Communities+:+A+Systematic+Literature+Review&hl=el&as_sdt=0&as_vis=1 (Accessed : February 2012).

Williams, M.A. and Hall, M.C. 2000. Guest editorial, *Tourism Geographies*, (2)1, 2–4

Wolfe, R.I. 1952. Wasaga Beach: The divorce from the Geographic Environment, *Canadian Geographer*, 1(2), 57–66.

WCED-World Commission on Environment and Development. 1987. *Our common future.* [Online] Available at: http://www.regjeringen.no/upload/SMK/Vedlegg/Taler%20og%20artikler%20av%20tidligere%20statsministre/Gro%20Harlem%20Brundtland/1987/Address_at_Eighth_WCED_Meeting.pdf (Accessed : February 2012)

WTO-World Tourism Organization. 1995. *Concepts, Definitions and Classifications for Tourism Statistics.* Madrid: WTO.

Chapter 10

Controversies of Second Homes and Residential Tourism in Portugal

José António de Oliveira

Introduction

Portugal has had a long history of housing shortage which recently, and in just a few years, turned into a surplus of supply (Oliveira 2011). The real meaning of this change in terms of urban and local development needs to be studied and incorporated in various legal and planning instruments for spatial organisation and local/regional development that the country has.

According to Portuguese Census, the number of people per dwelling fell from 3.9 in 1940 to only 1.8 in 2011 (INE 2012). This evolution, in addition to reflecting demographic change and the transformation of family structures, points to the need to grasp the actual characteristics of housing in general and of vacant dwellings in particular. Though this problematique merits being the focus of another study, it is addressed in this chapter because of its imbrications with the phenomenon of second homes in Portugal.

The National Institute for Statistics (INE http://metaweb.ine.pt/sim/conceitos/Pesquisa.aspx) developed a definition of second homes based on two statistical data collection concepts – 'secondary residence used for tourism purposes' and 'family dwelling unit for secondary residence'. The former concept, respecting the guidelines issued by the World Tourism Organization (WTO), INE defines as 'dwelling unit which is not the principal residence of the family and is used by one or more members of the family household for recreation, leisure and holiday or other activities that do not correspond to the exercise of an activity remunerated on that locality. Included are housing units rented by means of a timeshare contract', while the latter is defined just as 'family dwelling which is used only periodically and in which no one has permanent residence'. This definition is used in the Housing Censuses, and it will also be used as an operational definition for the discussion below.

In fact, any definition of second home can be broken down into two segments: first, type of property and, second, type of use. Transversal to these segments is the dimension of time, either in frequency or in intensity (number of visits and respective duration). In property terms, second home may be defined as one that belongs to someone who already has a first home that he/she considers to be primary, or permanent. From the point of view of the type and frequency of use, a second home can be used several times a year, e.g., for weekends, vacation,[1] or at any other time – but always predominantly for recreation and leisure.

We believe that the phenomenon of second homes should not be separated from two analytical frameworks that are closely interrelated: tourism and real estate market (Figure 10.1). The former is mainly marked by motives of recreation and leisure, and the latter by some kind of family investment, whether in a second home (new or used, purchased, or inherited) that may become a tradable good, or in a home that can enable, for instance, the enjoyment of an environment away from the daily routine, the mix of these goals also being a possibility. New or used, purchased or inherited second homes elicit different attitudes and practices, and have different impacts on the management of the territory in which they are located. This aspect should be given due attention by public authorities, especially those responsible for territorial development and spatial planning. In Portugal, this has practically not been the case so far.

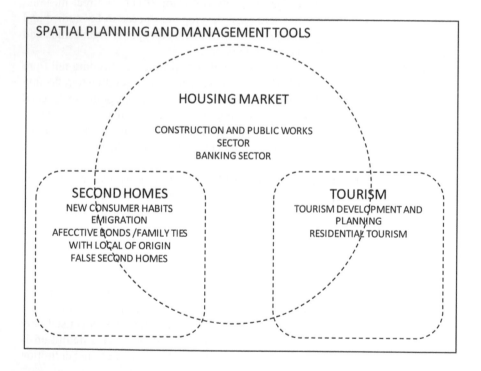

Figure 10.1   Relations between second homes, tourism and the housing market.

---

1   Some authors, such as M. Valenzuela Rubio (2003), distinguish second homes from holiday homes, associating the latter with the exclusive use for holiday periods so that they may therefore be located farther from the place of the first residence.

The phenomenon of the expansion of second homes and their relationships with residential tourism in Portugal has given rise to a wide range of controversies, and two of them are systematized and discussed in this chapter: the real significance of the second home phenomenon, and the relationship between second homes and tourism in the context of the Portuguese territorial planning policies. An attempt is made to clarify these controversies within a theoretical framework anchored in three analytical domains in which territorial management instruments can be applied effectively in regulating land occupation and use in Portugal:

- the **evolution of the housing market** and its relationship with a fragile economy that creates few opportunities for the investment of small savings; concurrently, the expansion of that market also ended as an opportunity for family savings by way of bank credits, which, regardless of the values of interest rates, until recently was guaranteed by the mortgaged property.
- the increased adoption of new **consumption habits** by a growing urban population ever more distanced from rural life, but with a large number of people living in big cities that still maintain family ties and affective relationships with their places of origin; emigration/outmigration from the countryside largely provoked the fall of the primary sector since the 1970s and facilitated the raise of the secondary sector, in which construction and civil engineering have had a higher share (number of firms and employees) than the manufacturing industry; another important factor is the improved road infrastructure, which has enabled significant reductions of distance/time and fostered the use of private automobile.
- the **actions of public authorities**, from municipal to central governments, which since the 1960s have put high stakes on tourism as an activity generating employment, wealth and the capitalization of public finances; in this context, as accommodation is bound to tourism, and given that recreation and leisure continue to be its primary motivation, it is safe to assume that in the interface of these overlapping linkages the expansion of second homes phenomenon has emerged; residential tourism arose as a business opportunity for the real estate component of tourism, which was already showing signs of saturation mainly reflected in the decline of some accommodation typologies, and it counted on strong international demand from countries like the United Kingdom, Ireland and Germany.

Aiming to clarify and systematize the complexity of issues related to the strong growth of second homes in Portugal since the 1990s, as well as to the controversies that underlie the interpretation of this phenomenon, a taxonomic analysis of all

278 counties[2] of the Portuguese mainland[3] are classified on the basis of a set of indicators of the interrelationships established in the context of those controversies. To this end, an analytical model was developed and the results of its application are presented in the following sections.

Given that in Portugal there is no quantitative, nor even qualitative, information systematically covering the country, or at least a significant set of counties, on the ways of use, physical characteristics, or the knowledge, attitudes and practice of second home owners and users, the conclusions resulting from this macroscopic analysis are, when appropriate, compared with the results of a survey of a characteristic sample of 163 owners of second homes located in six counties of the Oeste Region, a NUTS III NW of the Lisbon Metropolitan Area.[4]

Three Second Homes Realities

On the basis of the nature of economic and culture change and related socio-spatial processes in Portugal over the last several decades (Belyaev and Roca 2011, Medeiros 2005, Ferrão 1996, Gaspar 1989, 1993), a hypothesis can be put forward that a significant proportion of second homes in Portugal is not only a result of the adoption of new consumption patterns typically associated with postproductivist societies but is also largely a reflection of two interrelated factors: first, emigration and, second, out-migration from rural areas. The consequence has been that a sizable housing stock was left behind, abandoned, and that its fate in terms of ownership and use largely remains unknown nowadays. The Housing Census does not gather any physical data about such housing units, which only adds to the uncertainties about their true character.

Possible main explanatory domains of spatial distribution features and of expansion dynamics of second homes in Portugal are:

- as commonly accepted, the main use of second homes relates to recreation and leisure, and these are also, as widely recognized in the literature (and so far in the Portuguese statistical data available at www.ine.pt), the main motives for tourist mobility, so a positive correlation between their frequency of occurrence and the volume of tourism offer can be expected;

---

2   Although the parish is the basic administrative unit in Portugal, most statistical data are only available for counties (consisting of several parishes).

3   Second homes are analysed here on the Portuguese mainland only, i.e., excluding the Azores and Madeira Archipelagos. Thus, all references to 'national' actually refer to the country's continental part.

4   This survey was a part of a 2008–2012 research project 'SEGREX – Expansion of Second Homes in Portugal', coordinated by TERCUD – Territory Culture and Development Research Centre, Lusófona University, Lisbon, with financial support from FCT – Foundation for Science and Technology, Portugal.

- similarly, location of the second home is normally associated with the existence of specific amenities that function as pull factors, especially of an environmental nature, so some relation between that kind of occurrence and the frequency of second homes should also be expected;
- inversely, more urbanized environments can be obstacles to the expansion of second homes, so where population densities are high it can be expected to have, in relative terms, less second homes;
- but, as part of the Portuguese local planning experience it can also be expected that local policies to attract new residents, even part time ones, may result in an increasing number of second homes.[5]

All these domains are integrated as 'analytical dimensions' in the model for the taxonomic study of second homes presented below. Given that the above-mentioned hypothesis comprises different vectors of influence, in addition to the indicators developed in the realm of these dimensions, the model also includes:

- a vector that illustrates the intensity of the international emigration phenomenon;
- the occurrence of vacant housing units, which functions as a calibrator of the second home phenomenon in accordance with the stated hypothesis;
- aspects of the general framework of the abandonment of rural areas, taking into account the traditionally accepted dimensions, such as ageing, poverty and depopulation;
- in addition, and because the proliferation of housing buildings can also be the result of local strategies to expand sources of revenue, yet another dimension is added – local government finances.

As shown in detail in Table 10.1, the model developed for the study of second homes phenomenon at national level integrates the analytical dimensions and their operationalization by means of indicators derived from the official statistics provided by the National Institute of Statistics (INE, Statistics Portugal – www.ine. pt). It should be borne in mind that some indicators, although included in a specific analytical dimension, also contribute to the positive or negative characterization of other dimensions.

---

5   Spatial planning in Portugal was consolidated only after 1990 by the implementation of the Base Law of Territorial Organisation, which promoted the first generation of Municipal Master Plans. This was actually in direct relation with the country's membership in the EU since 1986 (Ferrão 2011).

Table 10.1 **Analytical model for the study of second homes and operational indicators.**

Analytical Dimension	Indicator	Measurement unit	Period / year of observation
Tourism demand and supply	Rate of change of accommodation capacity (beds) in all types of hotel establishments, 2002–2009	%	2002–2009
	Share of foreign tourists in all types of accommodation	%	2009
	Accommodation capacity	Nr. of beds per 1000 inhabitants	2010
	Number of guests in all types of hotel establishments per 100 inhabitants	Nr. per 100 inhabitants	2009
Environmental amenities	Share of area included in the Natura 2000 network	% of total county territory	2010
Urbanisation and economic power	Average amount of cash withdrawals from the automated tellers	1000 EUR per inhabitant	2005 and 2010
	Population density	pop/km2	2011
	Purchasing power index	index per inhabitant	2009
Land supply for tourism development	Share of territory allocated to tourism in the Municipal Master Plans	km2 / per 1000 of total county territory	2010 or 2008
Emigration	Average annual share of the value of bank deposits by emigrants in total value of bank deposits	%	2000–2010
Quality and nature of the housing constructions	Share of vacant dwellings per total number of dwellings	%	2001
	Share of second homes in total number of dwellings	%	2011
	Rate of change of second homes	%	1991–2011
	Number of all dwellings per number of families	Nr.	2001
	Number of primary family homes without electricity per number of families	Nr.	2001
Rural decline	Ageing index	Nr. of persons over 65 per 100 persons under 15 years of age	2010

Analytical Dimension	Indicator	Measurement unit	Period / year of observation
Local public finances	Total (current and capital) revenue of municipalities	1000 euro per inhabitant	2010
	Total (current and capital) expenses of municipalities	1000 euro per inhabitant	2010
	Local tax for onerous transfer of real estate and Local tax on real estate	% of current revenue	2010

Relationships amongst the indicators presented in Table 10.1, in addition to allowing a clarification of the real size and significance of second home distribution, should help understand other controversies outlined in detail in the following sections of this chapter. In order to make this possible, and in view of the difficulty in perceiving a data-table with over 5,000 observations (i.e., a data-table with 19 indicators for the 278 counties of mainland Portugal as units of analysis) a multivariate analysis method (Principal Component Analysis) was applied, together with a numerical taxonomy[6] that simply but effectively prevents the loss of useful information, in order to obtain the characterization of different counties.

The results clearly show that in general but sufficiently solid terms the country consists of three main groups of counties:

- the most urbanized and economically strongest counties;
- counties with the highest incidence of tourism activity, in which problems of environmental sustainability already replicate, or may come to replicate, albeit at another scale, what has taken place in Southern Spain; and,
- counties of the poorest municipalities with strongest emigration and with highest share of second homes in the total housing stock, together with dwellings of poor living standard (i.e., without electricity).

This model of indicators shows a high level of internal consistency as it reflects a system of relations that can be viewed as simple and obvious, but which also calls attention to other phenomena that seem less relevant, at least in the Portugal of the twenty-first century, as can be interpreted from the negative relation between highest levels of expense per inhabitant and the percent amount of tax collection over property in local public receipts. Taking into account only the values of Pearson's Correlation higher than 0.35 (although the Student's t-test could allow values above 0.19 for a confidence level of 98.95 per cent), this system of relations can be summarized as follows:

---

6   The software used was SPAD - Système Portable pour l'Analyse des Données, and an ascending hierarchical classification based on scores of the first 5 components was chosen using a clustering strategy based on the nearest neighbour method.

- On the one hand, the weight of second homes in total households is strongly related, in descending order, with the surplus of housing supply (i.e., higher number of dwelling units than families) and with the highest revenue per county inhabitant, as well as with counties' expenses, population aging, low purchasing power, incidence of emigration, low population densities and low importance of municipal revenues resulting from property taxes.
- On the other hand, the rate of change (1991–2011) of second homes and of tourist accommodation (beds), along with other indicators such as those related to environmental amenities or the incidence of vacant housing, are not relevant to the definition of the components. This is obviously due to spatial distribution patterns poorly correlated with some other indicators.

But the meaning of at least two indicators – current revenues and property tax collection in the counties – needs to be clarified since their contribution to the definition of the components may lead to false conclusions.

An important part of the counties' current revenues, especially in the case of the developmentally lagging counties, depends on the Financial Equilibrium Fund (Local Finances Law No. 2/2007) which enables transfers from the Central Administration, thus distorting the spatial distribution of this indicator (i.e., revenue figures may be higher than expected, especially in view of the fact that one of the criteria refers to the decentralization of some competences of the Central Administration, such as, for example, the maintenance of the national road network). As to the collection of property taxes, it is also clear that in Portugal the municipalities are free to choose the fee level within the limits set by Central Administration, but the tax on rural land property is generally higher than that on urban property (land and constructions).

Without going into great technical details, elements of the reliability of the used methodology need to be pointed out: the table with eigenvalues (Table 10.2) shows that three components explain 58 per cent of the variance of the initial set of indicators. This, alongside with the system of relationships described above, can be interpreted as proof of the coherence of the model of indicators used.

That coherence is again reflected in the loadings (correlation between the variables and each component) presented in Table 10.3 (below), which includes a preliminary qualitative characterisation of each of the identified components. Loadings tend to have ever more residual explanatory weights but they are not less interesting from the standpoint of understanding the richness of relations engaged in the explanation of the causes of the spatial distribution of second homes.

Based on the scores (relationship between statistical units and factorial axes) of the first five components, a numerical taxonomy was applied following the hierarchical ascending method with the nearest neighbours clustering strategy. The obtained results point that Portugal is a country that can be classified in two, three or six groups, where the division into three groups results in Fisher's test differential values of 0 per cent after five iterations. The following dendrogram cut is taken as a reference (Figure 10.2 below).

Table 10.2   Eigenvalues.

Component number	Eigenvalue	% of variance explanation	
		% for each component	Accumulated %
1	5.7572	30.30	30.30
2	3.6031	18.96	49.26
3	1.6595	8.73	58.00
4	1.1720	6.17	64.17
5	1.0313	5.43	69.60
6	0.9631	5.07	74.66
7	0.9080	4.78	79.4
8	0.7627	4.01	83.46
9	0.6336	3.33	86.79
10	0.5686	2.99	89.78
11	0.5499	2.89	92.68
Remaining components with an eigenvalue less than 0.5	-	7.32 (less than 1% per each remaining 9)	100.00

Table 10.3   Characterization of the components: loadings.

Component number	Positive or negative contribution	Main indicators for component characterization	Pearson's correlation coefficient with the component	Component description
1	+	Total (current and capital) revenue of the municipalities per inhabitant (1000 euro), 2010	0.77	Territorial contexts of low population density, where aging above the national average is already very onerous. The revenues of the counties do not reflect the weight of property taxes, because they depend on transfers from the Financial Balance Fund of the State Budget. The per capita expenditures of counties are high, and the purchasing power of their inhabitants is below the national average, which is also reflected in lower amounts of withdrawals from automated teller machines per person. The emigration phenomenon is more pronounced in these areas, which leads to high ratios of housing units per family, along with a significant share of dwellings without power supply. This is the portrait of the poorest rural part of the country, where second homes are, in relative terms, also most numerous, perhaps reflecting high incidence of the emigration phenomenon and, also, a methodological problem of the Housing Census as regards the classification of housing units.
		Share of second homes in total number of dwellings, 2011 (%)	0.76	
		Ageing index (number of persons over 64 years per 100 persons under 15 years), 2010	0.76	
		Total (current and capital) expenses per inhabitant (1000 euro), 2010	0.71	
		Number of all dwellings per number of families, 2001	0.68	
		Average annual share of the value of bank deposits by emigrants in total value of bank deposits, 2000–2010	0.52	
		Number of primary family homes without electricity per number of families, 2001	0.49	
	−	Per capita purchasing power index (Portugal = 100), 2009	−0.80	
		Local tax for onerous transfer of real estate plus Local tax on real estate in % of total current receipts, 2010	−0.80	
		Average amount of withdrawals from automated teller machines per person, 2005–2010 (1000 EUR)	−0.70	
		Population density (n° persons per km2), 2011	−0.56	
		Percent share of foreign tourists in all types of hotel establishments, 2009	−0.53	

Component number	Positive or negative contribution	Main indicators for component characterization	Pearson's correlation coefficient with the component	Component description
2	-	Number of guests in all types of hotel establishments per 100 inhabitants, 2009	−0.85	The developmentally lagging part of the country on the margins of any tourism development, and without surplus residential dwellings.
		Accommodation capacity (beds) per 1000 population, 2010	−0.79	
		Percent share of foreign tourists in all types of hotel establishments, 2009	−0.61	
		Number of all dwellings per number of families, 2001	−0.58	
		Share (‰) of territory (km2) allocated to tourism in the Municipal Master Plans in total municipal territory, 2010	−0.52	
		Average amount of withdrawals from automated teller machines per person, 2005–2010 (1000 EUR)	−0.51	
3	+	Share of vacant dwellings per total number of dwellings, 2001 (%)	0.70	The part of the country where the housing stock for purchase or rent ('vacant', according to the Census) and where the phenomenon of emigration has not been important.
	-	Average annual share of the value of bank deposits by emigrants in total value of bank deposits, 2000–2010	−0.51	
4	+	Population density (n° persons per km2), 2011	0.47	Second homes did not have high rates of growth between 1991 and 2011. Population density is above the national average, and dwellings without electricity are not many, but population ageing is a problem.
		Ageing index (number of persons over 64 years per 100 persons under 15 years), 2010	0.35	
	-	Rate of change of second homes, 1991–2011 (%)	−0.58	
		Number of primary family homes without electricity per number of families, 2001	−0.31	

continued

Table 10.3   *concluded*

Component number	Positive or negative contribution	Main indicators for component characterization	Pearson's correlation coefficient with the component	Component description
5	-	Rate of change of accommodation capacity (beds) in all types of hotel establishments, 2002–2009 (%)	-0.84	Sluggish in tourism offer, together with the apparent absence of significant emigration.
		Average annual share of the value of bank deposits by emigrants in total value of bank deposits, 2000–2010	-0.39	

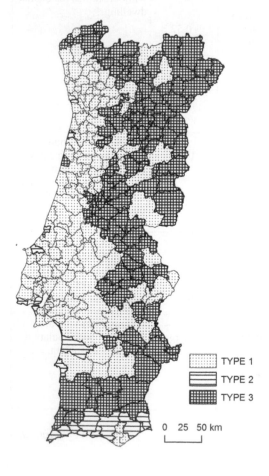

Figure 10.2

Three groups of counties according to weight of second homes, income disparities, emigration, ageing, and tourism development.

TYPE 1
TYPE 2
TYPE 3

0   25   50 km

As shown in Figure 10.2, Portugal is sharply divided between an extended coastal, more urban and developed part, and a more rural one with increasing problems of demographic and economic sustainability (Roca 2004). Within this latter part, whose share in surface area is 47.8 per cent of the national total, some exceptions can be observed: some counties with urban centres of regional importance (Guarda, Castelo Branco, Portalegre, and Évora), along with some counties close to the Spanish border (Valença, Chaves, Elvas, and Campo Maior), together with several economically more dynamic counties (Mira, Covilhã, Borba, Estremoz, and Vila Viçosa) are all exceptions to this cluster.

This dichotomous reality is overlapped by another pattern controlled by the intensity of tourism development which, covering most of the counties in the Algarve region, also includes others such as Lisbon and Oporto, Nazaré, Cascais, and Óbidos to the north of the Tagus River, and Grândola, and Sines in the south. The weight of second homes is relevant in all these counties, either as a part of urban developments created for that purpose, or within some of its urban agglomerations that are not necessarily county seats (e.g., Porto Covo in Sines, Praia d' El Rey and Vau in Óbidos, and Troia in Grândola).

Tables 10.4 to 10.6 below show the main statistical measures associated with each of the three groups retained for the analysis, which result in the characterization of each of the identified types.

Type 1 includes the counties where the share of second homes (see Figure 10.3 below), as well the number of dwelling units per family or those without electricity supply, is below the national average. These are counties where the relative share of the phenomena of ageing and emigration is less intense, and the purchasing power is above the national average.

It is implicit in the results that these counties have higher concentrations of resident population. Hence, although the value of revenues and county expenditures per capita are below the national average, suggesting the existence of higher levels of rates of return in the provision and use of public facilities and services, the dependence of local finances on property tax collection is relevant.

Type 2 includes the counties where, when compared with the national averages, the tourism phenomenon is more intense, both on the demand side (number of guests or share of foreign tourists) and supply (accommodation capacity). On the other hand, there is a strong positive correlation between the intensity of this phenomenon and the values of such indicators as the amount of withdrawals from automated teller machines and the purchasing power index.

But the most important feature related to the second homes phenomenon is the very high rates of change during the period 1991–2011, with average values close to 140 per cent, although its share in total housing stock has not been relevant to the clustering of this set of counties. This fact, coupled with the high importance of the property tax collection for local finances (average of 41 per cent for Type 2, as opposed to 18 per cent for the national total) clearly suggests that in these counties are located the second homes clearly associated with the tourism phenomenon, along with their use for leisure and vacations and more probably with a higher share of those owned by foreigners (see Figure 10.4 below).

Figure 10.3   Houses with coloured stripes inspired by the original wooden houses of the fishermen. Second homes in a Type 1 county (Aveiro, North-Central Portugal, on the Atlantic coast).

Figure 10.4   Advertisement for second homes and residential tourism development in the *Falésia d' El Rey* resort in a Type 2 county (Óbidos, South Central Portugal, on the Atlantic coast), located close to two other resorts in the same county – *Praia d' El Rey* and *Bom Sucesso.*

Table 10.4   TYPE 1 – Low relative representation of second homes, better housing standards, high purchasing power capacity, and a local government depending on property taxes slightly over the national average.

Indicator	Average		Standard deviation		Test value
	In the type	In the total of 278 counties	In the type	In the total of 278 counties	
Share of second homes in total number of dwellings, 2011 (%)	17.38	26.68	7.60	12.80	−12.73
Number of all dwellings per number of families, 2001	1.36	1.56	0.14	0.28	−11.95
Total (current and capital) revenue per inhabitant (1000 euro), 2010	0.69	1.00	0.21	0.47	−11.58
Total (current and capital) expenses per inhabitant (1000 euro), 2010	0.76	1.09	0.25	0.53	−11.02
Ageing index (number of persons over 64 years per 100 persons under 15 years), 2010	136.10	182.89	42.42	87.06	−9.41
Number of primary family homes without electricity per number of families, 2001	0.00	0.01	0.00	0.01	−6.79
Average annual share of the value of bank deposits by emigrants in total value of bank deposits, 2000–2010	5.61	7.97	4.31	6.87	−6.01
Local tax for onerous transfer of real estate plus Local tax on real estate in % of total current revenue, 2010	22.53	18.11	9.16	11.94	6.47
Per capita purchasing power index (Portugal = 100), 2009	84.44	76.03	21.01	24.22	6.08

Table 10.5   TYPE 2 – The touristic Portugal, with highest purchasing power, highest growth of second homes and a local government highly dependent on property taxes.

Indicator	Average		Standard deviation		Test value
	In the type	In the total of 278 counties	In the type	In the total of 278 counties	
Percent share of foreign tourists in all types of hotel establishments, 2009	50.34	13.52	16.07	17.07	9.72
Number of guests in all types of hotel establishments per 100 inhabitants, 2009	526.55	83.89	568.68	209.99	9.50
Share (‰) of territory (km2) allocated to tourism in the Municipal Master Plans in total municipal territory, 2010	26.35	3.39	31.26	11.70	8.85
Local tax for onerous transfer of real estate plus Local tax on real estate in % of total current revenue, 2010	41.21	18.11	10.65	11.94	8.72
Average amount of withdrawals from automated teller machines per person, 2005–2010 (1000 EUR)	3.07	1.79	0.94	0.69	8.35
Accommodation capacity (beds) per 1000 population, 2010	151.44	23.25	233.04	72.47	7.97
Per capita purchasing power (Portugal = 100), 2009	107.47	76.03	41.19	24.22	5.85
Rate of change of second homes, 1991–2011 (%)	138.48	83.05	83.74	47.69	5.24

Table 10.6   TYPE 3 – The depopulated, poor and aged rural countryside where second homes are more represented, along with a biased contribution of the emigration/out-migration phenomenon.

Indicator	Average		Standard deviation		Test value
	In the type	In the total of 278 counties	In the type	In the total of 278 counties	
Share of second homes in total number of dwellings, 2011 (%)	37.01	26.68	7.70	12.80	11.12
Total (current and capital) revenue per inhabitant (1000 euro), 2010	1.36	1.00	0.43	0.47	10.85
Ageing index (number of persons over 64 years per 100 persons under 15 years), 2010	250.20	182.89	91.54	87.06	10.65
Total (current and capital) expenses per inhabitant (1000 euro), 2010	1.47	1.09	0.51	0.53	9.90
Number of all dwellings per number of families, 2001	1.76	1.56	0.21	0.28	9.77
Average annual share of the value of bank deposits by emigrants in total value of bank deposits, 2000–2010	11.74	7.97	8.07	6.87	7.57
Number of primary family homes without electricity per number of families, 2001	0.01	0.01	0.01	0.01	6.93
Local tax for onerous transfer of real estate plus Local tax on real estate in % of total current revenue, 2010	8.53	18.11	4.06	11.94	−11.06
Per capita purchasing power index (Portugal = 100), 2009	59.88	76.03	8.31	24.22	−9.19
Average annual share of the value of bank deposits by emigrants in total value of bank deposits, 2000–2010	1.35	1.79	0.42	0.69	−8.64
Percent share of foreign tourists in all types of hotel establishments, 2009	5.34	13.52	9.13	17.07	−6.61

Type 3 includes the rural counties where the share of second homes in total housing stock is higher than in Types 1 and 2. Type 3 counties are the most depopulated and aged, as well as the poorest; this is also where the emigration/outmigration phenomenon was, at least in relative terms, more intense, and where public finances are most unsustainable (see Figure 10.5).

Figure 10.5 Traditional houses in the village of Montesinho, inside the Natural Park Montesinho. Second homes in a Type 3 county (Bragança, NW Portugal, mountainous area).

The described types, along with the relationships between the variables that best define each type, actually illustrate several controversial issues regarding how to tally and grasp the second home stock and residential tourism in Portugal which will be discussed below.

Controversy over the Methodology to Tally Second Homes: The Real Size of the Second Homes Stock and its Relationship to the Rural Exodus

From 1991 to 2011, the rate of change of dwellings considered as second homes in the Housing Censuses was 73 per cent. In 2011, the second homes stock reached a figure close to one million one hundred thousand units, which could mean that almost one third of the families owns such real-estate. This housing stock is a part and parcel of the Portuguese urban and rural landscapes, but does it actually fit appropriately into the second home concept?

This controversy has not been sufficiently explored and has remained accepted by common sense as a 'given fact'. Perhaps this is why it also reveals an important problem related to spatial planning in Portugal: the indolence with which central and local authorities face the second homes phenomenon by underestimating the empirical knowledge of the phenomenon itself (censuses, inventories) and the driving forces behind it (owners' origins, motivations, frequency of use, to name a few), as well as the potential effects on the management of infrastructure and public facilities (underuse vs. overuse).

Second homes in Portugal are registered by the Census of Housing, conducted at the same time as the Population Census, since 1960. Figure 10.6 shows the 2011 Census procedure used in identifying a second home.

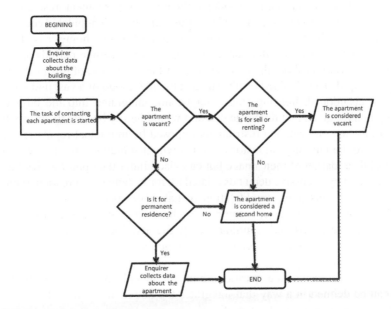

Figure 10.6 **Procedure chain for the definition of dwelling types and respective data collection by the 2011 Population and Housing Census.**

At the outset, the applied procedure reveals three types of problems for the study of second homes, as listed below in order of importance:

- the physical characteristics of the dwelling units considered as second homes are not known, much less those that refer to their owners or beneficiaries;

- any dwelling unit that is vacant and, at the same time, for sale or rent is not considered a second residence, a fact which calls into question their register and positioning in the market;
- any dwelling unit that is vacant and not for sale or for rent (even if on hold in terms of the real estate market) is considered a second home, which completely distorts the very meaning of the second home concept both in scientific terms and in terms of technical and management notions about this type of real estate.

As shown in the previous section, there is a strong correlation between the incidence of emigration and the relative frequency of existence of second homes (counties included in Type 3). Moreover, it was also found that there is a strong correlation between the dynamics of growth of such housing and a greater importance of the tourism phenomenon (counties included in Type 2).

From the viewpoint of the definition of second homes on the basis of ownership (the owner has another that is considered the main, or permanent, residence) they exist throughout the country, but in higher numbers in the depopulated countryside. However, from the point of view of the definition of second homes based on type and frequency of use (dwelling units used only during certain periods in a year), serious reservations about the real meaning of the expansion of this phenomenon must be raised. The obtained results suggest the likelihood of a vast underutilized, or even abandoned, housing stock, maybe not integrated into the buying and selling market given that domestic demand would hardly be able to absorb them.

To this fact one can add restrictions imposed by municipal regulations when it comes to the buildings in historic centres which sometimes tend to reverse the trend of degradation of their image but can also hinder the opportunities for their revitalization by excessive demands related to, for instance, the creation of modern housing conditions (e.g., amplification of habitable space and use of materials that are cheaper and/or easier to maintain). Thus, many historical centres continue to lose their residents and, simultaneously, the activities of retail and services, which, especially those of shorter range, cannot survive on a 'part-time resident population' only.

But how to answer the question about the actual number of second homes that can be defined in a way that integrates the above mentioned criteria related to the type of property and types of use? One way to arrive at a realistic number is by referring the only statistical source available in Portugal where the second homes emerge as an analytical category, i.e., the 'Travel Survey of Residents (IDR)', introduced in 2009 by the INE, whose results are published in the Tourism Statistics (INE 2011).

According to the IDR, 13,136 million overnight stays in second homes by Portuguese residents in Portugal who circulated within the country were recorded in 2010. This is the only available type of data. Given that the share of these overnights represented 24.3 per cent of the total number of overnights (53,964,000), in which 3,044 million people were involved, the use of the same proportion results in an estimation of 739,716 persons that spent at least one night in a second home. Taking into account that, on average, the Portuguese family size is about 3 persons,[7] second homes predictably occupied by Portuguese residents living in Portugal can be estimated at 246,572. Outside of this estimate are the second homes occupied by emigrants, foreign residents and foreign tourists in Portugal who also enjoy this form of housing/accommodation.

Since it is impossible to know the numbers, exact or approximate, involved in any of these categories,[8] it is safe to conclude that a good remaining portion of the housing units fall outside the scope of the second home concept commonly understood in the literature (Caldeira 1995, Casado-Diaz 2004), given that they are dwellings that have no use and are deteriorating, but census classifies them as second homes just because they are not on the market for sale or rent. This error is surely more frequent in the case of rural municipalities included in the previously described Type 3.

Controversy over Land Use and Spatial Planning: Second Homes and Tourism Development

Taking just the official statistics, 1,741 conventional hotel and all other accommodation establishments legally classified as providing tourism accommodations were registered in 2010 (INE 2011). Of this total, less than 2 per cent were 'holiday villages',[9] mostly (88 per cent) located in the Algarve Region. Generally, the evolution from 1995 to 2010 reveals the following:

* a significant and steady expansion of hotels and hotel-apartments;
* a tendency towards a decreasing number of the types of accommodation of lower added value which were the most frequent, and which in 2008 began to be classified as 'local accommodation', thus ceasing to be designated for 'tourism';

---

7   In the survey of second home owners in the Oeste Region (NUTS III in the NW of Lisbon Metropolitan Area), conducted as a case study in the aforementioned research project SEGREX, the average size of second home owners' family households was 3.2.

8   For example, as regards the number of Portuguese emigrants, in 2003 the INE stopped producing annually the 'Survey of Migratory Movements of Departure'. The only available records on emigrants are those provided by the Emigration Observatory (http://www.observatorioemigracao.secomunidades.pt), mostly referred to the 2008–2010 period. Despite shortcomings with data based on consular records only, this source points to a stock of 3,586,945 Portuguese emigrants. An analysis of official statistics and censuses of the main receiving countries of Portuguese emigrants made by the Emigration Observatory (data available online at http://www.observatorioemigracao.secomunidades.pt/np4/2661.html), pointed to 1,324,046 residents in those countries born in Portugal out of a total of 1,804,261 persons with Portuguese nationality.

9   Hotel type tourist residences, sometimes located outside urban settlements, which can offer services similar to a resort.

- after a period of strong growth, at least until 2005, tourist apartments started to register a negative growth rate as of 2008;
- as early as 1995 a stabilization in the number of 'holiday villages' was initiated despite their apparent densification (i.e., a progressive increase in its accommodation capacity).

In the context of the natural adjustment of supply to demand, these data draw attention to the following processes:

- the saturation of the 'holiday village' type, perhaps because of the competition from more modern planned tourist resorts, but most likely because of the similar offer outside the tourism industry norms (i.e., housing developments within urban or rural places);
- the reconfiguration of supply in types of accommodation with lower level of value added;
- the start of the fading of the 'holiday apartments' type, often based on a scheme of temporal fractional ownership (time-sharing), mainly located in sun and sea destinations, as they are close to the saturation of their carrying capacity, thus becoming less attractive, especially in view of the costs involved (sharing the annual maintenance) and the competition of the non-regulated (informal) lodging market;
- finally, second homes cannot be neglected as part of the residential tourism expansion, in the context of real estate market trends and of increased tourists' psychocentric behaviour, largely related to their ageing.

In Portugal, there is no specific regulation that applies to the promotion of second homes when they are outside tourism industry developments classified as 'tourist villages', 'holiday apartments', or 'tourist complexes', the latter being also commonly known and designated in the legislation as 'resorts'. When a second home does not fit into any of the types referred to above, general legislation on the acquisition of privately owned housing is applied. In fact, despite the only recent adoption of the concept of residential tourism, already since 1969 it has been possible to sell a second home legally classified as a type of tourist accommodation.

While admitting plural ownership (a concept that includes the horizontal property regime, in which exclusive property of one part of the building coexists with common ownership or joint ownership of parts used by all), current legislation stipulates that all accommodation units must always be available for tourist use, regardless of the type of ownership, through the mediation of a managing entity, which may be, or not, the owner-entity.[10]

---

10   In the case of tourist resort types current legislation stipulates that 'the housing units must be permanently allocated to tourist use, whose trade should always be in charge of an operating entity, and forbidding that right to individual property owners as a part of the plural property schemes' (DL No. 39/2008).

Part of the justification for this kind of 'legal approach' can be found in the land use and territorial development planning system in Portugal. First, in a process that is subject to two, sometimes conflicting, approaches, and that is also sometimes politically or policy-wise and economically justified, there is a strict control of the definition and extension of the perimeter of urban agglomerations.[11] Secondly, construction for residential purposes outside of those perimeters is facilitated by using two arguments with different implications (or impacts):

- the farmers' right to housing and the property of arable land (licensing casuistry is usually based on the criterion of average size of rural property, that is 7.5 hectares in some counties of the South of Portugal, but that can go to only 1,000 m2 in counties where average size of land properties is very small, such as in the Minho Region in the North); and
- the need for tourism development, which requires setting aside large areas of the county territory with specific environmental amenities, provided they are not in conflict with the values of the preservation of the natural environment either in the context of National Ecological Reserve (REN), Natura 2000 (RN2000), or any other type of territory classified as Special Protection Areas, resulting from the RN2000 or other EU Directives.

However, the differences between 'holiday villages' and 'tourist complexes' are not very clear. It seems that the intention of the law was to establish a major difference between the two types by way of the fact that 'tourist complexes' should include at least one hotel and other facilities or services, such as golf, tourist animation, or even a restaurant. Apparently, the legislators' idea was also that 'tourist complexes' could, given their more legitimate tourism orientation, be more easily developed outside urban perimeters.

In fact, the areas of the country with the most attractive amenities and/or best conditions for successful investment are outside the perimeters of urban areas, either in the neighbourhood of water surfaces (e.g., sea lagoons or dams), on the coasts with very low settlement densities (e.g. almost the entire coast between the towns of Sagres and Tróia, on the southern Atlantic shore), or in interface areas between the coast and the traditional and preserved rural countryside (i.e. a strip of about 60 km from the coast to the interior).

According to the National Registry of Tourism (https://rnt.turismodeportugal.pt/RNET), in 2011 a total of 40 tourism industry developments were classified as 'holiday villages' (32) and 'tourist complexes' (8). In the former type, 4 were projects and 5 corresponded only to future investment intentions. In the latter type, 1 was a project and 3 were intentions.[12] Taking the information available

---

11   This type of control is recommended as a way of minimizing the potential negative effects of second home expansion (Gurran 2011: 126).

12   Although the Law in force stipulates that 'Turismo de Portugal provide the National Register of Tourism Developments on its website (...) showing name, classification,

at *Turismo de Portugal* [i.e., the central government entity that supervises the tourism sector (www.turismodeportugal.pt)], large investments in 'integrated tourist resorts' were planned in 2006 (THR 2006). Although investment projects in areas below 70 hectares (enough to build, at levels of medium or low density, hundreds of home units) were not considered, the occupation of 44.7 thousand hectares, with 26.5 million m^2 of constructed area, and a total investment of 12.7 billion EUR, was predicted.

Most of those investment projects eventually collapsed with the deepening financial crisis, but this does not prevent questions about issues that illustrate the controversy between, on the one hand, strong regulation and control of the growth of residential tourism in the framework of tourist resorts, and, on the other hand, almost complete ignorance and neglect of the growth of the real estate classified as second homes, as some of them seem to correspond to housing stock without any use – but that gave rise to an initial investment in infrastructures which must now be maintained, thus contributing to the degradation of local government finances.

In fact, if residential tourism is understood, as it is by Portuguese tourism authorities, as something that combines housing with a complementary offer beyond simple support services for its maintenance (cleaning, laundry, ironing and small repairs), which implies the provision of other services, such as the ones previously referred to, then it is really crucial to assess its effects on local development. But, if the term 'residential tourism' were to be replaced by the term 'second home tourism',[13] apart from the inclusion of the resort developments already established (or near completion or in draft form) in the 1970s, the numbers would change dramatically, as more 1,098,336 homes classified as second homes in the 2011 Census would also be taken into account in this kind of assessment.

The negative perception that sometimes emerges as a reaction to the local/ regional development effects of real estate initiatives promoting residential tourism in 'integrated tourist resorts', inspired by the negative experience of, for example, southern Spain (Aledo 2009), eventually overlooks the existence of almost 1,100,000 second homes. On the other hand, all NUTS II regions in Portugal are protected by spatial planning instruments, such as Municipal Master Plans (Port.: PDM) and Regional Master Plans (Port: PROT), which define the maximum number of tourist beds, but unfortunately do not consider second homes in rural and urban areas at all.

---

capacity and location of the project (...)', it is actually not yet possible to obtain any data on their accommodation capacity. Most of these developments (70 per cent) are located in the Algarve Region.

13   In Spanish literature (Casado-Diaz 2004, Mazón et al. 2011, Aledo and Mazón 2004), residential tourism is closely related with second home tourism. As McWatters (2008:4) states, 'for at least the past decade Spanish scholars in the social sciences have utilized residential tourism to describe the permanent or seasonal residence of (mostly elderly) northern Europeans along the Costa del Sol region of southern Spain'.

Thus, from the viewpoint of effects on development at all scales, what would be preferable? To have plainly known 'n' developments legally framed and scrutinized by public and private authorities, or to have more than one million dwelling units (a number which tends to grow) considered as second homes, but about which little or nothing is known?

Moreover, the list of intended investments in the tourism sector anchored in the promotion of developments outside urban agglomerations shows a pattern almost opposite to the relative importance of second homes (Figure 10.7 below), which also illustrates the controversy between national policies on territorial planning (MAOTDR 2007)[14] and on sectorial development (MEI 2007).[15]

Conclusions

The second homes phenomenon and its expansion in Portugal is a topic which, although acknowledged as a problem that affects, for example, the quality of peri-urban landscapes, still escapes the attention of the State, not only as regards its simple and correct count, but also from the standpoint of understanding it at least as part of the tourism phenomenon, not to mention the effects on local public finances and development.

With reference to some of the development-related dimensions which affect, and can be affected by, the expansion of second homes, there is no doubt that despite their higher quantity in more urbanized and correspondingly more demographically dynamic coastal areas (Type 1 municipalities), they assume a greater relative weight in the rural counties of the interior of Portugal (Type 3 municipalities). This weight, strongly related to the rural exodus, made public investment efforts in the construction and modernization of equipment and infrastructure quite spurious – thus contributing to, *inter alia*, the permanence of unsustainability factors in local government finances.

The fact that a major source of revenue of the counties is property taxation, especially the ones on residential real estate, may partly explain the apathy with which the second homes phenomenon is (dis)regarded by public authorities. The

---

14   'The state of the landscape should concern all agents (...) The most critical situations are (...) those resulting from the destruction of the peri-urban areas, particularly in more urbanized areas and/or subject to pressure from tourism and second homes' (MAOTDR 2007:79). Despite this statement, in the Action Plan of the PNPOT (National Programme for Spatial Planning Policy) there is no specific measure for the correction of this effect of second homes.

15   In the National Strategic Plan for Tourism is declared that 'the main objective for Portugal should be to grow in terms of quality and not quantity, never confusing residential tourism with the real estate business. In all projects, tourism beds should be operating before any other aspects. The main focus should be on the creation of integrated resorts, with associated offers (e.g. golf, spa)' (MEI 2007:70).

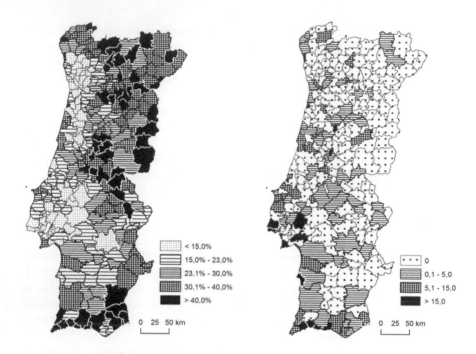

Figure 10.7   Left: Distribution of second homes in 2011 (% of the total number of dwellings); Right: territories allocated to tourism development in Municipal Master Plans (‰ km² of the total county area).

same does not apply to tourism real estate, which is sufficiently scrutinized by public authorities, resulting in a basic contradiction between the need to attract investment and the creation of obstacles to its realization. However, this problem arises, or at least is most acute, only when a construction takes place outside the perimeter of urban agglomerations and when no tourism development areas are already reserved in the Municipal Master Plans. Despite the predictions, or merely expectations, of growth for integrated tourist resorts, the allocation of these areas is not sufficiently represented at the level of these Municipal Master Plans, as their higher presence has been in the areas already marked by greater dynamism of the tourism sector (Type 2 municipalities).

The growth of real estate promotion associated with residential tourism (including international retiree migrations), either by means of building planned tourist resorts or, mainly, by using hundreds of thousands of vacant housing units within urban areas, can be envisaged as still possible in Portugal. The negative experience from southern Spain could hardly take place in Portugal, since the central government interferes in all planning processes related to urban developments by controlling both the expansion of the urban perimeters of the towns and the allocation of land usable by the tourism sector (mainly through accompanying measures included in the process of elaboration of the Municipal Master Plans) followed by the licensing of construction developments on those zoned spaces. Indeed, contrary to what is happening inside urban areas, construction and maintenance of infrastructure and facilities associated with residential tourism developments is the responsibility of private developers, so that they are not an additional burden for the national or local public finances.

In the above context, the aforementioned inquiry merits rhetoric repetition: what is more favourable – to be fully aware of all tourism developments that are legally framed and scrutinized by public and private authorities, or to have over one million dwelling units that are considered second homes, but about which too little is still clear? Answering this and similar questions calls for further policy research and instruments.

References

Aledo, A. 2009. Turismo residencial em Espanha: experiências e desafios. Paper presented at the International Conference 'Segundas Residências e Desenvolvimento Local', Óbidos [Online] Available at: http://tercud. ulusofona.pt/publicacoes/2009/Present_Aledo.pdf [accessed 1 May 2010].

Aledo, A. and Mazón, T. 2004. Impact of residential tourism and the destination life cycle theory. In *Sustainable Tourism*, edited by F.D. Pineda, C.A. Trebbia and M. Mugica. Southampton: WIT Press, 25–36.

Belayaev, D. and ROCA, Z. (eds.). 2011. *Portugal in the Era of the Knowledge Society*. Lisbon: Edições Lusófonas.

Caldeira, M.J. 1995. *Residência secundária na Área Metropolitana de Lisboa – Outros espaços outras vivências*. Lisbon: Faculdade de Letras da Universidade de Lisboa.

Casado-Diaz, M.A. 2004. Second homes in Spain. In: *Tourism, Mobility and Second Homes: Between Elite Landscape and Common Ground*, edited by C.M. Hall, and D.K. Müller. Clevedon: Channel View Publications, 215–32.

Ferrão, J. 1996. Três Décadas de Consolidação do Portugal Demográfico 'Moderno', in *A Situação Social em Portugal, 1960–1995*, edited by António Barreto. Lisboa: Instituto de Ciências Sociais, Universidade de Lisboa, 165–90.

Ferrão, J. 2011. *O Ordenamento do Território Enquanto Política Pública*. Lisboa: Fundação Caloust Gulbenkian.

Gaspar, J. 1993. *As Regiões Portuguesas*. Lisboa: MPAT/DGDR.

Gaspar, J. 1989. *Portugal Os Próximos 20 Anos. VI – Ocupação e Organiza*ção do Espaço – *uma Perspectiva*. Lisboa: Fundação Calouste Gulbenkian.

Gurran, N. 2011. Migración Residencial y Transformación Social en las Costas Australianas. In *Construir una nueva vida. Los espacios del turismo y la migración residencial*, edited by T. Mazón, R. Huete and A. Mantecón, Santander: Milrazones, 103–53.

INE, Statistics Portugal. 2011. *Estatísticas do Turismo 2010.* Lisboa: Instituto Nacional de Estatística, I.P.

INE, Statistics Portugal. 2012. *Database.* [Online] Available http://www.ine.pt/xportal/xmain?xpid=INE&xpgid=ine_base_dados&menuBOUI=13707095&contexto=bd&selTab=tab2&xlang=en [accessed May 2010].

MAOTDR – Ministério do Ambiente, do Ordenamento do Território e do Desenvolvimento Regional. 2007. *Programa Nacional da Política de Ordenamento do Território – Relatório.* Lisbon: MAOTDR [Online] Available at: http://www.territorioportugal.pt/pnpot/ [accessed 7 June 2012].

McWatters, M.R. 2008. *Residential Tourism: (De)constructing Paradise.* Bristol: Channel View Publications Ltd.

Medeiros, C.A. 2005. *Geografia de Portugal. Volume 2. Sociedade, Paisagens e Cidades.* Lisbon: Círculo de Leitores.

MEI – Ministry of Economy and Innovation. 2007. National Strategic Plan for Tourism. Fostering the Development of Tourism in Portugal. Lisbon: Turismo de Portugal, IP. [Online] Available at: http://www.turismodeportugal.pt/Portugu%C3%AAs/turismodeportugal/Pages/Publica%C3%A7%C3%B5es.aspx [accessed 7 June 2012].

Oliveira, J.A. 2011. The housing Problem and the Evolution of Homeownership Culture: from 'Clandestine' Neighbourhoods to Second Homes. In *Portugal in the Era of the Knowledge Society*, edited by D. Belayaev and Z. Roca. Lisbon: Edições Lusófonas, 305–39.

Roca, M.N.O., Silva, V. and Caldinhas, S. 2004. Demographic Sustainability and Regional Development: The Cases of Alto Minho and Alto Alentejo. Paper presented at the Congress of the European Regional Science Association, 41, Zagreb, 2004 – European Regional Development – Issues in the New Millennium and Their Impact on Economic Policy [Online] Available at: http://www-sre.wu-wien.ac.at/ersa/ersaconfs/ersa01/papers [accessed May 2012].

Valenzuela Rubio, M. 2003. La Residencia Secundaria en Ámbitos Metropolitanos: la Comunidad de Madrid. *Estudios Turísticos*, 155/156, 113–57.

Chapter 11

Lifestyles and Consumption of Do-it-Yourself Residential Tourists in Italy

Tullio Romita

Introduction

In 1999 the book *Il turismo che non appare* [The Undetected Tourism] was published and this was probably the first publication in Italy to address specifically the private home tourism (Romita 1999). The book was the result of years of research devoted to the study of tourism, using an approach which, from the sociological point of view, placed at its centre the question of sustainability in tourism.[1] Indeed, in an attempt to seek an explanation for the impacts on the environment caused by the tourist movement in some areas of southern Italy, the 'undetected tourism' was brought to light as very large, complex and widespread tourism practiced for decades through the use of private dwellings in a self-organized and do-it-yourself manner. This social phenomenon had played and continued to play an important role in the development of local communities and in the organization and management of the land, albeit little studied, or not at all, by the Italian scientific community and basically ignored by official statistics.

Over ten years after the publication of this work in Italy the state of things has not changed much. Even today private home tourism for vacation purposes is a form of tourism not officially surveyed either nationally or locally, and it remains basically impossible to determine its actual quantitative dimensions, both in terms of demand and of supply. Probably the largest changes are happening on the side of the scientific study of this social phenomenon, involving major national research institutes which are giving more space to research and publications on the subject.[2]

---

1   *Il turismo che non appare* is arguably one of the first publications ever to highlight in detail the existence of a strong correlation between informal and underground economy, and degradation of natural environment.

2   Here we would like to point out: the 'reports on tourism and services sector' of the Bank of Italy and the National Institute for Research on Tourism (ISNART); the publications of the research centre of the Touring Club of Italy (TCI); the research of the Italian National Institute of Statistics (ISTAT) on 'Travel and Holidays' of Italians and those on 'Second-Home Tourism' by some Italian Chambers of Commerce (among which the most recent and thorough was carried out in 2011 on tourists and on owners of holiday

As a main research topic at Centre for Research and Studies on Tourism (CReST), Department of Sociology and Political Science, University of Calabria, 'undetected tourism' has proven to be an important key to understanding the processes of development and change across society and local communities, as well as problems related to land use and consumption, particularly in areas of southern Italy (see http://www.sociologia.unical.it/crest/index.php). The knowledge gained on the subject has enabled to classify as 'touristic' hundreds of Italian municipalities which did not appear as such as a result of the presence of little or no conventional official accommodation structures. Also, research results provided an acceptable explanation for the fact that in hundreds of municipalities of almost all areas of the Italian territory, especially during summer, a population much higher than the resident one could be observed. Furthermore, it was possible to circumscribe some impacts on territory and on local communities, both negative and positive, produced by the presence of unorganized and unsurveyed tourism flows typical of undetected tourism. Likewise, it was possible to understand why in many Italian communes, even among the smaller ones, there was a higher number of 'empty or unoccupied' houses than the number of occupied dwellings, in some cases even in very substantial proportions.[3]

This chapter draws on the knowledge and experience gained at CReST with a focus on (i) the more recently studied so-called do-it-yourself tourists as a social figure generated as part of developing the local tourism industry based on the idea of the cottage (McCannel 1976), as well as on (ii) what happens in the local contexts in which tourism develops spontaneously and outside the official tourist market (i.e. in situations where the unofficial, do-it-yourself tourism is very diffused as practiced through private homes used for vacation purposes). Although this chapter is devoted to the features of what we have hitherto called 'undetected tourism' in Italy, in order to comply with the internationally established language, the term 'residential tourism' is used here.

Residential tourism in Italy: Evolution and Definition

In almost every country of the Western world, tourism became an important mass social phenomenon around the middle of the twentieth century. From those years onwards the tourist industry has expanded continuously, seeking and finding new markets, gradually imposing its rules, by providing standardized consumable products on a global scale. For over half a century, then, talking about tourism

---

homes by the Chamber of Commerce of Cosenza in collaboration with ISNART and Centre for Research and Studies on Tourism of the University of Calabria).

    3   The Census 'Population and Housing in Italy' conducted every ten years by the Italian National Institute for Statistics (ISTAT), classifies as either empty or unoccupied all the houses which at the time of the census are not occupied by the people who live there the whole year.

has meant commonly referring to a phenomenon in which a few people organize and plan how to travel and go on holiday for millions of others: touristic products created and marketed by tourism professionals are purchased and consumed by people who, thus, live an experience of travelling and living in controlled environments, and guaranteed by the tour operators and travel agencies – with the so-called 'environmental bubbles' (Boorstin 1961). In the 1950s and 1960s, the seaside and mountain areas, even those which could not boast a tradition of tourism, were open to tourism and were equipped to find their place in the national and/or international tourism market, encouraging and facilitating the construction of official accommodation structures and, simultaneously, of private houses for holiday use.

At the beginning, the path of development of resorts was substantially equal throughout the country but over the years, while some resorts have preferred conventional development models investing in the expansion of the tourist accommodation offer (not only through the creation from scratch of accommodations but also through the conversion of part of the existing private housing stock into hotels, pensions), other places have, however, preferred to give priority to tourism development models based on demand and supply of affordable private housing for holiday use. In any case, the general result of this process is that there are very few resorts now distributed throughout the Italian territory where one can track the significant presence of residential tourism. Residential tourism is therefore a phenomenon which has always existed in Italy, and which over time has also been transformed from an elite into a mass phenomenon.

Still, in the Italian case residential tourism is not and has never been just a 'form or type' of tourism, but is rather a social phenomenon which with conventional mass tourism only has in common the roots. In fact, important features of this phenomenon are that it has been essentially informal, undeclared and unregulated. In this context, the operation and behaviour of different actors (tourists, owners of holiday houses and host communities), unlike in the case of conventional residential tourism, have not been other-directed but primarily self-directed, i.e. involving do-it-yourself choices and decisions.

Residential tourism in Italy has always been a more important form of do-it-yourself tourism which through the years not only survived the explosive emergence of the tourism industry, but indeed became stronger and substantially different from the official rules of the tourist market. In fact, since the time when tourism became a mass social phenomenon, residential tourism in Italy has never ceased to exist and grow (Romita 2009a: 4–11). It can now count on a potential offer of more than three million vacation dwellings,[4] mostly built in the 1960s and 1970s.

Various reasons led to the emergence of residential tourism in Italy, whether they are related to the Italians' specific way of being and living, or are common to other countries. For one, private houses for vacation use reproduce conditions and lifestyles similar to those of 'every day' life, and also enable larger families

---

4   This figure was detected in the Housing Census of 1991 when the number of private homes for holiday use was specifically identified (ISTAT 1994).

to enjoy long tourism holidays, since they are cheaper than those offered by conventional tourism hotels and other non-hotel accommodation facilities. Moreover, ownership of a second residence in a resort as an element of social distinction becomes a financial investment; finally, a significant part of the stock of private houses for holiday use was built by emigrants, and for them the house is not only an opportunity for holidaying in their places of origin but also a means of forging strong links with the territory.

It is really rare to find cases of residential resorts where tourism has been developed thanks to defined development strategies. The supply of residential tourism is almost always a response to the spontaneous demand for tourism, in which local authorities and people benefit economically from the sale of land, the construction of dwellings, and the provision of services for the maintenance of homes for vacation and the residence of tourists.

As stated before, quantitative and qualitative characteristics of residential tourism in Italy have not been known, as monitoring instruments, national policies and legislation, which could make it more visible, have been lacking for quite some time. In addition, the theoretical and conceptual tools used to analyze the residential tourism phenomenon have been essentially those used to study and interpret the phenomenon of conventional tourism, thus largely inadequate. However, in recent years the situation has changed considerably, both in Italy and elsewhere. The rapid expansion of residential tourism worldwide spurred the interest of a growing number of scholars with a sizable new literature, especially on the Mediterranean Europe (Mazón and Aledo 2005, Huete 2008, 2009, Mantecón 2008, Mazón, Huete and Mantecón 2009, Hall and Müller 2004, Karayiannis, Iakovidou and Tsartas 2010, Roca, Oliveira and Roca 2010, Grzinic 2010, Romita 2010, among others). Nevertheless, the analytical difficulties and the absence of a consolidated set of scientific references in the literature on residential tourism are evident in the wide variety of terms used to indicate this phenomenon, ranging from undetected tourism, residential tourism and private homes tourism to second homes tourism, real estate tourism, holiday homes tourism, to name a few (Romita 1999, Mazón and Aledo 1997, ISNART 2011, Aledo and Mazón 2004, Karayiannis, Iakovidou and Tsartas 2010).

In the Italian case, the definition which seems to represent effectively this phenomenon is the following: 'Residential tourism is an informal and black economy phenomenon which is developed through private tourist accommodation, available in the area to accommodate the request from a self-directed, spontaneous, uncertain and unpredictable tourism demand which, in turn, organize and conduct their own choices through space and time of their tourism experience.' In this context, 'the residential tourist is a do-it-yourself tourist, who organizes the space and time of the holiday in total independence by staying in private dwellings and accepting only the daily life rules that are set by their hosts.' (Romita 2009b: 632).[5]

---

5 'Residential tourism is an economic activity resulting from the urbanization, building and selling of houses offered to the non-hotel tourism market almost always

The Spread of Residential Tourism in Italy

Although residential tourism in Italy has substantially been an undetected social phenomenon, and thus more difficult to quantify in terms of demand than supply, some local and national studies conducted in recent years have allowed a better understanding of its great importance and dissemination across the country, as well as of its dimension. On the supply side the only realistic data on private homes for holiday is from the 1991 Census of Population and Housing: 2,711,419 dwellings, with 9,762,086 rooms. Unfortunately, all other national censuses, both previous and subsequent, this type of data is not available since all such homes are included among those assessed by ISTAT as 'empty or unoccupied dwellings.' (Table 11.1)

Table 11.1   Occupied and unoccupied dwellings in the Italian Censuses from 1971 to 2001.

Census year	Occupied dwellings	Unoccupied or empty houses			Total dwellings
		Unoccupied and/or empty dwellings	Unoccupied dwellings for holiday use	Total unoccupied and/or empty dwellings	
1971	15,301,357	2,132,534		2,132,534	17,433,891
1981	17,541,752	4,395,471		4,395,471	21,937,223
1991	19,735,913	2,581,190	2,711,419	5,292,609	25,028,522
2001	21,967,516	5,324,477		5,324,477	27,291,993

Source: ISTAT, Population and Housing Census (Years: 1971, 1981, 1991, 2001).

Using the 1991 data on the number of homes and rooms as a reference, and assuming only 1–2 beds per room, it can be estimated that there have been about 10–20 million beds available in the segment of the residential tourism in Italy. Although this is probably an underestimation of its true dimension, these numbers are proof that the residential tourism supply is of prime importance in the Italian tourism industry, which until 2010 had about 146,000 units with 4,600,000 beds in its conventional official accommodation offer (i.e. hotels, tourist villages, camping, hostels, B&Bs).

---

outside official channels, whose beneficiaries live there permanently or semi-permanently, and that responds to a new form of residential mobility and developed society.' (Mazón and Aledo 2005: 18–19).

A remarkable consistency in the growing supply of residential tourism, especially in the southern regions (Romita 1999), was evidenced in different studies[6] and our estimate is that there were about 11 million residential tourists in 2001.

The Demand for Residential Tourism in Italy

The analysis of the demand for residential tourism in Italy is even more complicated. Since there are no data collected through censuses, one can only make estimates about the stock of holiday housing (and respective beds), while in evaluating flows generated by the residential tourism use can only be made of non-systematic studies that have tried to address this issue by proposing estimates of such flows and possible assessment models. For example, in some local studies, it was estimated that due to residential tourism in August alone the size of the present population can be up to fifteen times greater than the resident population (Romita 1999). According to another study made at national level, in 2003 residential tourism in Italy may have resulted in about 730 million stays (Mercury 2003). In a more articulate and prudent recent study, it is estimated that in 2010 Italian residential tourism generated 154 million overnight stays by international tourists and 296 million overnight stays by Italian tourists (ISNART 2011: 38).

In spite of the actual inability of these estimates to fully represent the situation of tourist flows attributable to residential tourism, and the need to create a system for monitoring the movement of residential tourists, it becomes quite obvious that tourist flows attributable to residential tourism in Italy carry great weight – with flows certainly more abundant than those conventional tourism is capable of producing.

The Choice of Private Homes in Italians' Travels and Vacations

According to the latest ISTAT national survey entitled 'Travel and Holidays of the Italians in Italy and Abroad' (ISTAT 2011), Italian tourists' interest in staying in private homes has prevailed and is growing, and holiday stays in such dwellings are on average longer than those at hotels and other conventional types of holiday accommodation. Comparing the 2010 and 2009 data (see Tables 11.2 and 11.3), the following can be concluded: 'in 2010, regarding stays in Italy, private homes for holiday are preferred, while in stays abroad more conventional accommodation facilities (hotels, tourist villages, camping, among others) are chosen. As regards Italian destinations, 53.9 per cent of trips and 63.6 per cent of overnight stays involve private homes for vacation, while 46.1 per cent of trips and 36.4 per cent of overnight stays involve conventional tourism accommodations. For trips abroad,

---

6   For example, in a study conducted by the Mercury company (Mercury 2003) which also used the 1991 census data as baseline, it was estimated that in 2001 the number of homes for holiday use grew to 2,917,172 units.

however, conventional facilities are used in 70.9 per cent of trips and 58 per cent of overnight stays, while private homes for vacation are chosen in 29.1 per cent of trips and 42 per cent of overnight stays.' (ISTAT 2011: 9).

Table 11.2   Italian trips by type of accommodation and main destination in 2010 (%).

Destination	Type of accommodation		Total
	Conventional accommodation	Private apartments for holiday	
Italy	46.1	53.9	100.0
North	50.2	49.8	100.0
Centre	47.3	52.7	100.0
South and Islands	39.1	60.9	100.0
Abroad	70.9	29.1	100.0
Total	50.7	49.3	100.0

Source: National Survey 'Travel and Holidays of the Italians in Italy and abroad', 2010 (ISTAT 2011).

Table 11.3   Total travel nights by Italian nationals, accommodation and main destination, in 2010 (%).

Type of accommodation	Destinations		Total overnight stays
	Italy	Abroad	
Conventional accommodation	36.4	58.0	41.0
Private apartments for holiday	63.6	42.0	59.0
Total	100.0	100.0	100.0

Source: National Survey 'Travel and Holidays of the Italians in Italy and abroad', 2010 (ISTAT 2011).

A similar conclusion is also highlighted in ISNART's 2011 annual report on tourism in Italy, in addition to interesting new information for the analysis of the behaviour of tourists who prefer to spend their vacation in private homes. For example, this preference has played a special role in the development of domestic tourism (see Table 11.4). In fact, amongst those who have stayed in private homes for vacation, only a small fraction travelled abroad (mainly France, the United Kingdom, Germany, Switzerland and Spain), while the Italian tourist destinations have been chosen most frequently (mostly Lombardy, Liguria, Toscana, Lazio, Piedmont, and Emilia Romagna). Moreover, it was recorded, always in the same survey, that the reasons which guide the choice of residential tourists when it comes to spend holidays in a private home are mainly the presence of friends and family, second home ownership and the willingness to revisit relatives and friends.

Table 11.4   Travel by Italian nationals, type of accommodation and main destination in 2010 (%).

Destinations	Type of accommodation		Total holidays
	Conventional accommodation	Private holiday apartments	
Italy	74.0	87.9	78.1
Abroad	26.0	12.1	21.9
Total	100.0	100.0	100.0

Source: Report on Italian tourism "Tourism Company 2011" (ISNART 2011).

Residential Tourism in Italy in Synthesis

The previously highlighted features of residential tourism in Italy can be summarized as a social phenomenon which:

- emerged with organized mass tourism, but then continued to grow over time quite independently;
- has developed on the basis of high demand and a spontaneously provided supply;
- has maintained the prevailing character of an undetected and informal social phenomenon;
- in consequence of its informal character, it is not possible to plan and to predict the economic, social and environmental issues at stake;
- has given rise to 'spontaneous local tourist contexts' in terms of social, economic and environmental effects wherever the residential tourism has developed in an unplanned manner, based on spontaneous agreements and do-it-yourself behaviour, including all relevant local stakeholders (homeowners, tourists, local people, civil servants, traders, etc.);
- has preserved the character of an undetected phenomenon due to the absence of a formal system of national observation and monitoring;

- because it is a largely non-formal phenomenon, the actual flow of tourists is not known;
- despite being an undetected phenomenon, some local studies and some national estimates lead to the conclusion that in most cases Italian residential tourism produces higher tourist flows than official (conventional) tourism;
- in consequence of its undetected status, destinations where official (conventional) tourism is not widespread do not appear in national tourism statistics, even when residential tourism is widely present in these destinations;
- in many national tourism areas coexists with conventional tourist accommodation facilities, which is sometimes complicated because of the self-direction nature of residential tourism and of hetero-direction nature of conventional official tourism;
- has been significantly influenced by the so-called 'roots tourism' (or, 'return tourism') practiced by the people who come back as tourists to their places of origin from which they emigrated in the past for economic reasons (i.e., employment in the most industrialized areas in northern Italy, or abroad).

Lifestyles and Consumption of Residential Tourist

Introduction

In the field-based studies of lifestyles and consumption patterns of the residential tourists carried out by CReST since 1997 it has generally been assumed that this kind of tourism is a social phenomenon which: (i) is essentially self-directed, thus avoiding approval and standardization processes that are typical for tourism industry and mass tourism; (ii) emerged and developed to meet the needs of a creative use of leisure time, because it feeds on self-directed choices; and (iii) respects the original cultures, promotes social relations and the creation of new forms of housing. The specific topics studied were the tourists' attachment to the holiday towns, vacation habits, motivations, behaviour and consumption patterns, as well as the terms and conditions of residence and of use of rental housing.[7] The main results of the research carried out by CReST in 2007 (over 600 interviews) and 2011 (around 3.000 interviews)[8] on lifestyles and consumption patterns

---

7 Field research at CReST was based on over six thousand interviews with residential tourists in seaside and mountain holiday destinations. For most significant results see Romita 1999, 2007, 2009a, 2009b; Romita and Perri 2006, 2009a, 2009b, 2011; Romita and Muoio 2009.

8 The research conducted in 2011 was commissioned by the Chamber of Commerce of Cosenza in collaboration with ISNART in Rome.

of residential tourists in Southern Italy (in northern Calabria, on the shores of the Tyrrhenian and Ionian Seas, and in the Sila mountain tourism towns) are presented below.

In the absence of official statistics on the profile of residential tourists at both national and regional levels, it is worth bringing here (Table 11.5) the most recent estimates made by ISNART (2011).

Table 11.5   Identikit of the tourist in private holiday homes, 2010 (%).

Gender	Italians	Foreigners	Total
Male	47.2	50.2	47.6
Female	52.8	49.8	52.4
Total	100.0	100.0	100.0

Age	Italians	Foreigners	Total
15–24 years	10.9	17.1	11.7
25–34 years	23.2	30.6	24.2
35–44 years	21.6	16.8	20.9
45 -54 years	17.3	12.2	16.6
55–64 years	13.0	13.4	13.1
65 years and over	13.9	9.9	13.4
Total	100.0	100.0	100.0

Education	Italians	Foreigners	Total
None / Primary school	3.1	1.3	2.9
Junior High School	10.9	13.0	11.2
High school diploma	50.3	42.3	49.3
Higher Education	35.6	43.4	36.7
Total	100,0	100,0	100.0

Profession	Italians	Foreigners	Total
Employed	60.1	62.3	60.4
Unemployed/looking for first job	7.4	7.9	7.5
Housewife/Retired	23.6	17.5	22.8
Student	8.9	12.3	9.3
Total	100.0	100.0	100.0

Social status	Italians	Foreigners	Total
Single with children	6.4	6.5	6.4
Single with no children	27.9	36.4	29.0
Couple with children	48.9	36.5	47.3
Couple with no children	16.8	20.6	17.3
Total	100.0	100.0	100.0

Source: Report on Italian tourism "Tourism Company 2011" (ISNART 2011).

Lifestyles of Residential Tourists

As has been seen, in the Italian residential tourism relations between tourism stakeholders and the social and economic relations and dynamics are the results of choices that only minimally depend on the institutional tourism public and private decision makers. Moreover, behaviour and choices are mostly determined by self-directed decisions of tourists in the do-it-yourself manner, as Italian residential tourists do not rely on anyone to organize and manage their tourism experience, since they arrange the space and time of the holiday in total autonomy, staying in private homes and respecting only the general rules of everyday life and, in particular, those in force in the host tourism town.

While there are several important studies on organized mass tourists,[9] knowledge on the residential do-it-yourself tourist has been quite limited. The first ever survey about the experience of Italian residential tourists, conducted in 2007 (Romita and Perri 2011),[10] enabled the characterization of their modes of behaviour, as described in considerable detail below.

---

9   See for example the contributions of McCannel (1976), Urry (1991), Cohen (1974), Augé (1999), and the Italian Savelli (1989), Nocifora (2008), and Dall'Ara (1995).

10   As part of the CReST's research project, a survey was carried out with more than 600 face-to-face interviews with residential tourists (owners and tenants) in seaside and mountain holiday towns in Southern Italy, with high-density private housing for holiday

First of all, residential tourists do not care for novelty, and perceive the holiday as an opportunity to break the routine of daily life, enjoying long stays (over three weeks) during which they attend public meeting places and participate in local tourist entertainment events; only rarely do they participate in visits and/or excursions to other places; their preferred activities are hiking and reading; during the holiday they rarely travel to other locations, preferring to stay as much as possible at the chosen place; they tend to explore the opportunities for recreation and entertainment made available by the host community, as well as enjoy the local natural and cultural resources; they endeavour to become accepted by the local community by means of opportunities for daily relationships and by establishing friendships.

Loyalty to the same places, including the housing developments for residential tourism (Figures 11.1 and 11.2) is another characteristic of residential tourists: they spend their holidays several times in the same location and return to it after a few years' break; as tenants, they often rent the same housing unit for years and leave their personal belongings there as they will be useful the following year; they frequently stay in the same place also for brief breaks (Easter, Christmas, weekends) throughout the year.

Although they are for the most part satisfied with the place chosen to spend their holiday, residential tourists also point out local problems and failings, and they are also inclined to give advice about improvements. This attitude does not depend only on holiday home ownership, travel accessibility, or proximity to friends and relatives, but rather on the fact that the choices made by the do-it-yourself tourists are totally self-directed, thus carefully weighed and taken to be compatible with expectations.

Furthermore, residential tourists go on holiday with their family, and tend to replicate their daily habits: they cluster with other families, relatives and/or friends, without giving up their own family independence; they organize everyday routines and set criteria that are similar to the usual ones, or try to build new, comforting routines, albeit within the limits of the objective conditions of the space and time available for each type of vacation (e.g., setting times for lunch and dinner compatible with 'holiday activities', and choosing traditionally favoured foods, but with more frugal meals in restaurants; organizing the use of home spaces and implements so as to replicate as much as possible those of the prime residence, including parking the car as close to the house as possible; keeping loyal to the same lido or beach spot, for example). The availability of a car is essential to ensure local mobility for the whole family (e.g., to go shopping).

An important characteristic of these do-it-yourself tourists is caring for the state of local tourism resources and evidencing a particular sensitivity to environmental resources in the tourism area selected for spending the holidays. Their typical

---

use. The interviews took place at the vacation residences and in outdoor places where tourists gather (such as beaches, picnic areas, sky lifts) during the month of August 2007.

Figure 11.1   Housing development for residential tourism, Tyrrhenian Coast, south Italy.

Figure 11.2   Residential tourism settlement, Tyrrhenian Coast, south Italy.

reasoning is: 'I chose to come to this town since the kind of natural and cultural resources available allow me to enjoy my favourite form of tourism. So if these are kept well I will find more satisfaction during the holiday. The local community ensures that resources are accessible and usable as a part of the efforts to cater for tourists, and I must assess the quality of the resources I'm using. In any case, as long as resources are of adequate quality and quantity to satisfy the reasons why I want to spend my vacation here, the chances that I come back to this town are very high.'

Residential tourists are very specific in the assessment of the tourism space: as attentive, interested and active observers of the environment where they wish to spend their holiday, they develop the knowledge that allows them to propose improvements in tourists' living conditions; if requested, they offer constructive suggestions and concrete ideas for progress regarding the local tourism offer, such as the low quality of the landscape and the environment, the shortage and/ or inadequacy of tourism facilities and infrastructure, to name just some examples (Figures 11.3 and 11.4).

Last by not least, residential tourists are so self-organized that they are very well integrated with the local setting. In the official (conventional) accommodation for vacations, tourists are not involved in any specific tasks, because all the services – from those that meet the most basic needs (food and lodging) to all others (sun umbrella on the beach, swimming pool, sports facilities and equipment) – are organized and provided by other agents, meaning the accommodation entity. In the do-it-yourself tourism, however, everything must be planned and in many cases managed by tourists individually: they have to ensure the necessary spaces for fundamental activities (as close as possible to the sea, to friends and to the home where they are staying); to temporarily or permanently modify the physical environment (e.g., planting the sun umbrella on the beach for the entire duration of the holiday); to build the 'space' for accomplishing specific tourist experiences without the need to follow a precisely defined course, but, rather, complying with the environment that is modified and used according to residential tourists' needs and objectives.[11]

Consumption Patterns of the Residential Tourist

The consumption patterns observed among the surveyed residential tourists are to a large existent very similar to those of their day-to-day lives: housing in a private home during the holiday and purchasing the food, cleaning stuff, and fuel; arranging parking for the car, or a beach place, clothing, medicines, and so forth. Moreover, depending on the lifestyle and propensity to consume, residential tourists have to bear the cost of souvenirs, restaurants, bars and pizzerias, as well as the tickets for shows, gyms and sports equipment, to name but a few.

When residential tourists are the owners of the house in which they stay during the holidays, they are responsible for a number of other expenses which also

---

11   For more on this topic see Beato 2007.

Figure 11.3   **Damaged residential tourism housing, Tyrrhenian Coast, south Italy.**

Figure 11.4   Residential tourism housing, Tyrrhenian Coast, south Italy.

mostly reflect the consumption style, such as for example, costs of swimming pool, gardens and private condominium maintenance and security, house and furniture repair, lighting, heating/cooling, telecommunications, water, as well as municipal property taxes, among others.

Consumption plays an important role in the place of residential tourism because it normally increases the overall wealth of the population given the great diversity of people the visitors must interact with. While the consumption style during their stay is influenced by the level of the general and specific quality of services offered, its selection is conditioned by the costs of travel, accommodation and additional expenses, as well as by the destination features, choice of residence, among others. Major findings from the 2007 survey on these and related features, are outlined below.

Travel and Stay

The majority of the surveyed Italian residential tourists spend their holidays with their families (four people on average), while foreigners do it mostly without children. This largely explains why the car is the preferred means of transport among Italians but not so among foreign residential tourists. Regarding the latter, most travel by plane, favouring the special low cost flights. In terms of expenditure, the total amount spent on transportation per person is around 100 EUR for Italians and 400 EUR for foreigners.

The length of stay of residential tourists ranges from one to three weeks. However, in many cases the stay is much longer, with a considerable share of respondents who stay on vacation for over two months.

If vacationing on a seaside area, residential tourists spend most of their time on the beaches, but this is much more typical of the Italians and less popular among foreigners, who prefer other activities. In fact, foreigners are more interested in, for example, participating in events of wine and food tasting, buying local crafts and other products, roaming the streets of historic urban nuclei, enjoying folk music shows or events. Italian residential tourists show a greater propensity for practicing dynamic sports, such as swimming, cycling and tennis, while a great majority of foreigners prefer the tranquillity of walking.

Destination Features

Residential tourists manifest a clear tendency to be regular guests of the place where their holiday house is located. In fact, a great majority, especially among Italians, had spent their holidays in the same location in the previous three years.

Opinions about the vacation destination are generally positive. The majority of respondents would recommend residing in the municipality where they are on vacation to friends and relatives. When this is not the case, it reflects a general dissatisfaction with respect to the available environmental resources.

Furthermore, the residential tourism home is usually located in the place where the climatic and environmental quality, rich landscape, overall tranquillity and relaxed atmosphere, local gastronomy and warm tourist reception can be appreciated. In fact, along with environmental resources, local products and acceptance by the resident population are the elements that are most interesting and highly valued among this type of tourists.

As regards the environment, a growing number of residential tourists pay increased attention to problems of pollution and poor waste management, and to the cleanliness of the sea and beaches; moreover, they recommend actions towards solving them.

Accommodation and Additional Expenses

Excluding the costs of travel and accommodation, the average daily expenditures borne by Italian residential tourists during their stay is on average 30 EUR vis-à-vis circa 28 EUR for foreigners. Most of these expenses are made on food and beverages: on average 10 EUR per day are spent on meals in restaurants and pizzerias, 5 EUR on consumption in bars, pastry shops and cafes and another 5 EUR on food products and wine, while the bulk of the average daily expenditure (about 20 EUR) comes from the purchase of food from supermarkets and shops.

In addition to restaurants, average daily costs of shopping for clothing comes to about 10 EUR per person (foreigners spend much more than Italians). In all other kinds of purchases expenses are less relevant but still remarkable, especially for leisure activities (almost exclusively by Italian customers) and for access to bathing facilities and equipment (daily average of about 10 EUR per person).

The range of rent prices of private holiday homes is between 350 EUR and about 600 EUR per week. Clearly, the price of renting is critical information needed for estimating the financial resources generated by residential tourism, but also one of the most difficult to obtain from the owners of these homes.

From the responses of the surveyed owners of residential tourism homes, there is a considerable difference in renting costs between the high and low seasons. More specifically, this difference is primarily evident for the largest and most expensive housing units. At any rate, a week-long tourist stay in a private home can cost anywhere between 100 and 5,000 EUR in high season, and between 50 and 4,000 EUR in low season.

Choice of Private Residence for Vacation

For Italian residential tourists, the choice of private houses to spend holidays is largely done through word-of-mouth among the tourists themselves and among the friends and relatives who live in the chosen location, for a large part usually based on the experiences of the convenience and costs of the previous holiday in the same house. In most cases, the chosen home is booked through the help of friends who either live in the destination area, or know the owners. Another popular booking method is direct contact with the owner, especially if the same home was previously used.

The Residential Roots Tourists
It is worth stressing that among the surveyed foreign residential tourists a majority were people whose families have roots (origins) in the place of vacation. In fact, the territory where this research was carried out is very strongly marked by the out-migration of the young from rural areas and small towns in the hinterland to the most industrialized Italian regions, as well as emigration to foreign countries in the 1950s and 1960s.

The emigrants and their descendants return every year, mostly during the summer vacation, to their places of origin, but the majority comes in the guise of tourists, including assuming behaviour and attitudes that are more proper of tourists than natives. Their roots, the emotional ties with the territory, the parental relationships as well as friendships and family possessions are the factors that attract them to the place. As a matter of fact, almost all surveyed foreign residential tourists chose their holiday area for their own desire to explore local traditions, or because it was the choice recommended by relatives and friends, and also because the latter in many cases offer free hospitality to these foreign tourists.

Operating Costs of Renting a House
To own a holiday house is a form of economic investment, and it presents some advantages (the house itself is a significant economic asset and renting it may enable accruing additional income), but as it is usually a second home used only for vacation for limited periods of time a year, it also involves a number of maintenance and management costs that tend to grow over time.

Survey records have provided a fairly accurate idea of what is involved in the ownership and economic exploration of a home for vacation in Italy. In one year, owners have to pay for fixed expenses (water, electricity, gas, heating, waste disposal) that on average amount to about 700 EUR. Apart from this, another 300 EUR may need to be paid for mandatory condominium expenses. About 2,000 EUR need to be paid to maintain sanitary facilities, walls, ceilings, floors and fittings, door and window frames. If there is a swimming pool, yet another 1,000 EUR must be added per year. Finally, there is the tax on second homes (in proportion to its dimension) to be paid to the municipality where the holiday home is located.

Most of the costs involved in the use of a second home are, of course, very similar to those of the first home (the one not primarily used for vacation), i.e., water, electricity, gas for cooking and heating. These represent a significant resource for local economies. Some costs may be reduced significantly but others cannot: for example, modern technology enables renouncing a fixed telephone line, but the owners must pay an annual fee for waste disposal although they produce daily waste, at best, only three months a year.

Conclusions

In a country that is so rich in cultural and environmental resources and that has for a long time been a preferred tourism destination the world over, it is hard to find places with no traces left by tourists. The demand is so widespread that even in the most remote villages in the hinterland of the Italian territory, though maybe for just a few days in the year, tourists look for resources, for the traditions, histories and cultures that Italian communities can express.

Residential tourism in Italy often embodies an effective response to this widespread demand. Its dominant feature is that it represents a largely informal and undetected social phenomenon with a wide range of effects and impacts on the land and on local communities. The negative effects are mostly related to the impossibility of governing the phenomenon from the point of view of its environmental and landscape-related impacts. However, the negative impacts have been offset by the positive ones. For example, there has been ample evidence of significant economic benefits for the local community as a whole, which tends to be much more balanced than the benefits from the organized tourism industry. Furthermore, positive impacts normally result from the considerable social interrelation between tourists and the local people, which is often contrary in the case of the conventionally organized tourism.

In fact, it is safe to claim that residential tourism is an encouraging social phenomenon with multiple positive implications for the economic and social growth of the country, even though it has generated negative effects on the environment and has interfered significantly with the urbanization process. In many parts of Italy it has overtaken official tourism, though it remains difficult to appreciate the phenomenon in terms of the tourism industry management logic because of its informal and undetected nature, which has spurred the development of many areas.

The Italian residential tourism has created spaces of 'spontaneous tourist contexts'. They do not have geographic and/or administrative dimension, or boundaries, but start and end wherever social, cultural and economic development is based on the ability to organize independently and beyond the formal rules, and to discover a shared space and time at the local community level. In this context, the main actors are the do-it-yourself tourists and other related actors who actually enable them to take on this role. They are special social figures, not just tourists. They are subjects in terms of the ability to decide for themselves, to foster the creation of tourism spaces and to influence their evolution.

Acknowledgments

The author would like to thank Dr. Lucia Groe and, especially, Prof. Isabel Canhoto, for their valuable assistance in the translation of the text into English.

References

Aledo, A. and Mazón, T. 2004. Impact of residential tourism and the destination life cycle theory. In *Sustainable Tourism*, edited by F.D. Pineda and C.A. Brebbia. Wessex: Witpress, 25–37.

Augé, M. 1999. *Disneyland e altri non luoghi*. Torino: Bollati Boringhieri.

Beato, F. (ed.). 1995. *La valutazione di impatto ambientale*. Milano: Franco Angeli.

Boorstin, D.J. 1961. *The Image. A guide to pseudo-events in America.* New York: Vintage Books.

Cohen, E. 1974. Who is a Tourist? A conceptual clarification. *The Sociological Review*, 22(4), 527–54.

Dall'Ara, G. 1995. *Perché le persone vanno in vacanza?*. Milano: Franco Angeli

Grzinic, J. 2010. Turismo residenziale e sostenibilità del turismo in Croazia. in *Il Turismo Residenziale. Nuovi stili di vita e di residenzialità, governance del territorio e sviluppo sostenibile del turismo in Europa*, edit by T. Romita. Milano: Franco Angeli.

Hall, C.M. and Müller, D.K. (eds.). 2004. *Tourism, Mobility and Second Homes: between Elite Landscape and Common Ground.* Clevedon: Channel View Publications.

Huete, R. 2009. *Turistas que llegan para quedarse. Una explicación sociológica sobre la movilidad residencial.* Alicante: Universidad de Alicante.

Huete, R. 2008. Tendencias del turismo residencial: el caso del Mediterráneo Español. *Universidad Autónoma del Estado de México*, 14.

ISNART. 2011. *Impresa Turismo 2011*. Roma: ISNART.

ISTAT. 2011. *Viaggi e vacanze degli italiani in Italia ed all'estero nell'anno 2010.* Roma: ISTAT.

ISTAT. 2004. *14° Censimento generale della popolazione e delle abitazioni.* Roma: ISTAT.

ISTAT. 1994. *13° Censimento generale della popolazione e delle abitazioni.* Roma: ISTAT.

Karayiannis, O., Iakovidou, O. and Tsartas, P. 2010. Il fenomeno dell'abitazione secondaria in Grecia e suoi rapporti con il turismo. in *Il Turismo Residenziale. Nuovi stili di vita e di residenzialità, governance del territorio e sviluppo sostenibile del turismo in Europa*, edited by T. Romita. Milano: Franco Angeli.

Mantecón, A. 2008. *La experiencia del turismo. Un estudio sociológico sobre el proceso turístico-residencial.* Barcelona: Icaria.

Mazón, T. and Aledo, A. (eds.). 2005. *Turismo residencial y cambio social. Nuevas perspectivas teóricas y empíricas.* Alicante: Aguaclara.

Mazón, T. and Aledo, A. 1997. *El turismo inmobiliario en la provincia de Alicante: análisis y propuestas.* Alicante: Diputación Provincial.

Mazón, T., Huete, R. and Mantecón A. (eds.). 2009. *Turismo, urbanización y estilos de vida. Las nuevas formas de movilidad residencial.* Barcelona: Icaria.

McCannel, D. 1976. *The tourist: a theory of the leisure class.* New York: Schocken.

Mercury. 2003. *Il turismo negli appartamenti per vacanza.* Firenze: Mercury.

Nocifora, E. 2008. *La società turistica,* Napoli: Scriptaweb.

Roca, M.N., Oliveira, J.A. and Roca, Z. 2010. Seconda casa e turismo della seconda casa in Portogallo. in *Il Turismo Residenziale. Nuovi stili di vita e di residenzialità, governance del territorio e sviluppo sostenibile del turismo in Europa,* edited by T. Romita. Milano: Franco Angeli.

Romita, T. (ed.). 2010. *Il Turismo Residenziale. Nuovi stili di vita e di residenzialità, governance del territorio e sviluppo sostenibile del turismo in Europa.* Milano: Franco Angeli.

Romita, T. 2009a. Il turismo c'è ma non si vede. *Rivista del Turismo del Touring Club Italiano,* 4, 4–11.

Romita, T. 2009b. Turisti per caso: ai margini o dentro il mercato?. In *XVI Rapporto sul Turismo Italiano,* edited by E. Becheri. Milano: Franco Angeli, 627–36.

Romita, T. 2007. Sustainable Tourism: the Environmental Impact of Undetected Tourism. *Tourismos: An International Multidisciplinary Journal of Tourism,* 2(1), 47–62.

Romita, T. 1999. *Il turismo che non appare.* Soveria Mannelli: Rubbettino Editore.

Romita, T. and Muoio, C. 2009. Turismo residencial: paisaje y consumo de lugares. in *Turismo, urbanizacion y estilos de vidas,* edited by T. Mazon, R. Huete and A. Mantecon. Barcelona: Icaria.

Romita, T. and Perri, A. 2006. La cura della risorsa ambientale come fattore di sviluppo del turismo. Il caso dei contesti turistici spontanei e del turismo fai-da-te. in *Atti del convegno: Turismo sostenibile. Trasformazioni recenti e prospettive future,* edited by E. Nocifora, O. Pieroni, T. Romita and C. Ruzza. Cosenza: Pronovis.

Romita, T. and Perri, A. 2009a. El turista fai-da-te. in *El turismo en el mediterraneo: posibilidades de desarrollo y cohesion,* edited by M. Latiesa Rodriguez. Madrid: Editorial Universitaria Ramon Areces.

Romita, T. and Perri, A. 2009b. Da emigranti a turisti. in *Atti del III Convegno Nazionale Turismo Sostenibile: ieri, oggi, domani,* edited by T. Romita, E. Ercole, E. Nocifora, M. Palumbo, O. Pieroni, C. Ruzza and A. Savelli. Cosenza: Pronovis.

Romita, T. and Perri, A. 2011. The D.I.Y. Tourist. *Tourismos: An International Multidisciplinary Journal of Tourism,* 6, 277–92.

Savelli, A. 1989. *Sociologia del Turismo.* Milano: Franco Angeli.

Urry, J. 1991. *The Tourist Gaze.* London: Sage.

Chapter 12

Policy Responses to the Evolution in Leisure Housing: From the Plain Cabin to the High Standard Second Home (The Norwegian Case)

Tor Arnesen and Birgitta Ericsson

Introduction

A policy response analysis related to the development of second homes[1] in Norway discussed in this chapter benefits from the option of following this phenomenon from its rise in the period between the two World Wars to the present day. The main concern here is to provide and consider a generalized overview of policy evolution, disregarding details.

By 'policy' is meant programmes and instruments/measures targeting a specific public issue through a course, or principle of action proposed and/or adopted by a government. Thus, a distinction is made between the concept of 'policy' and 'politics'. 'Politics' is understood as a process that enables the creation of a policy. Politics as a whole is situational and transitory, while policy as an outcome of politics lasts a significantly longer period of time. This makes politics the methods and tactics used to formulate and apply a policy (Cook and Scioli 1972).

The focus here is on the purpose-built leisure houses, excluding from the discussion family houses converted to leisure houses.

Second Homes and Public Policy in Norway: An Overview

The majority of purpose-built leisure houses in Norway are located in coastal and mountainous rural landscapes with natural amenities.[2] Keeping up with a general

---

1   Unless commented otherwise or following from the specific context, the concepts of 'second home', 'leisure house' and 'leisure cabin' are used synonymously.

2   By amenity is meant a feature that increases attractiveness or value (e.g. outdoor qualities as access to coast line, beaches, hiking areas, alpine slopes) of real estate or a geographical location. For a comprehensive discussion on rural amenities and economics, see Green, Deller and Marcouiller (2005).

growth in welfare throughout the post Second World War period, the standard of leisure houses has developed from plainly built cabins distributed in a dispersed settlement pattern towards increasingly higher quality second homes developed in agglomerations. Generally, these agglomerations are located separated from, but close to, 'the old' rural settlements.

Figure 12.1   Illustration of the spatial separation of and relation between second home agglomerations and rural settlements found in Norwegian mountain regions.

Note: The arrows underline the structure of the geographical layout found between the 'old' established rural settlements and the main road system found in the floor of the valleys (black arrows and markings of towns and rural settlements), and the newly established second home agglomerations in adjacent mountain areas (light grey arrows and markings of the agglomerations).

Source: Østlandsforskning.

Since 1970 the number of leisure houses in Norway (Statistics Norway 1972, 1976) has more than doubled by a linear trend to the figure of 410,289 purpose-built units in 2010, which is quite a lot in a country with 2.2 million private households.[3]

---

3   In addition to this, there are 32,238 units of non-purpose-built leisure homes registered in the cadastre. Of the purpose-built units, some are apartment buildings with more than one second home unit per building. These are not accounted for in Figure 12.1,

In this context, it seems pertinent to question the kind and effectiveness of policy responses, proactive or reactive, addressing such housing development, as well as what future directions these policies should take. The contention is that policy measures introduced over the years have succeeded fairly well in regulating land use and have recently included leisure properties in the general housing property tax regime. However, no policy measures have yet been formulated to address the evolution in (i) the interaction of second home agglomerations with 'old' rural centres, and (ii) the institution of 'home', since a growing number of households organize their life distributed by two houses, one domestic and one leisure home.

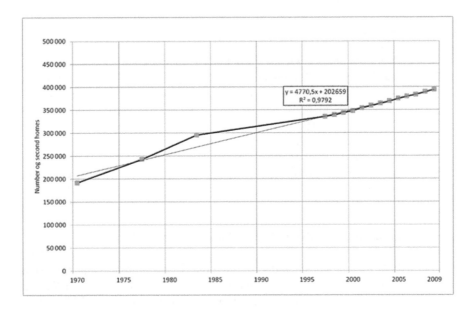

Figure 12.2 Growth of single-unit second homes in Norway since the 1970s.

Note: Numbers somewhat underreport the actual situation, but the assumption is that the underreporting is systematic thus mirroring the growth trend correctly.

Source: National cadastre database and Ericsson, 2005.

As shown in Figure 12.2, the linear trend growth rate[4] over the period 1970 – 2009 has been 4,770 new second homes per year.

---

and therefore the total number here is somewhat lower and only reflects single purpose-built second homes.

4   Linear trend line f(x) is calculated with the formula $f(x) = ax + b$, where a is the slope (growth rate) and b is the intercept.

Demarcation of Study Area

Of about 410,000 leisure houses in Norway today, approximately 50 per cent are located within a circle with a 200-km radius from the capital Oslo, henceforth called the Oslo Leisure Influence Region (OLIR). Over the years, developments in this region have been the main target for most second home policy initiatives. For the sake of the discussion in this chapter, illustrations and exemplifications refer mainly to this region. In fact, most of the major policy options related to second homes have been initiated as a response to developments in OLIR.

Agglomerations

If an agglomeration of leisure houses is defined as an area where neighbour leisure houses are no further than 200 meters apart, then approximately two thirds of the leisure houses in OLIR are nowadays located in agglomerations. There are some 30 large agglomerations with more than 500 units, 360 medium sized agglomerations with 100–499 units, and 1150 small agglomerations with 20–99 units. The number and size of agglomerations in this region increased significantly in the periods 1980–1989 and 1990–2004: of all new units, 65 and 76 per cent respectively were located in agglomerations (Overvåg and Arnesen 2007). This trend has continued to the present day.

Leisure House and Real Estate Policy Concerns before 1997

The year 1997 may be considered a turning point in leisure house and real estate policy initiatives in Norway, given that only since then have the cadastre data improved significantly (Ministry for the Environment 1999). Before 1997, policy initiatives were not based on cadastre data but on a geo-referenced inventory of ownership of leisure estates.

The policy areas that drew particular attention were:

i.   land use conflicts and challenges arising from the leisure house growth, including issues such as the privatization effects in congested coastlines, physical development patterns in the mountains, and disturbance and interference effects on the borders of nature protection areas in mountain regions;

ii. social policy issues and the quest for facilitating outdoor recreation activities 'for everybody', thus considering leisure houses a welfare good;

iii. policy instruments to deal with displacement effects of leisure housing in picturesque local communities with high demand in the real estate market.

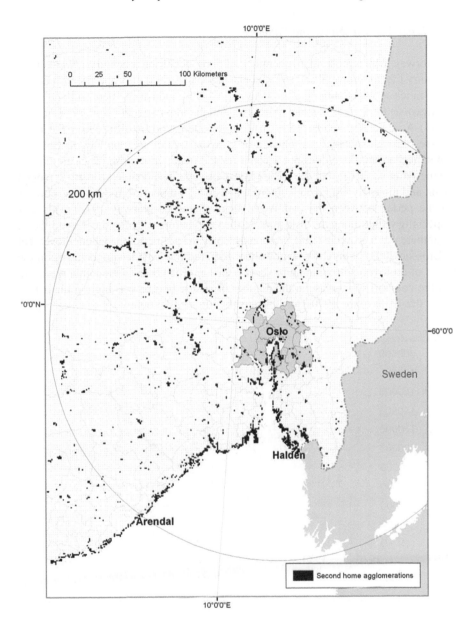

Figure 12.3   The Oslo Leisure Influence Region within a 200-km radius from Oslo.

Note: The grey area is the greater Oslo region. Second home areas, shown in black, are displayed with a minimum of 20 units per km².

Source: Norwegian Mapping Authority.

The Early Years: Figures and Patterns

In Norway, with almost no aristocratic traditions, acquiring a country *villa* to enjoy 'country life' for informal and casual social gatherings for family and friends became very much in vogue among a growing and wealthy industrial, merchant and other bourgeoisie throughout the late nineteenth and the mid twentieth century (Sørensen 2004). As argued by Veblen (1899), a lifestyle pattern tends to diffuse from the upper classes through society as a function of growing wealth creation and distribution. This mechanism alone does not account for the trends in the use of leisure houses (Arnesen et al. 2012), but the leisure house became fashionable property – though plain and modestly equipped – amongst a growing number of middle class citizens in the period between the two World Wars (Statistics Norway 1972). This was especially visible along the Oslo fiord. In the years before and after the Second World War there was a growing trend, promoted also by the politics of the labour movement (Tordsson 2003), whereby town dwellers acquired low cost leisure cottages on the basis of do-it-yourself building projects. This happened both in mountain areas and on the coasts. The increase in the number of leisure houses was high in the period from the 1930s to the 1970s, as illustrated by Figure 12.4 below.

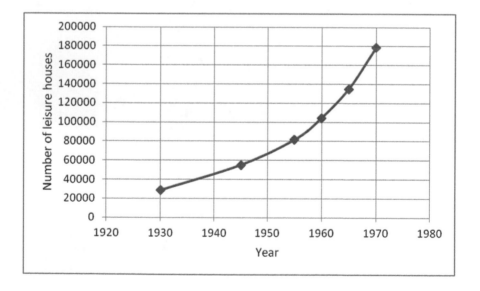

**Figure 12.4   The increase in number of leisure houses in Norway from the
              period between the World Wars to the 1970s.**

Source: Statistics Norway 1976: 19.

These numbers include single unit purpose-built leisure houses (79 per cent) and other domestic houses used for leisure (21 per cent). Although the available data do not allow for their quantification in separate figures, is it likely that the number of new purpose-built leisure units grew much more than the number of former domestic houses converted into leisure houses (Statistics Norway 1976: 20). In 1970, approximately 115,000 (60 per cent) of all leisure houses in Norway were located in OLIR, where also the highest annual rates of growth were registered (Statistics Norway 1976: 23). About 35 per cent of all leisure houses in OLIR[5] were located on the coast, 30 per cent in the mountains and 35 per cent in the forest belt between the coast and mountain areas. The share of purpose-built leisure houses increased from 30 per cent in 1935 to 92 per cent in 1965 (Statistics Norway 1976: 21). The surface area of 55 per cent of these houses was less than 50 m², but since 1970 a trend towards larger units was registered. Almost none were linked to a sewage system, and only some (30 per cent on the coast and 6 per cent in the mountains) were connected to the electrical power supply (Statistics Norway 1976: 42). Road access to these houses was uncommon.

Land Use Conflicts, Challenges and Policy Response: Coastal Issues

The leisure houses on the coast were extensively and almost exclusively used in the summer season,[6] i.e. 86 per cent had only summer use (Statistics Norway 1976: 55). They were generally located in the archipelago near a town, often reachable by boat, bicycle, public transport, etc., typically about one hour from home. A common family organization model was that mother and children would stay in the summer cabin during school holidays, while the father commuted to work. On average, leisure houses on the coast were used 65 days a year (Statistics Norway 1976: 55). This kind of household user model was so well established in the Oslo fiord archipelago that many small commercial ferryboats and buses frequently operating in the archipelago in the summer were nicknamed 'daddy-boats' and 'daddy-buses'.[7]

---

5   Throughout the late 1960s and the 1970s Statistics Norway operated with a geographical subdivision of '*handelsfelt*', or trade regions (Statistics Norway 1967: 3) that correspond quite well to OLIR as defined here, thus enabling early and recent trends in leisure house development to be compared.

6   In fact, in the period after the Second World War it was not allowed in Norway to allocate construction materials to the building of leisure houses, because it was all needed to 'rebuild the country'. This limitation contributed to the low-cost, low-standard (shack-ish) profile of these houses, usually built of 'left-over materials'.

7   One important policy progress that triggered this development were the Holiday Laws, introduced in 1920, by which all employees were granted 10 days of fully paid holiday every year. In 1937, this was expanded to 3 weeks. This promoted the attitude that holiday should be spent recreating away from the environment and worries of daily life. Access to outdoor life, fresh air and sun was encouraged by the health authorities and

Initial Restrictive Policy Responses at the coast

Given that coastlines are a limited asset and that the Oslo fiord region was and continues to be, by Norwegian standards, very populous, in the late 1930s concerns rose that private cabin owners were obstructing generalized access to beaches and other attractive parts of the coastline. 'No Admittance' signs became a political issue. This was an early case of what Halseth (2004) observed as the production of 'elite landscapes' in Canadian cabin tradition. As a policy reaction, an act was passed in the Parliament in 1937 aimed at securing general public access to the beaches along the coasts of Norway, not just the Oslo fiord archipelago. As another major policy initiative the government initiated an outdoor recreation act that was, however, delayed by the Second World War and adopted only in 1957 (Tordsson 2003). To this day, it is the most important act defining the highly cherished principle in Norwegian politics – open access and freedom to roam.

Initially the market for these summer coastal cabins was mostly local and for locals at the county level, but it gradually transgressed across counties. In 1970, about one half of all leisure houses in OLIR was owned by households from another county (Statistics Norway 1976: 23).The pressure on the coastal zone continued to increase in the post-Second World War period: between 1960 and 1970, 60 per cent of all new leisure houses built in Norway were located in OLIR (Statistics Norway 1976: 24). The beach stretches act of 1937 did aim at securing open access to the coast, but obviously became insufficient for halting the growth of cabins close to the sea, and conflicts between cabin owners and others continued to escalate.[8] Conflicts were also fuelled by a growing awareness of the value, in monetary terms, these barren skerries and stretches of coastline represented. As far as farmers and fishermen were concerned, this was barren land. As use of skerries and archipelagos became in vogue for recreational purpose, such marginal lands became extremely attractive. This process was repeated many years later in mountain areas with abundant recreational amenities (Overvaag 2009).

In 1970, the Ministry for the Environment was established, and took the lead in second home land use issues. With a few exceptions commented below, other ministries were conspicuous by their absence. Second home development was still very much seen as a policy area for land use regulation. Beyond that, they were seen as an *ad hoc* phenomenon. Second homes were not considered an interesting part of the evolution of civil society, with potential economic, social and functional

---

adopted as an ideal, and its materialization was the summer life in a shak-ish cabin in the archipelago. It was also important that the labour movement was very active in promoting physical well-being and outdoor life, thus securing the working class's access to coastal cabins. Many such cabins were built by labour unions or through worker welfare schemes in industrial companies.

8   It is worth noting that until the full implementation of this act in 1965, location and building of leisure houses in non-regulated areas was entirely a matter that concerned only the landowner and the builder.

influences in many communities – many of which rural and already feeling the heat of urbanization, accompanied by centralization and depopulation processes.

In 1972, the Ministry for the Environment developed an important policy instrument to deal with land use for recreational purposes – the so-called 'Skjærgårdsparken', or archipelago recreation park in OLIR. Through stepwise expansion from 1972 to 2009, the park area in OLIR covers approximately 52 km² of the islands and skerries of the archipelago area, stretching from Lindesnes to Bamble (see Figure 12.2). The park was established under a servitude agreement with a large number of landowners and the state along the coast. The main effect of the park delimitation is a strictly enforced building ban.

The growth pattern of coastal cabins and the parallel social development described above were not unique to Norway and the Oslo fiord archipelago. For example, Löfgren (2002) described a similar process along the Swedish southern coast with townspeople 'discovering' amenities and fostering the development of a coastal cabin culture and of an increasingly confident market.

Another step-up in the policy reaction to the coastal cabin development was taken in 1965, when the parliament passed a provisional coastal planning act that generally prohibited any building within 100 meters of the shoreline. This became a permanent law in 1971, and from 1985 the prohibition within the 100-meter belt from the coastline was included in the national planning and building act. Even though the growth of new coastal cabins in the Oslo fiord archipelago came almost to a halt during the 1970s, the general building ban in the 100-meter belt was to some extent sidestepped by local authorities, who had the competence to grant exemption to upgrade existing buildings, or allow former tool sheds, wharfside sheds and the like to be converted to cabins (Stokke et al. 2006). As a national policy reaction to this, in 1993 the parliament passed 'National Policy Guidelines for Planning in Coastal and Marine Areas in the Oslofjord Region' (Hvidsten 1996, Ministry for the Environment 1993).

While the number of coastal cabins grew on other stretches along the very long Norwegian coast, land use conflicts were still pretty much an Oslo fiord issue, gradually extending along the fiord all the way to the Swedish border in the east, and to the town of Arendal along the west wing of the fiord (Figure 12.3). Developments in this region drove policy making processes on coastal second home issues well into the 1990s:

> The Oslo fiord region is the part of the country with the greatest concentration of economic activity and settlement, while the fiord and the coastal zone comprise the most intensively used area of open-air recreation in the country and contain important natural and cultural values worthy of preserving. (...) It is also a primary political objective to increase the opportunities for outdoor activities in areas used for daily excursions and on holidays. Special priority is given to unbuilt areas of the shore that are suitable for bathing and open-air recreation. In the Oslo fiord in particular, these areas are such a scarce resource that, as a rule, re-allocation of such areas for other competing purposes should not take place. (Ministry for the Environment 1993)

One more issue should be mentioned regarding the pre-1997 policy reactions to land use issues in coastal areas. While the 100-meter coastal belt building ban made sense in OLIR, the Oslo fiord region, the same rationale was not that obvious along the long coastline stretching beyond 2,500 km into sparsely populated northern Norway. This issue has quite recently become part of the political debate (Stortinget 2010) and an amendment was adopted in the regulations of the planning and building act to allow for a more flexible and liberal practice in other stretches of the coast (Ministry for the Environment 2011).

Displacement Issues on the Coast

In relation to purpose built leisure cabins there have been little or no displacement conflicts, since the majority of these have been built in areas separated from permanent settlements (Figure 12.1). Rather than posing a displacement threat, they actually add to the economic activity and social life in the adjacent local communities during a few hectic summer months. That is not to say that displacement processes have not taken place at all. For example, along the coastline of OLIR, displacement of permanent residents has been very much present and still represents an important policy issue. This relates to the use of former houses for permanent settlement as summer residences in a number of coastal communities, typically smaller picturesque villages and outport communities connected to the mainland or on islands.[9] A main policy instrument, the concession act, to deal with displacement issues, was passed in parliament in 1974 and later amended in 1993.[10] The concession act regulates conditions for acquiring a property: local authorities have the competence to demand concession for all acquisitions of properties if the intention is to transform a domestic house into a leisure house, or transform a leisure house into a domestic house for permanent residence. Seventy-six municipalities, mostly in the Oslo fiord archipelago, have chosen to implement this act. According to the planning and building act,[11] local authorities are also given the competence to regulate location and extension of building areas, including permissible housing purposes. Local authorities may also provide regulations to limit the number of days per year a leisure house may be inhabited (typically 3

9   In attractive destinations, conversion of houses to second home use often creates inflated property values. It is often argued that this is causing a displacement of permanent residents, and undermining the foundation of municipality welfare and other services. However, there are also arguments that the current depopulation trend in attractive second home destinations is caused by a restructuring of the rural labour market (Marjavaara 2007a, 2007b).

10   Act of 28 October 1993 relating to concession in the acquisition of real estate property (The Concession Act).

11   Act No. 71 of 27 June 2008 relating to Planning and Processing of Building Applications (The Planning and Building Act) (the Planning section).

months a year). These acts are intended to be the prime policy instruments to prevent displacement effects from leisure housing.

Second Home Development in Mountain Areas before 1997

At the end of 1970s, 26 per cent of second homes in OLIR were located on the coast, at a maximum distance of 1 kilometre from the shoreline. The growth experienced on the coast during the 1970s was gradually reduced as the various policy measures described above became effective. For the next 20 years, there was a continued decline in the number of new units on the coast. Instead, the growth in the number of second homes moved to mountain areas.

Table 12.1   Growth in the number of second homes (SH) in OLIR in the periods 1980–2004 and 2000–2004 on the coast (land within 1 km from the shoreline), in the mountains (areas above 599 metres above sea level), and in the forests (areas between the coasts and mountains).

Increase in OLIR	Coast	Mountains	Forests	Total
Number of SH 1980 – 2004	4,740	18,606	5,024	28,370
Share of SH 1980 – 2004	16.7 %	65.6 %	17.7 %	100 %
Share of SH 2000 – 2004	10.4 %	73.5 %	16.0 %	100 %

As shown in Table 12.1, growth of second homes moved into the mountains sometime towards the end of the 1970s and from the 1980s onwards, and it developed into a new trend, characterized by:

- High technical standard of houses, with an increasing share of units linked to the electric power supply and sewage structures;
- Locations in agglomerations, a prerequisite for implementing high technical standard of the units;
- Good road access;
- Extensive use on a weekend basis, both in summer and winter;
- Often developed as an integral part of destination developments with alpine skiing slopes, among others.

In sum, a highly visible trend developed in mountain areas where second homes became almost equal to permanent homes in terms of technical standard (Arnesen et al. 2012, Overvaag 2009).

Two categories of policy measures aimed to regulate developments of second homes in mountain areas before 1997 should be pointed to. First, local and central authorities shifted positions in promoting policies and/or preparatory measures regarding planning and restrictive policies, though such measures and initiatives mostly fell within already established planning and policy instruments (Skjeggedal et al. 2009). Second, state authorities run most, if not all, of second home policies through the Ministry for the Environment, primarily with the intention of preventing conflict with national goals and measures in nature protection initiatives. This meant putting a ban on second home development in wild reindeer protection zones and other landscape protection areas, as well as in border areas ("buffer zones") of national parks. These policies have been effective, though not without skirmishes and conflicts with local interests. This has not been an obstacle to developing second home agglomerations in mountains, and certainly did not have a blocking effect, as was the case of measures adopted in coastal areas in OLIR. Nature conservation legislation does not have the explicit goal of curtailing second home developments as such. The possible regulative effect on the development pattern is rather a side effect than a prime goal.

It is worth noticing that as early as 1962 the Ministry for Government and Regional Development created the so-called 'mountain plan team' to study adequate development models for mountain areas (Sømme 1965). Placed high on the team's agenda, the team anticipated second home developments that much later would materialize as agglomerations of second homes. The team also addressed a broad range of issues related to second home development in mountains, including local economic development. However, their work did not result in any major policy proposals at the time. The initiative was again handed over to the Ministry for the Environment and its rather myopic focus on land use issues. In 1979, growth was more pronounced and visible in mountain areas and the Ministry for the Environment appointed a committee to 'consider the possibilities for further simplification that would contribute to a more swift and flexible processing of second home building cases' (NoU 1981). The committee was an expression of national policy ambitions to promote second home developments as a welfare benefit for all (Ericsson 2011). The Governement's policy program for 1978–81 stated the goal of making it less complicated to obtain permission to build second homes in order to 'give continually more people the opportunity to have access to their own second home' (Ministry of Finance 1977: 55).

Beyond issuing a state directive emphasising that local authorities should not be too restrictive in allowing second home developments (Ministry for the Environment 1982), and including second home development areas as a new physical development category in the revised planning and building act of 1985, these initiatives did not result in any major policy changes. Nevertheless, it is worth noticing that even the Ministry for the Environment supported second home development in mountain areas at that time, and only in the 1990s did become more restrictive and reserved towards this phenomenon (Skjeggedal et al. 2009).

As for local and regional authorities in the mountain areas of OLIR, attitudes towards second home development varied and shifted over time from the 1970s onwards. Some counties and municipalities in the mountains took a proactive stance, and policy initiatives were taken to promote second home developments. This was often seen as integrating second home development as an element of local tourism development policies related to the establishment of alpine skiing destinations or general winter tourism development. Others took a more sceptical approach, which was probably a reason for issuing the above mentioned 1982 state directive.

Development of New Second Home Policy Dimensions after 1997: Coasts and Mountains

In the period 1995–1997 all-out efforts were made to improve cadastre data, including leisure-related real estates. Until then, even basic data on leisure houses were not systematically collected and updated, and were thus unreliable. This is why prior to 1997 there was almost no national leisure house policy based on information about such units regarding their standard and monetary value as well as fiscal potential. In taxation, leisure houses were simply subsumed under the general personal taxation of capital. The valuation of the leisure houses did not reflect their market value. Though a socio-economically effective taxation regime should reflect market values (Skatteutvalget 2003), leisure houses and estates did not comply with that norm and were grossly undervalued.[12]

While until the end of 1990s leisure houses were not an integral element of the fiscal and taxation regimes in Norway, but rather an appendix to them, in retrospect this had to change for two reasons: (i) a comprehensive shift towards agglomerations of second homes and away from dispersed second home settlements; (ii) swift growth in tangible fixed assets represented by newly built second homes in parallel with the establishment of second homes as an autonomous credit objects. These points are discussed below.

Since the mid-1990s there has been an ongoing, albeit hesitating and partly embryonic, process to integrate second homes into core policy areas such as fiscal and taxation, national transportation, social welfare and rural development. Some of these policy areas are more advanced than others, but there has been a continuing debate about addressing second home and leisure estates on a much broader scale of policies than before.

Agglomerations

Increasing development of second homes in agglomerations, with units with high technical standard and connected to public infrastructure (electricity, sewage,

---

12 Some economists argue that this situation is still applicable to most Norwegian real estate properties (Van den Noord 2000).

roads) has been a pan-national trend, though with some specificities, namely: the trend diffused from east to west and from south to north of the country (Overvåg and Arnesen 2007); an uneven development pattern was recorded regarding a small number of major winter tourism municipalities that attract a disproportionately great share of second home growth (Ericsson 2011). Figures 12.5 and 12.6 illustrate such agglomerations in a mountain area and on the coast of Norway.

Figure 12.5   A typical second home agglomeration in mountain landscapes, winter season, in Norway.

Figure 12.6   A recently developed second home agglomeration on the coast in Norway.

These specificities are important in terms of local regional development in rural areas. But in terms of policy instruments, they unfold within the established national second home policy regime discussed above. Though cadastre data are still unreliable regarding the technical standard of the units, it is reasonable to assume that the standard of all units built since 1997 is high. If this assumption somewhat overestimates the development since 1997, it is compensated for by uncounted existing buildings in agglomerations that were renovated and upgraded, connected to the electric power grid and to sewage infrastructure. At the current growth rate, a viable assumption seems to be that (i) today, approximately one quarter of the entire second home stock is of high technical standard, and (ii) all newly built second homes belong to that category.

Assets and Acquisition

Investments in second homes have substantially increased and should nowadays be compared to investments in a residential or domestic home. This tendency is well illustrated by comparing free marked transfers of second homes with domestic houses (see Figure 12.7 below)

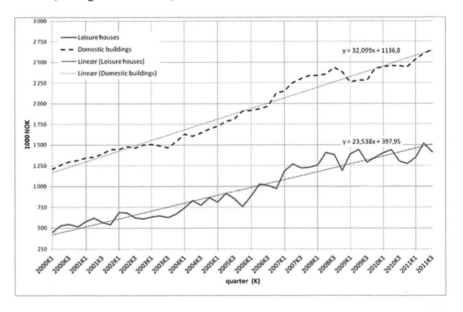

Figure 12.7   The evolution of free market sales of domestic houses and leisure houses since 2000.

Source: Statistics Norway (2011).

On average, the prices of second homes have practically developed almost on par with free market prices of domestic houses, at least since 2000. The increase in price in the second homes market has been on average approximately 23,500 NOK per sold unit per quarter. This is on average circa 75 per cent of the increase in price of the domestic house market.

This is consistent with another important element in second home development: they are now autonomous credit objects. Commercial banks today offer financing of second homes on terms similar to those of domestic houses, with mortgage in the second home itself. This financial development has consolidated the second home market, and opened options for more household actors to move into the market without having to submit a guarantee in other assets they might possess, or the need for a guarantor. This again represents a potential for younger households to enter the second home market at an earlier stage in the household formation path. They are not necessarily dependant on testamentary inheritance, which generally comes later in life, or on waiting for the time when their own capital accumulation potential is sufficiently high to afford down payments on domestic house loans, for instance. The owner's age at acquisition of a second home is not consistently recorded in the cadastre, but market analysts cited in the press reported an increasing share of younger households as actors in the second house market (Mikalsen 2008, Statistics Norway 2011).

Approximately one fifth of all Norwegian households own a second home today, just as it was in in 1970. This means that the number of second homes, like the number of households, has just about doubled since 1970. According to Statistics Norway (2011), 38 per cent of the households consist of only one person and they are probably less frequent in second home ownership than the 62 per cent of households consisting of couples. It can thus be assumed that the share of family households owning a second home has increased over the years.

Conflicts and the Debate on Sustainability

From a physical development policy perspective, the trend described above is well served by established policy instruments and regulations. However, it has not been without conflicts. The modern second home development represents a considerable resource allocation into the function of leisure. From many quarters, both in politics (Bjerke 2001, Lier Hansen 2001) and research (Aall et al. 2005, Gansmo, Berker and Jørgensen 2011, Williams and Kaltenborn 1999), issues of sustainability in modern second home developments have been raised. Harsh criticism was formulated of this trend as diverting from the so called 'traditional Norwegian values' marked by moderate consumption and a simple cabin life with fewer means, in favour of high comfort, luxury consumption and questionable sustainability in energy use, among other aspects. Though this discussion may be important, so far it has not resulted in any new policy instruments or regulations regarding consumption. The exception are building regulations also for second homes that have been tightened over the years as thermal insulation is concerned (Norwegian Building Authority 2011).

This Norwegian discussion of a development away from 'the traditional cottage' towards high standard, homely comfortable leisure houses should also be assessed in a dynamic perspective. The romantic idolatry of simple conditions should not only be seen as it manifests in the chosen level of quality in leisure houses, but also in relation to the standard level of domestic residences in similar time periods, as well as economic restraints. Thus much of today's disputed luxury fittings are not to be perceived as raised standard, but rather as maintenance of quality level according to modern requirements, and as an expression of increased personal wealth.

There is also another aspect of sustainability to be mentioned here: it has been well documented that high standard leisure houses are significantly more used than less equipped ones (Ericsson and Grefsrud 2005, Flognfeldt 2004, 2007, Velvin 2000). The sustainability perspective makes sense only if it relates to how resources are used. E.g. given two leisure houses, it is less (rather than more) sustainable to use one less than the other, which again implies the above mentioned unit upgrades. A side benefit of this, as regards sustainability, is seen in an extensive operation to connect leisure house agglomerations to the sewage grid. The ongoing tightening of the building regulation policy for leisure houses seems to incorporate current physical development trends with a demand for high standard leisure housing. As discussed by Aall et al. (2011), authorities have not been proactive in this respect and have been slow to promote a more sustainable practice in terms of this policy.

Property Tax Introduced

Work on amending the property tax law had already started in 1995 (Ministry of Finance 1996). An issue in this process with regard to the topic under discussion was the expressed intention that 'leisure properties should be assessed in the same way as domestic houses' (Ministry of Finance 1999: chap. 9.4).[13] The property tax law was finally amended in 2006 to give local authorities the power, as an option, to issue property tax on leisure houses on par with domestic houses (Ministry of Finance 2008) by removing the former condition for local property tax stating it could only be applied to properties in urbanized areas of a municipality. In 2011, 145 of the 429 municipalities in Norway had taken advantage of this option (Statistics Norway 2012). Beyond the obvious fiscal importance of this policy change, this should be interpreted as a major change in local authority's approach to second homes. From a theoretical perspective, the fiscal inclusion of second homes can be seen as a first step to redefining leisure houses from an appendix to being a equal part of the housing stock in the community.

---

13   For comparison, this brought Norway to the level of Sweden. For a number of years second homes and permanent homes have been assessed in the same way by the Swedish national tax board (Marjavaara and Müller 2007).

Future Policy Issues

In sum, from a land use policy perspective and from a property tax policy perspective, leisure house policy seems to be fairly well developed in Norway. The question now is: what challenges lie ahead calling for new policy initiatives in the years to come? In lieu of conclusion of the above discussion, this issue is briefly considered by pointing to one societal development which leisure house policy should start addressing.

As illustrated in Figures 12.2 and 12.3, agglomerations of second homes with a rapidly increasing share of high standard units prevail in OLIR. The current trend – especially relating to these high standard second homes – is one of frequent stays over weekends and prolonged stays in the second home (Arnesen et al. 2012). If this trend continues to develop and becomes widespread, it seems reasonable to interpret this as an evolution in the institution of 'home'. One could even argue, as we have done in a recent article (Arnesen et al. 2012), that what is called 'second home' should more properly be called 'second house in a home' – or a 'multi-house home'. In a multi-house home, integrated houses fulfill different or overlapping household functions in a coherent functional unit called home, including the leisure house that primarily fulfils leisure functions. To the extent that this may be a suitable conceptualization of current trends in the organization of home for a substantial part of households, it implies a new relation between the urban core and its rural hinterland. It implies that the idea of a household with only one address to designate its home in the registry office increasingly deviates from the empirical reality. A household can be at home both in the core and in the periphery, so to speak. Moreover, since data from the registry office have fiscal implications (households pay income tax to the municipality in which they have their address), this also has an effect on, for example, local authority finances. These are meant to mirror local population, and in fact they do not if there are homes not registered. The situation is especially disadvantageous for affected rural municipalities. This question has already been introduced as a political issue promoted by mayors of municipalities with a high number of second homes. It can be expected that in the years to come new policy response discussions will address this and other topics related to local authority finances as the multi-house phenomenon develops further.

References

Aall, C., Høyer, K., Hall, C. and Higham, J. 2005. Tourism and climate change adaptation: the Norwegian case. In *Tourism, recreation and climate change*, edited by C.M. Hall and J. Higham, Aspects of Tourism 22, Clevedon: Channel View Publications, 209–21.

Aall, C., Klepp, I.G., Engeset, A.B., Skuland, S.E. and Støa, E. 2011. Leisure and sustainable development in Norway: part of the solution and the problem. *Leisure Studies*, *30*(4), 453–76. doi:10.1080/02614367.2011.589863.

Aanesland, N. and Holm, O. 2002. Boplikt–drøm og virkelighet. Oslo: Kommuneforlaget.

Arnesen, T., Overvåg, K., Skjeggedal, T. and Ericsson, B. 2012. Transcending Orthodoxy: Leisure and the Transformation of Core – Periphery Relations, in *Regional Development in Northern Europe. Peripherality, Marginality and Border Issues in Northern Europe*, Regions and Cities, edited by M. Danson and P. de Souza, Oxon: Routledge, 182–95.

Bjerke, S. 2001. Energieffektivt hytteliv. Presented at the Hyttekonferansen, Hellerudsletta: Ministry of the Environment. [Online] Available at http://www.regjeringen.no/nn/dokumentarkiv/Regjeringen-Stoltenberg-I/md/Taler-og-artikler-arkivert-individuelt/2001/energieffektivt_hytteliv.html?id=264809 [accessed: February 2012].

Cook, T. and Scioli, F. 1972. Policy analysis in political science: Trends and issues in empirical research. *Policy Studies Journal*, 1(1), 6–11. doi:10.1111/j.1541-0072.1972.tb00051.x.

Ericsson, B. 2005. *Fra hyttefolk til sekundærbosatte: et forprosjekt* (Vol. 04/2005). Lillehammer: Østlandsforskning.

Ericsson, B. 2011. *Second Homes i Norge. Bidrag til en nordisk utredning* (ØF-rapport No. 1/2011). Lillehammer: Østlandsforskning. [Online] Available at http://www.distriktssenteret.no/filearchive/second-homes.pdf [accessed: January 2012]

Ericsson, B., and Grefsrud, R. 2005. *Fritidshus i innlandet: Bruk og lokaløkonomiske effekter.*(Vol. 06/2005). Lillehammer: Østlandsforskning.

Flognfeldt, T. 2004. 'Second Homes as a Part of a New Rural Lifestyle in Norway'. In *Tourism, mobility and second homes. Between Elite Landscapes and Common Ground*, edited by C.M. Hall and D.K. Müller, Clevedon/Buffalo/Toronto: Channel View Publications, 233–43.

Flognfeldt, T. 2007. Developing Tourism Products in the Primary Attraction Shadow. *Tourism Culture & Communication*, 7(13), 133–45.

Gansmo, H.J., Berker, T. and Jørgensen, F.A. (eds.). 2011. *Norske hytter i endring. Om bærekraft og behag.* Trondheim: Tapir Akademisk Forlag.

Green, G.P., Deller, S.C. and Marcouiller, K. (eds.). 2005. *Amenities and Rural Development. Theory, Methods and Public Policy.* New horizons in environmental economics. Cheltenham, UK: Edward Elgar Publ. Ltd. [Online] Available at http://www.bpatc.org.bd/elibrary/files/12713234901845421264.pdf [accessed: January 2012].

Halseth, G. 2004. The'cottage'privilege: increasingly elite landscapes of second homes in Canada. In *Tourism, mobility, and second homes: between elite landscape and common ground*, edited by C. Hall & D. Müller. Buffalo: Channel View Books, 35–54.

Heggem, R., Flø, B.E., and Rye, J.F. 2003. Drømmen om bopliktens endelikt. *Nationen*. Oslo. [Online] Available at http://www.nationen.no/meninger/ Kronikk/article710766.ece [accessed: January 2012].

Hvidsten, H.Ø. 1996. *Kampen om strendene. En studie av rikspolitiske retningslinjer for planlegging i Oslofjorden som virkemiddel i miljøvernpolitikken* (Master thesis). Universitetet i Oslo | University of Oslo, Oslo.

Jaffe, A.J. 1989. Concepts of property, theories of housing, and the choice of housing policy. *The Netherlands Journal of Housing and Environmental Research*, 4(4), 311–20. doi:10.1007/BF02503099

Lier Hansen, S. 2001. Bærekraftig hyttebygging. Presented at the Fagseminar om framtidsrettet hyttebygging. Geilo 28. mars 2001, Oslo: Ministry of the Environment. [Online] Avaliable at http://www.regjeringen.no/en/ dokumentarkiv/Regjeringen-Stoltenberg-I/md/Taler-og-artikler-arkivert-individuelt/2001/baerekraftig_hyttebygging_av_statssekret.html?id=265191 [accessed: December 2011].

Löfgren, O. 2002. *On holiday: A history of vacationing* (Vol. 6). Los Angeles: University of California Press.

Marjavaara, R. 2007a. The displacement myth: Second home tourism in the Stockholm Archipelago. *Tourism Geographies*, 9(3), 296–317.

Marjavaara, R. 2007b. Route to destruction? Second home tourism in small island communities. *Island Studies Journal*, 2(1), 27–46.

Marjavaara, R. and Müller, D.K. 2007. The Development of Second Homes' Assessed Property Values in Sweden 1991–2001. *Scandinavian Journal of Hospitality and Tourism*, 7(3), 202–22.

Mikalsen, B.-E. 2008. Slik blir hyttemarkedet. Boligprisene skal fortsatt opp, dermed kommer hytteprisene fortsatt til å stige, mener Bjørn Erik Øye. *DN.no*. Available at http://www.dn.no/privatokonomi/article1323784.ece [accessed: 6 March 2008].

Ministry of Finance. 1977. *Langtidsprogrammet 1978-1981*. [Stortingsmelding No. St.meld. nr. 75 (1976-77)]. Oslo: Ministry of Finance.

Ministry of Finance. 1996. *Boligtaksering og prinsipper for boligbeskatning*. [Odelstingsproposisjon No. St meld nr 45 (1995–96)] (p. 78). Oslo. [Online] Available at http://www.regjeringen.no/nb/dep/fin/dok/regpubl/ stmeld/19951996/st-meld-nr-45_1995-96.html?id=133629 [accessed: April 2008]

Ministry of Finance. 1999. *Ny lov om eiendomsskatt*. (Norges Offentlige Utredninger No. NOU 1996: 20). Oslo: Ministry of Finance. [Online] Available at http://www.regjeringen.no/nb/dep/fin/dok/nouer/1996/nou-1996-20/1.html?id=116121 [accessed November 2011].

Ministry of Finance. 2008. *Om endringer i skatte- og avgiftslovgivningen mv. Tilråding fra Finansdepartementet av 1. februar 2008, godkjent i statsråd samme dag. (Regjeringen Stoltenberg II)* [Odelstingsproposisjon No. Ot.prp. nr. 31 (2007–2008)] (p. 78). Oslo. [Online] Available at http://www.regjeringen.

no/nb/dep/fin/dok/regpubl/otprp/2007-2008/otprp-nr-31-2007-2008-.html?id=498751 [accessed: December 2011].

Ministry of the Environment. 1982. Rundskriv T-13/82. Kommunenes behandling av hyttesaker. Oslo: Ministry of the Environment.

Ministry of the Environment. 1993. T-1048 National Policy Guidelines for planning in coastal and marine areas in the Oslofjord region. Supplementary comments on the National Policy Guidelines for planning in coastal and marine areas in the Oslofjord region. Oslo: Ministry of the Environment. [Online] Available at http://www.regjeringen.no/en/dep/md/documents-and-publications/Circulars/1993/National-Policy-Guidelines-for-planning-.html?id=107847 [accessed: January 2012].

Ministry of the Environment. 1999. *Lov om eiendoms-registrering. Om et forbedret eiendomsregister og forslag til ny lov om eiendomsregistrering til erstatning for delingsloven. Utredning fra et lovutvalg oppnevnt 15. januar 1996 Avgitt til Miljøverndepartementet 25. januar 1999* (Norsk Offentlig Utredning – NoU No. NOU 1999: 1). Oslo: Ministry of the Environment. [Online] Available at http://www.regjeringen.no/nb/dep/md/dok/nou-er/1999/nou-1999-01.html?id=375578 [accessed: January 2012].

Ministry of the Environment. 2011. FOR-2011-03-25-335. Statlige planretningslinjer for differensiert forvaltning av strandsonen langs sjøen. [Online] Available at http://www.lovdata.no/for/sf/md/md-20110325-0335.html [accessed: January 2012].

Mønness, E. N. and Arnesen, T. 2012. A Domicile Principle in Farmland Policy? On Farm Settlement Policy and Experience in Norway. In *Regional Development in Northern Europe. Peripherality, Marginality and Border Issues in Northern Europe*, edited by M. Danson and P. de Souza. Regions and Cities. Oxon: Routledge, 212–31.

Norwegian Building Authority. 2011. Hvilke energikrav gjelder for fritidsboliger og for tilbygg på fritidsboliger? [Online] Available at http://www.dibk.no/no/Tema/Energi/Ofte-stilte-sporsmal/Hvilke-energikrav-gjelder-for-fritidsboliger-og-for-tilbygg-pa-fritidsboliger-/ [accessed: 15 March 2012].

NoU. 1981. *Hytter og fritidshus* (Norsk Offentlig Utredning – NoU No. 1981:21). Oslo: Universitetsforlaget.

Overvaag, K. 2009. *Second Homes in Eastern Norway. From Marginal Land to Commodity* (Doctoral Thesis). NTNU, Trondheim.

Overvåg, K. and Arnesen, T. 2007. *Fritidsboliger og fritidseiendommer i omland til Oslo, Trondheim og Tromsø.* (Report No. 4/2007). ØF-notat. Lillehammer: Østlandsforskning.

Skatteutvalget. 2003. *Skatteutvalget. Forslag til endringer i skattesystemet. Utredning fra et utvalg oppnevnt ved kongelig resolusjon 11. januar 2002. Avgitt til Finansdepartementet 6. februar 2003.* (Norges Offentlige Utredninger No. NoU 2003: 9). Oslo: Finansdepartementet. [Online] Available at http://www.regjeringen.no/nb/dep/fin/dok/nouer/2003/nou-2003-9.html?id=381734 [accessed: November 2011].

Skjeggedal, T., Overvag, K., Arnesen, T. and Ericsson, B. 2009. Hytteliv i endring. *Plan, 6,* 42–49.

Statistics Norway. 1967. *Statistisk måndeshefte. Monthly Bulletin of Statistics.* (Statistisk måndeshefte No. Nr. 2, 1967) (p. 94). Oslo: Statistics Norway. [Online] Available at http://www.ssb.no/histstat/sm/sm_196702.pdf [accessed: December 2008].

Statistics Norway. 1972. *Fritidshusundersøkelsen 1970 | Holiday House Survey 1970* (p. 185). Oslo: Statistics Norway. [Online] Available at http://www.ssb. no/histstat/nos/nos_a509.pdf [accessed: December 2008].

Statistics Norway. 1976. *Fritidshus 1970* (Report) (p. 95). Oslo: Statistics Norway. [Online] Available at http://www.ssb.no/histstat/sagml/sagml_20.pdf [accessed: December 2011].

Statistics Norway. 2011. Population and Housing Census 2001 Steady increase of small households. *Statistics Norway.* [Online] Available at http://www.ssb.no/ english/subjects/02/01/fobhushold_en/ [accessed: 6 March 2012].

Statistics Norway. 2012. Transfer of properties. Preliminary figures, 4th quarter 2011 Record high property transfers. *Transfer of properties.* [Online] Available at http://www.ssb.no/english/subjects/10/14/10/eiendomsoms_en/ [accessed: 2 March 2012].

Stokke, K.B., Anker, M., Omland, A., Skogheim, R., Skår, M. and Vindenes, E. 2006. *Planlegging og forvaltning av urbane friluftsområder.* (Notatserie No. 2006/133). Oslo og Lillehammer: NIBR, NIKU, NINA, 70.

Stortinget. 2010. Innst. 284 S (2009–2010): Innstilling fra energi- og miljøkomiteen om representantforslag fra stortingsrepresentantene Ketil Solvik-Olsen, Henning Skumsvoll, Oskar Jarle Grimstad og Torkil Åmland om å styrke lokal forvaltning av arealer i strandsonen. Stortinget. [Online] Available at http:// www.stortinget.no/no/Saker-og-publikasjoner/Publikasjoner/Innstillinger/ Stortinget/2009-2010/inns-200910-284/?lvl=0 [accessed January 2012].

Sømme, A. 1965. *Fjellbygd og feriefjell.* Oslo: JW Cappelen.

Sørensen, E. (ed.) 2004. *Gulskogen og landlivets gullalder.* Drammen: Drammens Museum.

Tordsson, B. 2003. *Å svare på naturens åpne tiltale.* (Doctor Scientiarum). Norges Idrettshøgskole, Oslo.

Van den Noord, P. 2000. The tax system in Norway: past reforms and future challenges. *OECD Economics Department Working Papers.*

Veblen, T. 1899. *The Theory of the Leisure Class: An Economic Study of Institutions.* New York: Penguin.

Velvin, J. 2000. *En kartlegging av hytteturisme som ledd i utvikling av bærekraftige bygdesamfunn. En rapport fra hytteundersøkelsen i Sigdal og Krødsherad kommuner.* (No. 17). Kongsberg: Høyskolen i Buskerud, 112.

Williams, D.R. and Kaltenborn, B.P. 1999. Leisure places and modernity: The use and meaning of recreational cottages in Norway and the USA. *Leisure practices and geographic knowledge,* 214–30.

PART IV
Conclusion

Chapter 13

Evolving Forms of Mobility and Settlement: Second Homes and Tourism in Europe

Paul Claval

Second Homes in Europe in Early Modern Times

Second Homes and Agricultural Activity

Second homes have existed in Europe for a long time: they first grew out of the forms of mobility which prevailed in rural areas. Transhumant herders, in the Swiss Valais for instance (Brunhes and Girardin 1906), had their first homes in their village. In May, at the time of haymaking, they stood in their *mayens*, higher on the slopes; from June to September, when their herds grazed the high altitude grassland of Alps, they lived in their *chalets* where they produced cheese. They had also *cabanons* in the small vineyards they often owned and cultivated in the Rhone Valley. Visitors, who discovered Alpine landscapes at the end of the eighteenth century or the beginning of the nineteenth, Ramon de Carbonnières (1781–1782) for instance, were struck by the high number of buildings in areas where the population density was low: most of them where occupied only a few days, a few weeks or a few months during the year.

Second homes reflected the mobility of at least a part of rural workers in mountainous or Nordic regions, where people had to capitalize on the diversity of local environments in order to feed themselves and their families.

Second Homes and Urban Elites in Early Modern Times

Second homes were also linked to the development of urban centres. The feudal society had a peculiarity: its lords did not live in cities, but in castles built in the estates they owned, among their serves; the source of their military force, administrative power and economic wealth was there. But increasingly, they had to spend spells of time in cities, where they sold a part of their crops, bought what was not produced in their farms, and where their sons attended schools. Double residence became the rule: wintertime in the city, and a move back to the castle in summer, at the time when it was important to have an eye on the serves, sharecroppers or farmers to get the part of crops or income entitled to the lords or landlords. What was then the main home and the second one? It was not at all clear.

In Italy, the evolution went so far that the cities soon gained the feudal control of the adjacent rural areas: they were surrounded by lands they ruled upon, and which formed a county – hence the name of *contado* given in Italian to rural areas.

In medieval cities, in Italy or elsewhere, higher classes were mixed: there were feudal lords commuting between their estates and their urban mansions, and merchants, lawyers, judges, whose activity was urban. These well-to-do permanent urban dwellers began, too, to buy lands and build second homes in the vicinity of the city in order to escape its heat and epidemics in the summertime. The fresco of the Good Government painted by Ambrogio Lorenzetti in the City Hall of Sienna, showed clearly that this evolution was well engaged by mid fourteenth century, as underscored by Emilio Sereni (1962). By the sixteenth century, Florence and most Italian cities were surrounded by *Bel Paese*: Sereni named thus this ring of suburban beautified countryside where second homes and gardens were numerous, as in the Venetian *Terraferma*, where Palladio designed many wonderful *villas* for the Venetian aristocracy (Cosgrove 1984).

The movement expanded all over Europe in the sixteenth and seventeenth centuries. In England, it gained momentum during the eighteenth century, thanks to Lord Burlington, the importation of the Palladian *villa* and the invention of the landscape garden (Cosgrove 1984).

The second homes of the new aristocracy were frequently suburban, since men often had to commute between their two homes in order to run their business in the city. But some institutions managed to develop rather long summer holidays, such as universities, courts or parliaments: hence the integration of long rural stays in the agenda of a part of upper classes.

In a way, these modern elites followed in the Roman aristocrats' footsteps, which devoted to *otium* a good part of their time, and lived for long stretches of time in their second homes along the bay of Naples, especially around Herculanum or Baia, or in Capri, or closer to Rome, near Ostia, as Pliny the Younger in the *villa Laurentiana* he owned and described in his letters (Baridon 2006).

In the sixteenth century, British aristocracy had the feeling of being somewhat behind the movement of Arts, Letters and Science that the Renaissance had launched in Italy. Hence the habit of sending young men for a *tour* of continental Europe: there were bound for Venice, Florence, Rome or Naples, and passed through France or the Netherlands, Germany and Switzerland. *Tourism* was born in this way.

The mobility of Western well-to-do people in the early modern time was expressed in two ways: they often owned second homes; their sons began to practice *cultural tourism*. During the eighteenth century, a growing number of wealthy families in Britain owned three homes: one in their rural estates, one in London where they spent the 'Season', and the third one in a spa, Bath or Cheltenham, for instance, and later, in a sea-resort like Brighton. In Bath, the crescents built by John Wood the Younger offered attractive housing for families that did not wish to invest too much in this third home (Johns 1965, Penoyre and Ryan 1958).

Second Homes and the Industrial and Transport Revolutions

In the Nineteenth and Early Twentieth Centuries

The revolutions in transport and industry had a deep impact on Western societies, their individual mobility and their settlement patterns. Thanks to the railways and steamship, it began to be increasingly easy to feed cities. Because of the industrial revolution, the urban centres which benefited from the cheap energy of waterfalls or coal attracted a growing number of workers. At the same time, and because of the emergence of entrepreneurs and the development of services, the higher class was complemented by middle *strata*, which copied the way of life of the former aristocracy.

Three main forms of mobility developed: (i) from overpopulated areas, the migration of landless peasants or craftsmen ruined by the new manufactures was bound to cities and industrial areas; (ii) many businessmen or tradesmen frequently visited major urban centres in order to meet bankers, wholesale dealers, importers or exporters; (iii) high and middle classes travelled for tourism and visited spas, sea- or mountain-resorts, or major foreign cultural centres, like Florence, Rome, Venice, or Paris.

A part of the identity of the lower classes was still rooted in the rural areas they migrated from. Whenever possible, they tried to visit back their home country; they sent their young children there to be raised by their grandparents; many of them tried to retire to their villages, where they had inherited a family house or bought one. *Roots tourism* developed in this way all over Europe – but the cost of travelling and the lack of holidays limited its expansion for a long time. Depending on the structure of families in the rural areas of origin, on the characteristics of their property markets, and on the existence, or not, of minor rural buildings, roots tourism took widely differing forms, as shown in the chapters of this book.

Business was the force behind the development of second homes in major cities – *pied-à-terre,* as they are called in French and often in English. Some of these second homes were located in major cultural centres, where artists, musicians or intellectuals liked to have short or long stays.

A growing number of middle-class families managed to have a second home where they stood for the summer holidays: some were located in the regions or localities where these families had their roots; many were built or bought in places where the environment was attractive, along the sea or around the lakes, which enjoyed both a wonderful environment and a mild climate, in the Swiss, Bavarian, Austrian, French or Italian Alps. Some families chose to have second homes in, or close to, a historic city, especially in Italy. Mountains remained less attractive before the diffusion of skiing and winter sports. Only a few people practiced mountaineering: hotels, inns or beds and lodgings there provided accommodation.

In tropical countries, European colonizers often managed to spend the hottest and wettest part of the year in a mountain resort where they frequently owned or rented a second home: Darjeeling or Simla in India, for instance.

Twentieth Century Changes

With affluence, the patterns of mobility began to change in the interwar period, because of the use of cars, the progressive enlargement of middle classes, the first forms of the Welfare State, and the emergence of mass societies. After the Second World War, the pace of transformation accelerated thanks to the social policies that insured a minimum income for all, holidays for every worker and the generalization or retirement pensions, thanks also to the transport and communication revolutions, which made long distance travel quicker and cheaper and improved the accessibility of remote places.

The evolution accelerated almost everywhere from the seventies or eighties, when the effects of these changes began to build up. From then on, second homes became testimonies of world-scale mobility, with (i) businessmen travelling all over the planet in search of the best partners and markets, (ii) workers converging towards countries with employment opportunities, high wages and efficient social security systems, and (iii) tourists attracted by the sun and beach of Mediterranean or tropical sea-resorts, by the new forms of spas, and by the high quality facilities offered by many touristic regions. In such a context, tourism became a major economic activity. Its incidence on the price of land and housing grew rapidly.

What were the effects on second homes of such rapid evolution?

Businessmen and the members of the jet set increasingly bought *pied-à-terre* second homes in global cities (London, Paris, New York, Los Angeles…), major touristic cities (Miami, Nice and Monaco, Malaga, Punta del Este…) or touristic centres (Saint-Tropez, Ibiza, Capri…). Some owned three, four or five homes all over the world. These second homes were used both for business and tourism.

Mass tourism had generated a massive increase of the number of visits to the places the families of the new urbanites came from – never was roots tourism as important. This tourism was, however, relatively short-lived: the links with the place of origin were already weaker for the second generation. They often disappeared for the third one. At the same time, rural depopulation reduced the number of relatives still living there. Root tourism was progressively replaced by residential tourism, where the availability of cheap houses played a major role – either old or newly built ones.

Mass tourism had created a massive demand for hotels, camping sites and all forms of entertainment and leisure activities in attractive areas, i. e. the vicinity of major cities for weekends, coastal areas for bathing, surfing and other water sports, high mountains for skiing, mountaineering, and lower mountains and picturesque hills or rural areas for hiking.

In the areas of touristic concentration, the construction of second homes had become a major economic activity. For about two decades, it emerged as one of the main growth factors along the Mediterranean coasts in Spain, France, Italy, Croatia, Greece, Cyprus, Tunisia, and up to a point, in Turkey or along the Atlantic coast, in Morocco, and Portugal. It explained the economic growth of these areas in the 80s and 90s, and because of the speculative bubble it generated in Greece,

Spain and Portugal, it was partly responsible for the economic crisis that is currently striking these countries.

Behind Second Homes: Dreams, Geographical Imagination and Ideologies

Human mobility does not result only from economic forces, or the existence of transport facilities. It reflects the dreams and geographical imagination of people, and the ideologies that shape them.

The Influence of Antiquity: Real, but Limited

Roman people's quest for *otium* was completely forgotten by the end of the Middle Ages. During the Renaissance, it influenced the small groups of humanists. Their ideas did not result in a resurrection of the Roman pattern of second homes, but was certainly important in many British aristocratic families' decision to complete the education of their sons by means of a Grand Tour to Italy.

The humanists' dreams also had an important influence in the way second homes were designed. They had to offer an urban look: the form Palladio gave them drew its inspiration from Roman ruins and monuments; their gardens were conceived as green architectures and peopled with statues of ancient gods and goddesses.

The Christian Tradition: Pilgrimage and Retreat into the Desert

Christianity introduced two forms of travels: pilgrimages, and retreats into the desert for praying, meditating and repenting.

There were different motivations for a pilgrimage: the love of the Lord, Mary or a Saint and the faith in his/her power; the will to atone for a mortal sin; the wish to thank God or his intercessors for His (or their) favour. The journey itself was seen as punishment and a means of purification before meeting the Lord, Mary or the Saint: it was all the more efficient if it was uncomfortable. The visit to the sanctuary was generally short. Pilgrimage had nothing to do with second homes.

Retreating into the desert in order to pray, meditate and repent had a different character. Christians spoke of 'desert', but desert only meant for them isolation, quietness and nature. It was generally made of a stretch of wild moor or forest, or isolated islands. In order to be efficient, a retreat had to be long enough: Jesus withdrew to the desert for forty days! Because of the will to repent, accommodation was generally minimum – a cave, a hut. It was certainly not a true second home, but it conferred a high status to long stays in nature.

For humanists, second homes had to exhibit some features of urbanity even if they were located in the countryside. For Christians, wild nature had a salutary effect. In this way, Christianity developed a taste for the simplicity of natural settings, where they would stay for rather long periods in order to do some soul-searching.

The Naturalistic Drive

For the Ancient Greeks or Romans, Nature was sacred. For most Christians, it appeared more as the hideout of Satan and his devils than as a holy place. It exposed them to temptation. Resisting it, like Saint Anthony, was one of the reasons for which a retreat into the wilderness had a salutary virtue for the soul.

Attitudes changed in the eighteenth century. The experience Europeans had in the New World was significant in this respect (Chinard 1911). Amerindians were wild warriors. They ignored pity, and could inflict terrific sufferings to their prisoners, but as noted by Jesuit missionaries as early as in the sixteenth century, they ignored some forms of mortal sin – more peculiarly greed and miserliness. To live in a natural environment had certainly something to do with such a situation. By the beginning of the eighteenth century, this line of thought gave birth to the myth of the Good Savage (Lahontan 1704).

When far from civilization, like Robinson Crusoe in his island, people had to make a considerable effort in order to humanize the natural environment in which the hazards of life had placed them. For Jean-Jacques Rousseau, many of the virtues of the primitive man were preserved in the remote mountains of Europe, as explained in the *Vicaire Savoyard* (Rousseau 1762). Staying in nature was conducive to fantastic experiences of fusion with the natural elements, as in the tenth *Rêverie d'un promeneur solitaire* (1782).

From that time on, the image of second homes and the practice of tourism began to be closely related to the idea of nature.

Nature and Nineteenth Century Domestic Feminism

This new conception of nature had important consequences for the way people conceived family life. In North America and, up to a point, in protestant Northern Europe, it was responsible for the domestic feminism as defined by Harriet Beecher-Stowe (Ghorra-Gobin 1987). Family was the basic social institution. In order to be efficient, familial responsibilities had to be shared by the father and the mother: the father had to work and earn enough for the whole family; the mother had to educate the children and to look after their physical and moral health. In order to succeed, the family had to settle out of the central part of the city, where pollution and moral depravation were high. Hence the value given to a suburban home, and, whenever it was impossible to purchase one, to a second home in a healthy area, along a fresh seacoast for instance, where kids would bathe in rather cold water and breathe iodized air.

Confronting Nature as a Factor of Self-development

By the end of the nineteenth century, the virtues attributed to nature became more numerous. In an urbanized and industrialized world, women and men had much less opportunities than in the past to develop their physical strength and aptitudes:

to reach physical self-fulfilment was impossible in everyday environments. In order to achieve it, people had to benefit from holidays where they could hike, run, swim, climb, develop their strength and build a harmonious and beautiful body: hence the new significance of second homes. They provided women and men with the possibility to fortify their bodies and strengthen their will.

The Utopian Model: The City in the Countryside

The longing for nature, which developed in Western societies from the eighteenth century onwards, was a part of a more general dream: the wish to combine the advantages of city and nature. Such a dream was present in Eastern cultures, as shown by Augustin Berque (2010). It became the basis of one of the most pregnant form of modern utopia during the nineteenth and early twentieth century in Western Europe and the United States. Ebenezer Howard gave to this ideal the form of garden cities. Frank Lloyd Wright expressed it another way in *Broadacre* (1932).

Such a utopia was conducive to the fragmentation of the city, an overall suburbanization, and a situation where second homes ceased to be useful, since the main home possessed at the same time the virtues of an urban residence and of a rural one.

In such settings, the opposition between first and second homes tends to be increasingly blurred.

Sustained Development and New Views over Nature

The conception of nature changed substantially during the past generation. Ecologists were responsible for a new attitude about human presence in nature: this presence was increasingly considered to be harmful to the resilience of ecosystems.

The benefits that nature was supposed to offer women and men appear at the same time less evident. It is no longer necessary to walk, hike, ski, climb or swim in natural environments in order to achieve well-balanced physical development: fitness centres, swimming pools, climbing walls are as efficient. They can now be found in every urban setting. Practicing sports in true nature is increasingly a privilege of wealthy persons. As a result, the demand for second homes has changed: they may be situated everywhere, provided that they are numerous enough to justify huge investments in sport and entertainment facilities.

We are living in a period where the ideologies of holiday and leisure time are changing: until a generation ago, they were mainly conducive to the dispersion of tourism and second homes. Today, they are balanced by the growth of new forms of touristic attitudes and behaviours, new types of infrastructures, and the idea that humanity has to reduce its impact on nature, and to restore it whenever it has been damaged during the past generations.

The Distribution of Second Homes in Europe: Heritage and Today's Processes

Second homes are one of the facets of mobility and reflect its transformation in Europe and the world throughout the past five centuries: from an agrarian civilization to an industrial, commercial and urban one; from a time when travellers mainly walked, rode horses, mules or camels, or sailed, to a time when they use cars, trains, planes and all forms of motorboats and ships.

The evolution of mobility cannot be only explained by technical progress alone. It reflects the social legacy of times when agriculture was more important and societies were more hierarchical.

The Distribution of Second Homes and the Legacy of Pre-industrial Europe

The rural legacy explained the presence in many parts of Europe of buildings temporarily used for culture, sheep or cattle rearing, but which could easily be transformed into second homes. The rural legacy was also responsible for the strong economic links which characterized roots tourism: working in the family farm during the holidays, using the food produced there to supplement what was bought by the migrants in their new urban residence. The legacy of the rural past was mainly social and cultural: the migrants still saw themselves as members of their community of origin: they did not visit the place where it was located, but rather their parents or friends. They did not come primarily for leisure and entertainment, but for solidarity and the reaffirmation of their identities: as shown in this book by Maria de Nazaré Oliveira Roca, they are not interested only in the local amenities (they do not display only *topophilia*). They are glad to participate in the development of their community, which is what this author calls *terraphilia*.

As far as second homes were linked to the rural origin of their users, their location was not linked to amenities, the proximity of the sea, mountainous areas, or picturesque landscapes: they were scattered all over rural areas where out-migration towards urban areas was high.

The aristocratic legacy was different. For the nobility, to own rural estates had, in the Middle Ages, a power dimension; its economic significance remained important for much longer. There were, however, other reasons for wealthy and powerful people to move back to their castle, manor or mansion every year. They experienced real pleasure in living in a place where their authority was accepted, where they enjoyed a high status and benefited from everyone's respect. The social pre-eminence of aristocracy was more directly lived in their country residences that in their urban ones.

This high statute was expressed in different ways. In many parts of Europe, hunting was a privilege of the landed aristocracy. To participate in a hunt with hounds was an honour. In nineteenth century France, wealthy industrialists and businessmen living in Paris purchased expensive second homes with big shoots in the forested area of Sologne: a good part of high level economic contacts and negotiating was performed there.

As a result, and until the end of the nineteenth century and sometimes well into the twentieth, the second homes of well-to-do people did not have to be located in amenity-rich areas: their attractiveness lied in the presence of lower class clients and protégés, and of woods or forests for hunting. Aristocratic second homes were scattered all over the country.

The Distribution of Second Homes, Industrialization and Modernization

The legacy of the nineteenth century phase of the industrial revolution was different. Because of the pollution of major urban centres and industrial areas, upper and middle classes began to be much concerned with the health of their children, and looked for the fresh and safe atmosphere of Brittany in France, the West Coast in England, or New England in the United States. For elderly persons, countries with a mild winter climate were favoured; young men often patronized mountainous environments. Most of travelling was done by train, which explained the concentration of second homes in areas well deserved by the main railway lines, and their concentration in a few areas.

From the beginning of the twentieth century, new factors ruled the location of second homes. Shorelines were still sought, but with the new sun cult, the Mediterranean sea-resorts became ever more attractive for summer holidays – in the past, they had been selected only for winter stays. Because of the new favour for winter sports, high mountains became increasingly attractive. With the generalized use of cars for travel, second homes spilled more evenly over the areas offering amenities.

Contemporary Forces behind the Location of Second Homes

In societies where people enjoy much leisure time, the attitudes towards holidays, as well as the forms of mobility they induce and the distribution of the second homes people use, change. These new trends are in a way contradictory: in a time of urban congestion and generalized pollution, people are dreaming of nature, quietness, safe and pure environments – hence a drive towards dispersion. At the same time, people wish to be more active during their holidays: the proportion of those who make do with sunbathing is declining. People wish to bathe, surf, boat, hike, climb, raft, play badminton, tennis or golf, attend concerts, plays, operas, frequent restaurants and nightclubs. This means that tourism requires huge investments; investors can only receive good return on them if they are made in dense enough areas. It is increasingly difficult to have houses as second homes: in many sea- or mountain-resorts, the majority of them are now apartments. Building second homes has ceased to result mainly from individual initiatives and the work of local craftsmen. It is increasingly the job of property developers.

Travelling has been profoundly modified by the revolution of jet planes: mass tourism has ceased to be confined to a radius of a few hundred kilometres around the areas of departure. Two forms of travel now coexist, depending on the distance: when the journey is relatively short (200 or 300 km for weekends, up to 1,000 km for summer holidays), cars are widely used, which favours the dispersion of second homes; houses represent a fair proportion of them. For longer distances, air travel is dominant, which is conducive to the concentration of second homes in more restricted areas; and most of these are apartments.

Many people in Northern Europe have winter vacations in the Canary Islands, Costa del Sol, Majorca, Morocco, Tunisia, Greece, or Cyprus. The development of this form of tourism is linked to the activity of low-cost airline companies. Well-to-do European people visit Florida, the Bahamas, the Caribbean, Mauritius, the Seychelles, the Maldives, Sri Lanka, Thailand, Bali, or even the Australian Golden Coast. At the same time, wealthy Arabs of the Middle East buy *pied-à-terre* in London, Paris, Zurich, Berlin, Vienna, or Rome, or in Nice and the Riviera: they use them for business; their wives go shopping in luxury shops; they enjoy the high quality of services of these large cities.

The proportion of second home owned by foreigners is growing: it is possible for middle class Englishmen, Germans, Dutch or Scandinavians to buy second homes in France, Spain, Portugal, Italy, Croatia, Greece, as well as in Scandinavia or Ireland.

Second Home Tourism in Europe Today

The Diversity and Complementarity of Approaches

The twelve chapters of this book use different scales and adopt different perspectives on second homes and tourism in Europe. The phenomenon is analysed at the national scale in the British Isles and Ireland (Chris Paris), Norway (Tor Arnesen and Birgitta Ericsson), Finland (Mervi J. Hiltunen et al.), Russia (Tatyana Nefedova and Judith Pallot), France (Jean-Marc Zaninetti), one of the chapters focuses on Italy (TullioRomita) and another on Portugal (José António de Oliveira). The focus is on regional problems in Greece (the Cyclades Islands) (Olga Karayanis et al.), Italy (Calabria) (Antonella Peri), Spain (the region of Alicante) (TomásMazón et al.) and partly for Portugal (the Oeste Region) (Maria de Nazaré Oliveira Roca). The approach is predominantly economic for the British Isles, ecologic for Sweden (Dieter K. Müller). In Norway, emphasis is on the management of land use, especially in the areas where there is a concentration of second homes. In Finland, the interest is focused on the dynamics of spatial change and its eco-social impact in the last fifty years. The role of roots tourism is central in the papers on Portugal, and Italy. In Portugal as well as in Italy, the role of second home tourism in the rural areas where emigration was high in a recent past remained for long ignored (TullioRomita) or undervalued (José António de Olveira). The analysis is more historical, social and cultural for Russia.

Does this variety of interests and coverage constitute a weakness for this book? Not at all: each chapter sheds light on a different facet of second home tourism; it does it in a particular country, but helps to explain what occurs in the others. It results in a deeper global diagnostic about the problems and evolution of second homes at the European scale.

Two facts are stressed by all the contributors: (i) the difficult assessment of the number and spatial distribution of second homes all over Europe; (ii) and the accelerated growth in their number for the past twenty of thirty years, until the economic crisis of 2007.

Problems of Definition and Measurement

For all the contributors, assessing the number of second homes was difficult. There were different reasons for that. The first one stemmed from the rapid changes that have affected mobility in the last fifty years. Thanks to the revolutions in transport (the democratization of car use, high speed trains, cheap air flights for long distances) and communication (telephone, cell telephone, television, internet), it has become easier to visit distant places and enjoy the facilities of modern life in formerly isolated locations.

Second homes are increasingly comfortable. Fifty years ago, in Nordic countries or Russia, many of them were rustic wood cabins, or even mere shelters: no running water, no sewers, no flush toilets, no electric power, no telephone lines, no heating! They could only be used in summertime. The equipment of second homes is today increasingly similar to that of first ones. They may be inhabited year-round.

This evolution is parallel with the lengthening of human life: people have more years to live after retirement. As a result, many couples transform at that time their second homes into main ones. With the development of teleworking, an increasing number of persons stay permanently in homes which had been otherwise second ones. In big cities, many businessmen or wealthy foreigners buy *pied-à-terre*, which they use for business, shopping, entertainment or other forms of activity. Here, what is difficult is not to know whether they are first or second homes, but whether they are used for tourism, or not.

In the regions where rural depopulation was important, many houses remained empty. Others were transformed into second homes by the out-migrants, but remained closed most of the year. As a result many second homes were recorded as vacant houses or apartments. In Italy the category of second home was ignored: hence the lack of adequate statistics (Tullio Romita). Nobody was able to assess the possibilities of accommodation they offered until two or three decades ago. The situation has in a way been similar in Portugal: as a result, the role of tourism in the poorest rural regions of the country has been neglected (José António de Oliveira).

The quality of the chapters in this book results largely from the attention given to the definition of second homes and the careful assessment of the available statistics.

The Growing Role of Amenities

The distribution of second homes first reflects both the proximity of the places or areas the tourists come from, and the geography of amenities. Seashores, mountainous areas, picturesque regions, numerous rivers and lakes insure high attractiveness to some areas, and explain the lack of significant second homes development elsewhere. In Nordic countries, like Finland, where tourism is nature-oriented, the omnipresence of forested areas and the number of lakes explains the overall presence of second homes (Mervi J. Hiltunen et al.), but also their higher concentration in the Saupeselka Region, where the surface of inland lakes is the highest. In England, second homes congregate mainly along the coasts and in the Lake District, Yorkshire, East England, South and Southwest England, the Cotswolds, and the Welsh borderland (Chris Paris). In France, in 1968, the density of second homes was already high along the coasts, around Paris, in Lower Burgundy, central Normandy, the Loire Valley, Northern Alps, Eastern and Southern Central Massif, and Provence. Second homes are today twice as numerous, with high densities along the coasts, in all mountainous areas, in Brittany, and in most of Southern France, except for the plains of Aquitaine (Jean-Marc Zaninetti).

The qualities that contribute to the amenity of a place or an area depend, however, on time and countries. The attractiveness of seashores and mountains is general, but today people are more attracted to warm seas than in the late nineteenth or early twentieth centuries. Italian or French people have a taste for picturesque rural areas, whereas Nordic people prefer the solitudes of Northern forests and lakes. There is a cultural dimension to amenities.

Historical and Cultural Dimensions and the Role of Heritage

The distribution, age and use of second homes do not reflect only the geography of amenities. They result from a complex history and have social and cultural components.

The omnipresence of second homes in Nordic countries partly reflects the difficulties of rural life in regions where people had to rely on the cultivation of a few fields, cattle rearing – with often some forms of transhumance – and forestry. It explained the high number of wooden constructions, which could be easily transformed into second homes.

In France and the Mediterranean countries, rural overpopulation was omnipresent until modernization, i.e. the second half of the nineteenth century in France, the beginning of the twentieth in Northern Italy, the middle of the twentieth in Southern Italy, Greece, Spain, or Portugal. The rural exodus was very important and left many vacant houses. The significance of this rural heritage in the development of second homes is clear. In 1982, for instance, more than 40 per cent of all second homes in Burgundy had been built before 1871 (Jean-Marc Zaninetti)!

Because of the widespread former overpopulation of rural areas, second homes are present everywhere, even if their densities vary according to the specificities of regions. The open field areas of North-Eastern France, with their naked landscapes and small villages, appeared less attractive than the picturesque landscapes of dispersed farming, with their profusion of woods, meadows and hedges. In Italy, the big capitalist farms and their poor workers' houses in the Po valley are repulsive.

The case of Russia is fascinating (Tatyana Nefedova and Judith Pallot). There, second homes, dachas, are a fundamental institution. They originated in the eighteenth century out of a social situation which was at the same time similar to and different from those observed in Western and Central Europe. Dachas were rural estates given to the nobility, which did not owe its power to its rural roots, but to the favour of the Tsar: Russian nobility was not rooted in feudal *seigniory*, but in the fulfillment of civil or military services in the Tsar's name. It certainly explains the higher density of *dachas* around the two capital cities, the former and religious one, Moscow, and the modern and political one, Saint Petersburg.

In the nineteenth century, the evolution became similar to that observed at that time in other European countries: the new bourgeoisie adopted aristocratic attitudes and acquired dachas in the same area as the *boyars*. The initial policy of the Soviet Regime did not differ, initially, from the Tsarist one: dachas were granted to major institutions of the USSR, academies for instance. The inability of the Soviet economy to feed urban centres correctly, especially regarding vegetables and fruits, created a demand for plots of lands where to cultivate potatoes and plant apple trees. In this way, dachas became a mass phenomenon after the Second World War. When flying over the Saint Petersburg or Moscow areas, the often large allotments created in this way are prominent features in the settlement pattern.

After the collapse of the Soviet Regime, it became legal to own dachas. The demand for dachas grew. According to familial income, they began to differentiate, many of them being transformed into high quality houses by the wealthiest part of the population.

The Naturalistic Utopia

The reasons for owning a second home differ widely. They establish links with the rural societies from which many urbanites originated. They allowed the development of small-scale subsistence farming during wars and in socialist regimes. For most people, they introduced alternative ways of life thanks to their proximity to nature, the informal character of their social life, the early disappearance of most of the signs of social hierarchy – in clothing, for instance. In this way, second homes were an expression of the democratic and naturalistic utopias of late nineteenth and early twentieth century (Mervi J. Hitunen et al.; Tor Arnesen and Birgitta Ericsson; Dieter K. Müller).

This form of utopia displaced an older aristocratic one: that of a society where the power of the upper classes could be expressed in a patriarchal atmosphere, and where some pleasures were the privilege of the ruling minority.

In the contemporary world, the democratic dimension of the second home utopia has disappeared with the growing disparity of incomes at nation- and world-scales. It is increasingly difficult for those who have low incomes to purchase second homes, be they houses or apartments, or a plot of land to build one. Social stratification is back, even if expressed in a different way, through the presence or absence of costly and prestigious sports grounds, tennis courts or golf links.

The dream to live close to nature has not disappeared, but it does not have the same significance and the same attractiveness as in the past. It does not have the same significance, since, in a consumption society, tourism is just another possibility to spend one's money and to affirm one's economic status: a second home is just another item to buy. It does not have the same attractiveness as in the past, because of our time's growing ecological awareness: when living in second homes, are not tourists increasing their ecological footprint (Dieter K. Müller; Tor Arnesen and Birgitta Ericsson; Mervi J. Hiltunen et al.)? Are they not disturbing natural ecosystems and reducing their resilience?

The holiday utopias of today are no more naturalistic. They are green ones, with a growing concern for their eco-social impact (Mervi J. Hiltunen et al.): hence the emphasis on higher densities, lower energy consumption, and the ban of cars or flush toilets in many cases – particularly in the Nordic countries. The landscapes of second homes are changing rapidly. It is still difficult to imagine the forms they will take during the next twenty or thirty years.

The Problems of Second Home Tourism

Second Home Tourism and Social Justice

As long as tourism was just a marginal activity of well-to-do people, it expressed the social inequalities present in society. The problem was to build, through reform or revolution, a more equalitarian social system.

Tourism has progressively ceased, however, to be practiced only by privileged upper classes. It has become a middle class pursuit, before being transformed into a mass phenomenon. The problem was to open it to all the components of society. Everyone was entitled to a right to tourism, just as Henri Lefebvre (1968) professed that everyone was entitled to a right to the city! This theme appeared more particularly important in Nordic countries.

The will to reduce social injustices in the field of tourism was expressed in different ways: it involved the development of physical planning and land use regulations, in order to offer an access to amenities to the lower classes. These actions were particularly important along shorelines. The concern with social justice also involved offering more land to build second homes, in order to counter the speculation on land prices. It finally involved the provision of public infrastructures at low cost in touristic areas. These measures were significant components of the planning regulations in Norway, or of the eco-social conscious policy developed in Finland.

This social dimension of second home tourism might be also partly achieved through the mass construction of second homes by property developers, as in the new sea-resorts along the coastlines of Spain (Tomás Mazón et al.), Portugal, France, Italy, or the Greek Islands, and in some mountain resorts of the Alps, particularly in France.

Second home tourism generates, in the areas where it is especially important, another social problem: since it consumes much land, it is responsible for the rapid rise of land and housing prices. As a result, a part of the native population of touristic areas does not have enough money to buy houses or apartments, or rent them. Many of them have to settle in places where landscapes are dreary, or in the periphery of touristic areas, which often involves long distance commuting. As a result, public authorities have to plan low-cost housing for them.

With the growing differentiation of incomes which characterizes contemporary European societies, social equity problems are becoming more acute. Second homes as a form of naturalistic utopia belong to the past.

Second Home Tourism and Economy

One of the main results of the contributions gathered in this book is to emphasize the economic significance of second home tourism. It was already and important sector of activity in the 60s and 70s. Second homes have multiplied since them: their building has mobilized the energy of many property developers, employed many workers in the building trade, and created a lot of commercial and other service activities. Many areas were, in this way, transformed into residential economies (Jean-Marc Zaninetti): they no longer participate directly in the production of goods; they recycle money earned elsewhere and transferred there by tourists. Assessing the importance of these new residential economies may be one of the targets of research on second home tourism, as shown in the chapter on France.

A second perspective may be chosen in order to shed light on the economic dimension of second home tourism: tourists speculate on the price of the land, house or apartment they buy in order to increase their assets. This perspective is thoroughly explored in the case of Britain (Chris Paris), where second home owners increasingly appear as speculators operating both on the national and international property market – British people buying second homes in the British Isles or in continental Europe depending on the respective strengths of Pound and Euro, foreigners investing in London or in some prestigious second home areas, the Cotswolds for instance, depending on the economic perspectives of their own country and the UK.

In order to acquire a second home, many people borrow money. From the beginning of the 90s, banks have become increasingly active in this sector of activity.

Many of them have made risky loans to low-income persons. Europe participated in the subprime crisis, i.e., the granting of real property loans to insolvent persons, but differing from the United States, this affected perhaps more the second home market than the first one. It was particularly true in some Mediterranean countries, Greece, Spain, or Portugal. Tomás Mazón et al. present here a wonderful monograph of the municipalities of Santa Pola, Guardamardel Segura, and Torrevieja, along the Southern coastline of the province of Alicante: the majority of new housing was bought by Spanish customers. For the majority of them, it is hard to repay their loans: the whole sector is struck by a crisis, without any hope of a rapid solution. The situation in the Cyclades Islands is not so different.

These case studies ask a fundamental question. The development of tourism, and more specifically of second home tourism, had a curious effect in European countries: a good part of the savings of the working population was not invested to improve the infrastructures, collective facilities and spatial organization of the areas where this people lived, but in the places where they spent their holidays. It transformed the economies of many parts of these countries into residential economies. As long as the flows of tourists and money from urban and industrial areas towards touristic ones were growing, residential economies were buoyant. As soon as they were reduced, these economic areas experienced deep economic depression. Because of their specialization in residential economy, they have no means of solving their problems.

In this way, the study of the economy of second home tourism helps to understand the economic problems of European countries and regions which specialized too much in residential economy – Greece, Spain, or Portugal for instance.

Second Home Tourism and Landscape Preservation

Second home tourism often relies on the beauty and harmony of landscapes – this is one of the main amenities people are looking for. As long as the bulk of second homes were made of the former houses of farmers or craftsmen, their presence did not alter the visual environment. In fact, it improved it, since second homes were carefully looked after and restored.

The situation is now different: a growing proportion of second homes were purposely built. In some areas, they are scattered all over the area, which transforms the general outlook and is conducive to what French physical planners call *mitage,* i.e. the intensive – and abusive – building of houses in the countryside. Elsewhere, property developers have created massive allotments of houses or apartment houses, well exemplified in this book by 'the horizon of bricks in Santa Pola' photographed by Tomás Mazón.

Thirty years ago, landscape preservation was considered a major problem in touristic areas. It was the main motivation behind the European Landscape Convention, or Florence Convention, signed by the Council of Europe in 2000. Its preamble explains that:

'The member States of the Council of Europe:

Acknowledge that the landscape is an important part of the quality of life for people everywhere: in urban areas and in the countryside, in degraded areas as well as in areas of high quality, in areas recognized as being of outstanding beauty as well as everyday areas;

[Are] aware that the landscape contributes to the formation of local cultures and that it is a basic component of the European natural and cultural heritage, contributing to human well-being and consolidation of the European identity.' (conventions.coe.int/Treaty/EN/Treaties/Html/176.htm).

At that time, the main problem created by landscapes was that of preservation. A new concern was present, but the Convention paid only lip service to it:

[The European States are] concerned to achieve sustainable development based on a balanced and harmonious relationship between social needs, economic activity and the environment." (Conventions.coe.int/Treaty/EN/Treaties/ Html/176.htm).

The situation is quite different today.

Second Home Tourism and Ecology

Until the 90s, the major concern about the use of nature by second home tourism was landscape preservation. People had a right to enjoy forest, river-, lake- or sea-shores, mountains or rural areas. The role of planners was to help comply with their demand for building sites. Today, the attitude is utterly different. Social aims have to be combined with ecological ones, as expressed by the Finnish policy: to meet 'eco-social' targets. The purpose of Mervi J. Hiltunen et al. is 'to identify and discuss the environmental and social impacts of second home tourism in Finland', with the main problem being linked to energy consumption:

"The increasing popularity and amenity demands for second home living have initiated concerns on the environmental consequences of such development [...]. In general, the increase in living space, equipment rate and year-round use of second homes also increases energy consumption and natural resource use [...]."

The energy consumption linked to second home tourism results from the improved quality of second homes, the more frequent travels to visit them since they are increasingly used year-round, and the locally generated mobility, mainly by car, they induce.

Using more energy efficient techniques of building and heating may reduce the ecological impact of second homes. Substituting tourist centres to dispersed forms of settlement could also improve the situation: apartments instead of private housing, sewerage systems, closer commercial facilities, water processing, to name but a few. It means that new land use regulations have to be issued, and that second home developments have to be increasingly integrated into rather dense settlements.

This book shows that second home tourism has become such an important sector of the economy that it is no longer possible to let it develop freely: it is the source of new forms of social deprivation; it generates residential economies that are particularly sensitive to the economic cycle; it often impairs beautiful landscapes and increases human pressure on natural environments. As a result, it is one of the major physical planning stakes of touristic areas.

References

Baridon, M. 2006. *Naissance et renaissance du paysage*, Arles: Actes Sud.

Berque, A. 2010. *Histoire de l'habitat idéal: de l'Orient vers l'Occident*. Paris: Éd. du Félin.

Brunhes, J. and Girardin, P. 1906. Les groupes d'habitation du Val d'Anniviers comme types d'établissements humains, *Annales de Géographie*, 15: 329–52.

Chinard, G. 1911. *L'Exotisme américain dans la littérature française XVIe siècle*. Paris: Hachette.

Cosgrove, D. 1984. *Social Formation and Symbolic Landscape*. London: Croom Helm.

Ghorra-Gobin, C. 1987. *Les Américain setleurs territoires. Mythes et réalités*. Paris: La Documentation Française.

Johns, E. 1965. *British Townscapes*. London: Arnold.

La Hontan, Louis-Armand de Lom d'Arce. 1704. *Dialogues eu entretiens entre un sauvage et le baron de Lahontan*. Amsterdam: Vve de Boeteman.

Lefebvre, H. 1968. *Le Droit à la ville*. Paris: Anthropos.

Penoyre, J. and Ryan, M. 1958. *The Observer's Book of Architecture*. London: Frederick Warne.

Ramond de Carbonnières, L.F. 1781–1782. *Lettres sur l'état civil, naturel et politique de la Suisse*. Paris: Belin.

Rousseau, J.-J. 2011. *Profession de foi du Vicaire Savoyard*. Genève: Slatkine reprints; first published as Book Four of *Emile*, 1762.

Rousseau, J.-J. 1782. *Les Rêveries du promeneur solitaire* (faisant suite aux *Confessions*). Lausanne: F. Grasset.

Sereni, E. 1961. *Storia del Paesaggio agrario italiano*. Bari: Laterza; French translation: *Histoire du paysage rural italien*. Paris: Juillard, 1964.

Wright, F.L. 1932. *The Disappearing City?* New York: W.F. Payson.

Index